Ernst Schering Research Foundation Workshop 7
Basic Mechanisms Controlling Term
and Preterm Birth

W0227960

Ernst Schering Research Foundation Workshop

Editors: Günter Stock
Ursula-F. Habenicht

Ernst Schering Research Foundation
Workshop 7

Basic Mechanisms Controlling Term and Preterm Birth

K. Chwalisz, R. E. Garfield,
Editors

With 86 Figures

Springer-Verlag
Berlin Heidelberg GmbH

ISBN 978-3-662-21662-0   ISBN 978-3-662-21660-6 (eBook)
DOI 10.1007/978-3-662-21660-6

© Springer-Verlag Berlin Heidelberg 1994

Originally published by Springer-Verlag Berlin Heidelberg New York in 1994.

Softcover reprint of the hardcover 1st edition 1994

Typesetting: Data conversion by Springer-Verlag

21/3130–5 4 3 2 1 0 – Printed on acid-free paper

Preface

The mechanisms that initiate labour (i.e., the conversion of the quiescent uterus to an active and reactive organ at term) are poorly understood. Of considerable importance are the factors that control preterm labour and preterm birth, with their devastating effects on society. This problem is the leading obstetrical issue. Preterm labour affects approximately 10% of all pregnant women (even higher numbers are re-

Abb. I. The participants of the workshop

ported for less developed countries or where prenatal healthcare is a low priority) and it is the leading cause of morbidity and mortality of babies. Current practices to arrest preterm labour, once initiated, are ineffective. In addition, procedures to stimulate labour and dilate the cervix at term may not be used effectively.

On October 19th – 21st, 1992, a Schering Foundation Workshop took place in Berlin to discuss the "Basic Mechanisms Controlling Term and Preterm Labour". Leading scientists from Europe and North America were assembled to consider the fundamental systems which regulate the uterus and cervix during pregnancy. The topics in the symposium ranged from key cellular events governing uterine contractility and cervical dilatation during labour to clinical advances and applications. This book contains the proceedings of the workshop.

We were pleased to have been part of the workshop and to have been responsible for its organization. We gratefully acknowledge the contributions of the authors of the chapters in this book and the assistance provided by the Schering Research Foundation, in particular Dr. Ursula Habenicht.

K. Chwalisz
R. E. Garfield

Contents

List of Contributors

Assal Meliani, Aines
Biologie Cellulaire et Moleculaire de la Relation Materno-Fetale,
92140 Clamart, France

Baulieu, Etienne-Emile
INSERM U 33, Lab. Hormones, 80 rue du General Leclerc,
94276 Le Kremlin-Bicetre, Cedex, France

Baumann, Peter
Wayne State University/Hutzel Hospital, 4707 St. Antoine Blvd.,
Detroit, Michigan 48201, USA

Calder, Andrew A
Department of Obstetrics and Gynaecology, Centre for Reproductive Biology,
University of Edinburgh, Chalmers Street, Edinburgh, EH3 9EW, UK

Chaouat, Gérard
Biologie Cellulaire et Moleculaire de la Relation Materno-Fetale,
92140 Clamart, France

Chwalisz, Kristof
Schering AG, Postfach 650311, 13353 Berlin, Germany

Cotton, David
Wayne State University/Hutzel Hospital, 4707 St. Antoine Blvd.,
Detroit, Michigan 48201, USA

Dang, Duc-Can
Biologie Cellulaire et Moleculaire de la Relation Materno-Fetale,
92140 Clamart, France

Delage, Genevieve
Biologie Cellulaire et Moleculaire de la Relation Materno-Fetale,
92140 Clamart, France

Djian, Valentino
Biologie Cellulaire et Moleculaire de la Relation Materno-Fetale,
92140 Clamart, France

Do Khac, Lien
Endocrinologie et Régulations Cellulaires, C.N.R.S., URA 1131, Bt 432,
Université Paris Sud, 91405 Orsay Cedex, France

Dudenhausen, Joachim
Department of Obstetrics, University Hospital Rudolf Virchow,
Free University of Berlin, Pulsstraße 4, 14059 Berlin, Germany

Garfield, Robert E.
Division of Reproductive Sciences, Department of Obstetrics
and Gynecology, 301 University Blvd. Rt. J-62, The University of Texas,
Medical Branch, Galveston, Texas 77555-1062, USA

Gomez, Ricardo
Wayne State University/Hutzel Hospital, 4707 St. Antoine Blvd.,
Detroit, Michigan 48201, USA

Goureau, Olivier
Endocrinologie et Régulations Cellulaires, C.N.R.S., URA 1131, Bt 432,
Université Paris Sud, 91405 Orsay Cedex, France

Grover, Ashok K.
Department of Biomedical Sciences, McMaster University,
1200 Main Street West, Hamilton, Ontario L8N 3Z5 Canada

Haluska, George J.
Department of Obstetrics and Gynecology, Oregon Health Sciences
University, Portland, Oregon 97201, USA

Harbon, Simone
Endocrinologie et Régulations Cellulaires, C.N.R.S., URA 1131, Bt 432,
Université Paris Sud, 91405 Orsay Cedex, France

Jones, Colin T
Laboratory of Cellular and Developmental Physiology, Institute of Molecular
Medicine and Nuffield Department of Clinical Medicine,
John Radcliffe Hospital, University of Oxford, Oxford OX3 9DU, UK

Khan, Islam
Intestinal Diseases Research Unit, McMaster University,
1200 Main Street West, Hamilton, Ontario L8N 3Z5 Canada

Khouja, Ahmad
Laboratory of Cellular and Developmental Physiology, Institute of Molecular
Medicine and Nuffield Department of Clinical Medicine,
John Radcliffe Hospital, University of Oxford, Oxford OX3 9DU, UK

Leiber, Denis
Endocrinologie et Régulations Cellulaires, C.N.R.S., URA 1131, Bt 432,
Université Paris Sud, 91405 Orsay Cedex, France

Martal, Jacques
Inra, 78350 Jouyen Josas, France

Mazor, Moshe
Ben-Gurion University of the Negev, Beer-Sheva, Israel

Menu, Elisabeth
Biologie Cellulaire et Moleculaire de la Relation Materno-Fetale,
92140 Clamart, France

Mironneau, Jean
Laboratoire de Physiologie Cellulaire et Pharmacologie Moléculaire,
URA CNRS 1489, Université de Bordeaux II, 3 place de la Victoire,
33076 Bordeaux, France

Novy, Miles J.
Division of Reproductive Sciences, Oregon Regional Primate Research
Center, 505 NW 185th Avenue, Beaverton, Oregon 97006, USA

Ropert, Sylvie
Biologie Cellulaire et Moleculaire de la Relation Materno-Fetale,
92140 Clamart, France

Romero, Roberto
Wayne State University/Hutzel Hospital, 4707 St. Antoine Blvd.,
Detroit, Michigan 48201, USA

Tabb, Thomas
Department of Reproductive Sciences, University of Texas,
Medical Branch, Galveston,Galveston, TX 77550-2776, USA

Tanfin, Zahra
Endocrinologie et Régulations Cellulaires, C.N.R.S., URA 1131, Bt 432,
Université Paris Sud, 91405 Orsay Cedex, France

Warsop, Helen
Laboratory of Cellular and Developmental Physiology, Institute of Molecular
Medicine and Nuffield Department of Clinical Medicine,
John Radcliffe Hospital, University of Oxford, Oxford OX3 9DU, UK

Wichelhaus, Daniel
Laboratory of Cellular and Developmental Physiology, Institute of Molecular
Medicine and Nuffield Department of Clinical Medicine,
John Radcliffe Hospital, University of Oxford, Oxford OX3 9DU, UK

Yallampalli, Chandra
Division of Reproductive Sciences, Department of Obstetrics and Gyneco-
logy, 301 University Blvd. Rt. J-62, The University of Texas,
Medical Branch, Galveston, Texas 77555-1062, USA

1 Control of Myometrial Contractility and Labor

Robert E. Garfield and Chandra Yallampalli

1.1 Introduction

At the end of gestation, the uterus becomes increasingly contractile and reactive to excitatory agents to eventually reach a state in which it forcefully expels the fetus and other products of conception. This contractile state (labor) is achieved when contractions of different regions of the uterine wall become stronger, more frequent, and synchronous. The coordination of these contractions is believed to be required for the normal progression of parturition, and the absence of this activity throughout pregnancy is thought to be essential for adequate nourishment of the developing fetus (Csapo 1981).

In a significant proportion of pregnant women (ca. 5–10%) the uterus begins to contract early. Some preterm labor cases can be re-

lated to pathological conditions, while others have no apparent underlying causes. Spontaneous preterm labor ranks as the number one problem confronting the obstetrician. However, thus far, treatment to arrest
preterm labor has escaped medical science. Part of the reason for the
inability to control preterm labor may be attributed to the lack of understanding of uterine contractility. However, another part of the problem can also be related to the inability to diagnose term or preterm
labor. In both situations the physician is sometimes faced with the de-
cision to either inhibit or induce/augment labor. Unfortunately there is
no objective manner in which to evaluate the labor state. Frequent contractions and/or state of the cervix are used as indicators of labor.
However, neither is an adequate objective parameter since contraction
frequency gives no information about synchrony or force of contractility and cervical dilation or effacement sometimes occurs independently of uterine contractions.

Contractions of the uterus are contingent upon the underlying electrical activity. The frequency, duration, and magnitude of uterine contractions are recognized as being dependent on the frequency of action
potential discharge, the duration of the train of action potentials within
each muscle cell, and the total number of cells simultaneously and synchronously active (Marshall 1962). Therefore, the propagation of
potentials from pacemaker regions or from areas influenced by stimulatory agents to adjacent and distant cells is of fundamental importance
in events controlling excitability and contractility. The observation that
gap junctions appear between myometrial cells during the onset and
progression of labor was a major breakthrough in our understanding of
circumstances that control this process.

In this review, we will briefly summarize studies of the presence,
function and control of gap junctions in the myometrium. These
studies suggest that the synthesis, permeability, and degradation of the
junctions are physiologically regulated and that either pathological
conditions or pharmacological intervention might alter their normal
progression and function to modify labor and delivery. We will also
discuss recent evidence that a nitric oxide system is present in the
uterus and that it acts to maintain quiescence. Finally, we will discuss
possible models of labor and predicted steps in the control of this phenomenon with implications for the inhibition or stimulation of labor.

1.2 Intercellular Communication in the Myometrium

1.2.1 Basis for Contraction

The sequence of contraction and relaxation of the myometrium results from the cyclic depolarization and repolarization of the membranes of the muscle cells. The spontaneous electrical discharges in longitudinal muscle from the uteri consist of intermittent bursts of spike-action potentials (Kuriyama 1961; Csapo 1962; Marshall 1962; Kuriyama and Suzuki 1976; Kanda and Kuriyama 1980). In contrast discharges of the circular muscle consist of single plateau-type (that is, spike followed by a slow, sustained depolarization) action potentials in non-pregnant, early and midpregnant uteri, changing to repetitive spike-shaped action potentials superimposed on a plateau that gradually diminishes in amplitude towards term, to spike bursts at delivery (Osa and Fujino 1978; Anderson and Ramon 1976).

Contractile activity in uterine smooth muscle is initiated by a rise in the intracellular concentration of free ionized Ca^{2+} to approximately $10^{-5}M$ from a resting level of about $10^{-7}M$ (see Chap. 4, this volume). The source of this activator Ca^{2+} is extracellular (Ca^{2+} ions which flow into the cell down their electrochemical gradient in response to a change in membrane permeability) or intracellular (that is, Ca^{2+} ions released from intracellular storage sites) or a combination of both (Mironneau 1973; Kuriyama 1961; Mironneau et al. 1984). Conversely, a reduction of intracellular free Ca^{2+} (either as a result of efflux into the extracellular space or reuptake into intracellular storage sites) terminates contraction (Marshall 1974; Kuriyama 1961; Grover 1986). In the myometrium, the inward Ca^{2+} current (through voltage-dependent transmembrane Ca^{2+} channels) during action potentials initiates contractions (Mironneau 1976; Kawarabayashi and Osa 1976; Bengtsson et al. 1981; Kao 1989).

1.2.2 Propagation

Some specialized mechanism of conduction must be present between the cells to coordinate their activity because the myometrium is composed of billions of small muscle cells (Csapo 1981). Bozler (1938) was one of the first to indicate that rhythmic contractions can only be understood by postulating some mechanism of conduction which coordinates the activity of the numerous elements. The innervation of the uterus was not considered to be of importance in regulating myometrial activity associated with the birth process by Reynolds (1949), because procedures which interfere with nerve impulses fail to change the outcome. Furthermore, there is no recognized specialized conduction pathway comparable to the Purkinje fiber system of the heart present in the myometrium. Additionally, the effects of stimulants on contractility cannot account for synchronous patterns of contractility during labor. For these reasons, it is thought that electrical activity must propagate from cell-to-cell to coordinate mechanical events among the individual myometrial muscle cells (Abe 1968; Finn and Porter 1975; Csapo 1981). Thus, it is evident that the manner by which this activity propagates between the smooth muscle cells and the factors which regulate this process during gestation and labor are of singular importance to an understanding of the mechanisms which maintain pregnancy and initiate parturition.

1.2.3 Gap Junctions as Sites of Propagation and the Basis for Synchrony During Labor

Gap junctions are intercellular channels that link cells to their neighbors by allowing the passage of inorganic ions and small molecules. They have been found between cells in every tissue and organ examined and are essentially ubiquitous in the animal kingdom. In the electron microscope they appear in regions of close apposition between cells as zones of paired, paralleled membranes of unusually smooth outline separated by a narrow space of constant width – the "gap" (Figs. 1,2).

Gap junctions consist of pores which connect the interiors of two cells. The pores are composed of proteins, termed connexins, which span the plasma membranes to form a channel. Each gap junction can be made up from a few to thousands of channels and each channel is constructed from six connexin proteins in one cell aligned symmetrically with six connexins in the adjacent connected cell. The gap junction proteins have been cloned and antibodies to the connexins have

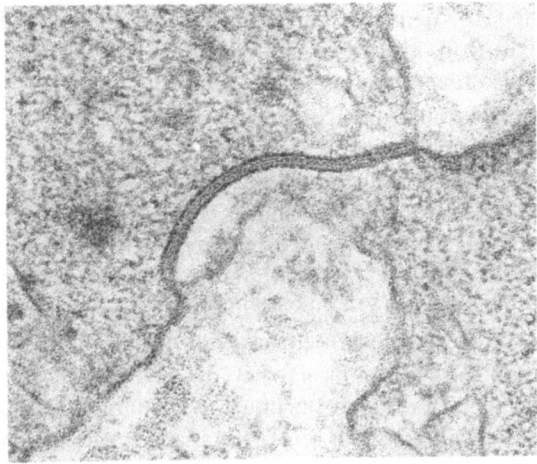

Fig. 1. High magnification electron micrograph showing gap junction between two myometrial cells (× 146000)

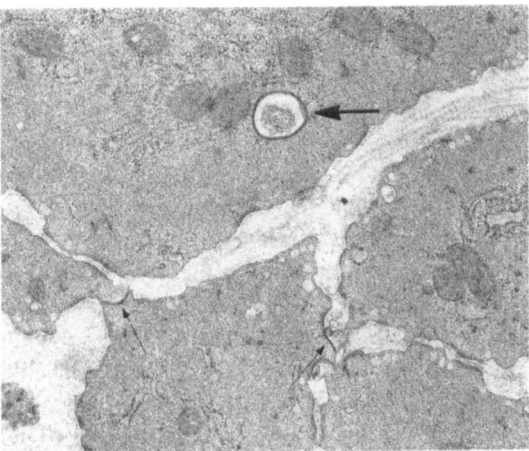

Fig. 2. Intermediate magnification electron micrograph demonstrating numerous gap junctions (*small arrows*) between myometrial cells from rat uterus during delivery. Note internalized gap junction in one of the muscle cells (*large arrow*) (× 24500)

been prepared. In the myometrium a 43-kDa protein, termed connexin 43, is thought to be the major component of the gap junction. Connexin 43 is also found in other tissues, including between muscle cells of the heart, where it is thought to be required for synchronizing cardiac contractility (Beyer et al. 1989).

In all species studied, the onset and progression of labor contractile activity during term or preterm labor is invariably associated with the presence of large numbers of gap junctions between the myometrial cells (Garfield et al. 1977, 1978). Moreover, improved electrical (Sims et al. 1982) and metabolic (Cole et al. 1985) communication between uterine smooth muscle cells is associated with the formation of junctions, supporting the hypothesis that the gap junctions permit the myometrium to behave as a functional syncytium during parturition. These alterations in the extent of structural and functional coupling are significant, therefore, in that the presence of gap junctions and cell-to-cell communication probably represent the biophysical basic for synchronous and effective uterine contractile activity during labor.

Evidence from our studies and that of others suggests the presence of specific physiological mechanisms for regulating and producing alterations in structural and functional coupling in the myometrium during pregnancy and parturition. Below we describe the possible mechanisms involved in the control of coupling in uterine smooth muscle: (1) the presence of the gap junctions and hence the extent of structural coupling; (2) the permeability of the gap junctions, and hence the extent of functional coupling in the myometrium; and (3) the degradation of the gap junctions. The integrated function of these control mechanisms presumably operates to ensure the appropriate activation and maintenance of coordinated activity in the myometrium and effective delivery of the fetus(es).

It is now established that gap junctions occupy a significant percentage of the area of the uterine smooth muscle cell plasma membrane (ca. 0.1–0.4%) only during term or preterm labor. That is, when the muscle is functionally active during pregnancy (Garfield et al. 1982, 1985). Gap junctions are consistently absent or present in low frequency and small size in nonpregnant, as well as preterm and postpartum animals (Garfield et al. 1977, 1978, 1979). In pregnant animals, the junctions begin to form about 1 day prior to the onset of labor. Gap junctions are always present in large number (ca. 1000 per cell) and in-

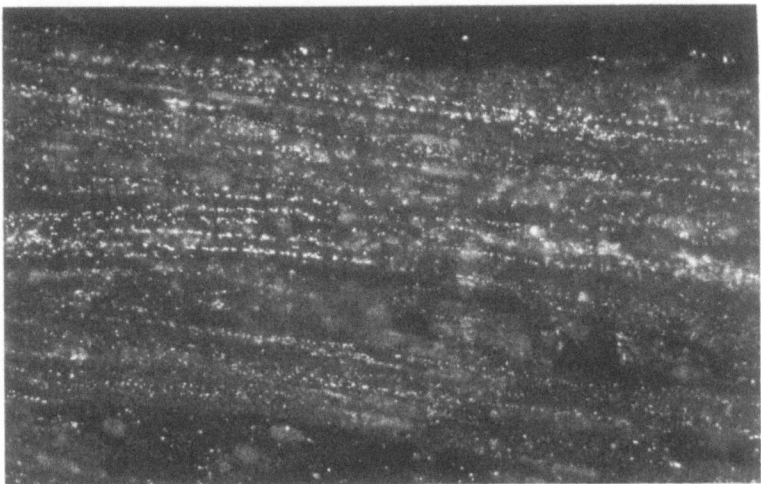

Fig. 3. Myometrial gap junctions in guinea pig uterus demonstrated with connexin 43 antibody during delivery. Each bright fluorescent spot represents a gap junction as an aggregation of connexin 43 proteins. From Chwalisz et al. 1991 (× 1000)

creased size (ca. 250 nm) during normal delivery of the fetuses, but disappear within 24 h after parturition (see Fig. 3; Garfield et al. 1977). This pattern of altered structural coupling in the myometrium is particularly prominent in rats and rabbits (Garfield et al. 1978; Puri and Garfield 1982; Demianczuk et al. 1984). Guinea pigs and sheep (Garfield et al. 1979, 1982) appear to differ slightly in that they demonstrate higher numbers of gap junctions prior to term. The gap junction profile in women during pregnancy and normal, spontaneous vaginal delivery is not known, but junctions are present in greater numbers in tissues from women undergoing cesarean section and in labor compared with those not in labor (Garfield and Hayashi 1981). It is also significant that gap junctions are invariably present in myometrial tissues from animals undergoing premature labor either as a result of experimental manipulation or pathology (Garfield et al. 1978, 1980a,b, 1982; MacKenzie and Garfield 1985a,b, 1986). Moreover, if the development of gap junctions is delayed, then pregnancy is prolonged (Garfield et al. 1978). Thus, myometrial gap junctions are dynamic and transitory structures whose presence is clearly associated with the conversion of the uterus into an active organ just prior to parturition. There is no known exception to this phenomenon and for this reason gap junctions appear to be necessary for effective labor.

1.2.3.1 Electrical Coupling

Studies to define the electrical role of the gap junctions between myometrial cells are not easily accomplished because of the small sizes of cells and their complex arrangement. Direct measurement of electronic spread of current is difficult because injected current rapidly dissipates in all directions. However, the following indirect studies indicate that myometrial cells are better coupled electrically during labor at term or preterm when gap junctions are elevated:

1. Correlation of gap junctions, labor and electrical/contractile activity (Garfield et al. 1988; Demianczuk et al. 1984; Verhoeff et al. 1985; Miller et al. 1989)
2. Increased space constant (Sims et al. 1982)
3. Decreased impedance (Sims et al. 1982)
4. Increased propagation distance of spontaneous and evoked potentials (Miller et al. 1989)
5. Increased conduction velocity (Miller et al. 1989)
6. Increased synchrony of burst of action potentials (Miller et al. 1989)
7. Decrease in input resistance (Blennerhassett and Garfield 1991; Sakai et al. 1992)
8. Increased responsiveness to contractile agonists (Garfield and Beier 1969; Chwalisz et al. 1991)

Moreover, the above studies show that when the junctions are decreased or closed coupling is similarly reduced. These multiple correlation studies provide strong support for the proposal that gap junctions control electrical coupling between myometrial cells.

1.2.3.2 Metabolic Coupling

We have also evaluated coupling by comparing diffusion of radioactive and fluorescent dyes. Because the probes to study this aspect are considerably larger than electrolytes, measurement of diffusion between cells is thought to represent metabolic cooperation between cells. Our studies indicate that when the gap junctions are elevated during term or preterm labor there are dramatic increases in:

1. Diffusion of deoxyglucose between myometrial cells (Cole et al. 1985; Cole and Garfield 1986)
2. Diffusion of Lucifer yellow and carboxyfluorescein (Blennerhassett and Garfield 1991)

Also, studies of cultured myometrial cells show that when gap junctions are present, fluorescent dyes can rapidly diffuse between cells (Sakai et al. 1992). The same studies demonstrate that when the gap junctions are closed, diffusion between cells is inhibited.

From the above studies we conclude that the myometrium is better coupled electrically and metabolically when the gap junctions are elevated and that these contribute to the ability of the uterus to contract synchronously during labor.

1.2.4 Gap Junctions as Sites of Hormonal Control of Labor

1.2.4.1 Hormones Control the Presence of Myometrial Gap Junctions

Our studies and those of others (see Garfield et al. 1988) suggest that the changes in steroid hormones and prostaglandins that precede or accompany labor, long recognized as controlling labor (Csapo 1981), regulate the presence of the gap junctions between the myometrial cells.

The changes in the steroid hormones, as reflected in tissue and plasma levels, that occur before labor are thought to be responsible for activating the synthesis of the myometrial gap junctions through genomic mechanisms (Garfield et al. 1978; 1980a,b; Fig. 4). Evidence that progesterone suppresses the junctions is suggested from the following observations: (1) progesterone normally declines in rats and

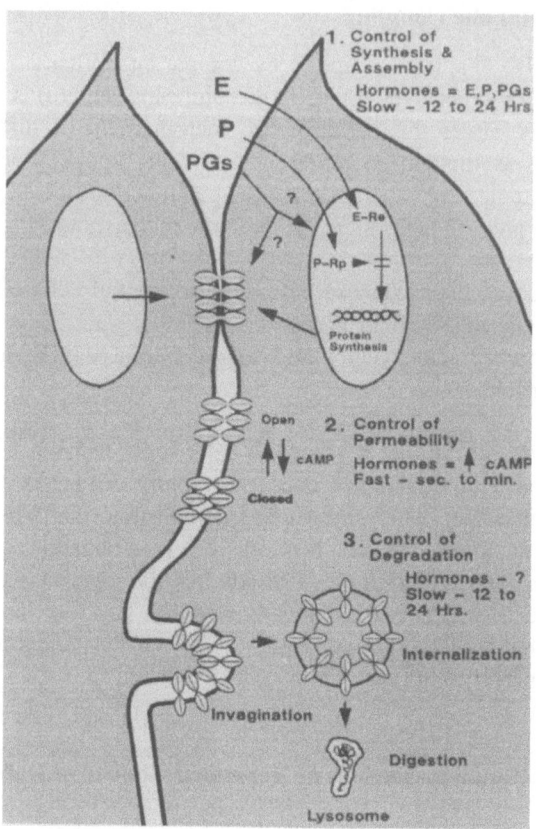

Fig. 4. Two myometrial cells showing sites for the control of the synthesis, permeability, and degradation of the junctions. Estrogen (*E*), progesterone (*P*), and the prostaglandins (*PGs*) regulate site 1. An increase in cAMP closes the junctions (site 2). Degradation of the junctions involves invagination, internalization, and digestion (site 3). *Re,*estrogen receptor; *Rp,*progesterone receptor

rabbits (Csapo 1981) at term prior to the development of gap junctions and labor. If progesterone levels are maintained by injections of the hormone, animals do not develop junctions and they do not go into labor (Garfield et al. 1980a,b). (2) Ovariectomy leads to premature progesterone withdrawal, appearance of junctions and labor. Progesterone treatment after ovariectomy prevents all three effects (Garfield et al. 1982). (3) Progesterone will inhibit the estrogen induced development of the junctions in nonpregnant animals (MacKenzie and Garfield 1985a,b). (4) Antiprogesterone compounds induce preterm development of the junctions and progesterone receptor agonists with high affinity will prevent this effect (Garfield et al. 1987). (5) Progesterone will suppress gap junction development in vitro, providing estrogen is present (Garfield et al. 1980a,b).

Estrogens promote the synthesis of gap junctions in the myometrium and in other target tissues which possess steroid receptors (MacKenzie and Garfield 1985a,b). That estrogens stimulate the formation of gap junctions is provided by the following observations: (1) Estrogens rise prior to labor (Csapo 1981) and injections of estrogens into pregnant animals stimulates the premature appearance of gap junctions and labor (MacKenzie and Garfield 1985a,b). (2) Injections of estrogens into immature, mature and ovariectomized rats induces gap junctions (MacKenzie and Garfield 1985a,b). (3) Estrogens promote gap junctions in tissues incubated in vitro (Garfield et al. 1980a,b). (4) Antiestrogen compounds such as tamoxifen prevent the estrogens from stimulating the increase in gap junctions (MacKenzie and Garfield 1985a,b). (5) mRNA extracted from estrogen treated rats induces the formation of gap junctions in cells which normally lack gap junctions (Dahl et al. 1980). We contend that estrogen may stimulate the synthesis of gap junctions by interacting with its receptor and stimulating the specific genome responsible for coding for the gap junction protein (Fig. 4).

Prostaglandin synthesis inhibitors such as indomethacin and meclofenamate alter the area of gap junctions in the myometrium indicating that the prostaglandins and/or leukotrienes are also in some way involved in the control of the junctions. The manner in which the metabolites of arachidonic acid influence the junctions must be complex since in some conditions prostaglandin inhibition reduces gap junctions (Garfield et al. 1980a,b) and in other conditions (MacKenzie and Garfield 1985a,b) it stimulates their presence.

1.2.4.2 Hormones Control the Permeability of Myometrial Gap Junctions

Changes in junctional permeability may be obtained through alterations in the gap junctional connexons in a variety of cells, such as an all-or-none closure or dilation of the cell-to-cell channel, leading to a state of either decreased or enhanced coupling, respectively. Modulation of junctional permeability by hormones or neurotransmitters have been described (Peracchia 1980). That the permeability of the junctions and, therefore, the extent of functional coupling in the myometrium may be regulated by endogenous mechanisms has been demonstrated (see below). Improved functional coupling would be expected to promote greater electrical and contractile synchrony in the uterine wall and lead to an enhanced rate of intrauterine pressure development and more effective labor. Alternatively, gap junction closure would reduce synchrony and lead to ineffective labor and a prolongation of pregnancy. We have used the 2-deoxyglucose (2DG) diffusion technique to study the influence of several agents on coupling in the myometrium of parturient rats (Cole et al. 1985; Cole and Garfield 1986). Our studies suggest that the permeability of the junctions in the parturient myometrium is decreased by intracellular Ca^{2+}, pH and cAMP. Elevated cAMP produced by treatment with dibutyryl- or 8-bromo-cAMP reduces cell–cell diffusion of 2DG in the myometrium, and this can be mimicked by inhibiting phosphodiesterase activity with theophylline or by stimulating adenylate cyclase with forskolin. That cAMP may play a role in regulating functional coupling in the myometrium is significant in that relaxin, prostacyclin (PI_2) and B_2-adrenoceptor agonists appear to influence labor and parturition and exert inhibitory effects on the myometrium by elevating intracellular cAMP. Moreover, in experiments with porcine relaxin, carbacyclin (a stable PGI_2 analog) and isoproterenol (a nonspecific β-adrenoceptor agonist) the diffusivity of 2DG was found to be significantly lower in treated than in control tissues. These data indicate that there may be specific receptor and secondary messenger-mediated physiological mechanisms for controlling cell–cell communication in the myometrium independent of the systems controlling structural coupling (Fig. 4).

A cAMP-mediated uncoupling mechanism may be involved in maintaining pregnancy in instances of premature junction formation and/or in species such as the guinea pig, sheep and, possibly, human, in which low, but significant numbers of gap junctions are present throughout pregnancy (see above). Perhaps the high levels of relaxin and prostacyclin observed in preterm pregnant animals act to elevate intracellular cAMP and, alternatively, prevent synchronous activity in the myometrium. A decline in these hormones or their receptors or an antagonism of their action by oxytocin or stimulatory prostaglandins (e.g., PGF2 or PGE2) may facilitate a shift to patent gap junction channels and the development of syncytial behavior. Additional studies are needed to fully understand the action of various substances on the permeability of the junctions.

1.2.4.3 Control of the Decline of Myometrial Gap Junctions Following Parturition

It should be noted that many of our studies have implied that changes in the synthesis of the gap junctions are the important determinants in regulating the presence of the junctions in the myometrium. However, the destruction or degradation of the junctions may also play a significant role in controlling their numbers and sizes. Gap junctions rapidly decline from the muscle cell surface following delivery. We have suggested that immediately postpartum the signal for synthesis may be withdrawn (i.e., decrease in estrogen and prostaglandins) and junction degradation may proceed by the formation of annular contacts followed by internalization and endocytosis (Garfield et al. 1980a,b; Figs. 2,4). It remains to be demonstrated what substances, if any, are involved in stimulating or inhibiting this process. This is clearly an important aspect as an inhibition of the breakdown of gap junctions may have the same effect as stimulating their formation, and a stimulation of degradation should have the opposite effect. The system of degradation is not easy to quantify with morphometric methods as there are so few internalized junctions.

1.2.4.4 Gap Junctions as Sites for the Coordination of Action of Agents Which Either Stimulate or Inhibit Labor

Agents that stimulate or inhibit the myometrium to contract or relax, respectively, can influence the muscle cells in a number of different

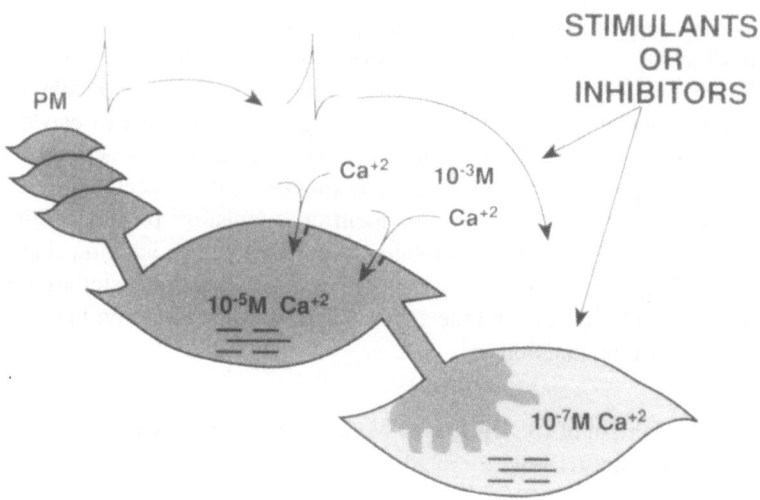

Fig. 5. A group of myometrial cells interconnected by gap junctions. As potentials propagate from pacemaker regions *(PM)* they initiate the influx of Ca^{2+} from the extracellular space to produce a contraction. Action of agents which stimulate or inhibit myometrial contractility are superimposed upon the electrical events to either augment or suppress them, respectively

ways that may or may not directly affect the gap junctions. However, these agents do not work independently of the gap junctions and the propagation of action potentials. As noted above, the influx and efflux of Ca^{2+} ions in the muscle cells during the passage of the action potentials and the opening of voltage-dependent Ca^{2+} channels are responsible for the cyclical increases and decreases in tension. This mechanism can be modified by endogenous or exogenous hormones, neurotransmitters, and other agonists and antagonists to either increase or decrease the excitability of the muscle and thereby raise or lower the ability of action potentials to propagate. Thus, the effects of excitatory and inhibitory agents are superimposed upon the driving force that is supplied by the propagation of action potentials (Fig. 5). Examples of the role of gap junctions in the action of stimulants and inhibitors are illustrated below.

The uterus is thought to become increasingly sensitive and reactive to contractile agonists such as oxytocin prior to or during parturition

(Csapo 1981; Soloff et al. 1979). The concentration of oxytocin receptors in the myometrium is controlled in the same way as the gap junctions by steroid hormones, i.e., stimulated by estrogens and inhibited by progesterone. However, oxytocin does not by itself produce contractions independently of the underlying electrical activity. Oxytocin increases the frequency and the force of spontaneous contractions of the uterus. This effect has been correlated with an increase in Ca^{2+} influx by (1) suppression of -ATPase, thus inhibiting Ca^{2+} extrusion, (2) opening Ca^{2+} channels and (3) stimulating IP3 and thus releasing internally stored Ca^{2+} ions (Marshall 1962). Oxytocin acts by increasing the Ca^{2+} content of myometrial cells and thereby depolarizes the cells bringing the resting membrane potential closer to threshold. This effect will increase the action-potential propagation in the presence of gap junctions. Thus, oxytocin action is superimposed upon the underlying electrical activity and is dependent upon gap junctions.

Carbachol, another stimulant, depolarizes myometrial cells similarly to oxytocin (Marshall 1962). Eicosanoids may also operate through the same mechanisms but they have varied effects, some producing relaxation and others contraction of the myometrium, and their effects may be immediate or delayed. Their immediate action may be similar to that of oxytocin, while the delayed effects could be mediated through the stimulation of gap junction synthesis (Garfield et al. 1980a,b), as prostaglandins seem to be involved in the formation of gap junctions (see above).

Gap junctions are also involved in regulation of relaxation (inactivity) of the myometrium. As mentioned above, Ca^{2+} ions are the key to contraction–relaxation. The efflux of Ca^{2+} and consequent lowering of intracellular Ca^{2+} to promote relaxation is regulated by several mechanisms: extrusion of Ca^{2+} by the Ca^{2+} pump, an Na^+/Ca^{2+} exchange system, and uptake into the sarcoplasmic reticulum. Many substances can indirectly affect these systems in a number of ways but like contractile agonists they do not function independently of gap junctions and the propagation of action potentials.

Lack of conducted action potentials and inactivity of the myometrium are promoted by decreases in (1) pacemaker activity, (2) excitability of the muscle cells, and (3) cell-to-cell coupling. Any agent that suppresses the generation of action potentials, closes gap junctions, or decreases their number inhibits the active events and thereby affects contractility. These contractile antagonists may act to prevent activity not by causing relaxation of the myometrium but by inhibition of the propagation of action potentials. Generally, β-noradrenergic agonists act by increasing Ca^{2+} efflux and K^+ conductance, resulting in a hyperpolarization of the myometrium and thus taking the resting membrane potential further from the threshold for excitation. Their net action is thus to inhibit the ability of action potential propagation throughout the tissue. However, the most important action of β-agonists and similar substances which increase cAMP levels is probably to close myometrial gap junctions to prevent propagation and not to change the membrane potential or the efflux of Ca^{2+}. β-Agonists have been shown to close gap junctions, thereby decreasing the ability of action potentials to propagate (see above). These agents therefore promote the inactivity of the myometrium rather than causing direct relaxation. These conclusions are also based on the observations that β-agonists at low concentrations cause a cessation of spike activity without an appreciable hyperpolarization (Kao 1989). Thus, the primary mechanism of action of β-agonists may be to decrease junctional permeability. Since the action of agonists and antagonists, either inhibitory (hyperpolarizing the myometrium) or stimulatory (depolarizing the muscle), are dependent upon the status of gap junctions and the presence of the junctions, and their functional state will dramatically influence the contractile patterns the agent produces.

1.2.4.5 Summary of Gap Junction Studies

I have briefly attempted to review the evidence that gap junctions play a necessary role in the gradual evolution of uterine contractility during labor. Our studies suggest that the synthesis, function, degradation, and permeability of myometrial junctions are physiologically regulated by hormones (Fig. 4). Thus, there are several sites for the control of the junctions and the management of term or preterm labor.

1.3 Nitric Oxide Inhibition of Myometrial Contractility

One of the most exciting and significant recent advances in biology and medicine is the discovery that nitric oxide is involved in controlling relaxation of various smooth muscles, including vascular (Furch-gott and Vanhoutte 1989; Ignarro and Kodowitz 1985; Moncado et al. 1991), intestinal (Sanders and Ward 1992), tracheal (Li and Rand 1991), and corpus cavernous muscle (Pickard et al. 1991). However, there are no studies that indicate that nitric oxide might regulate uterine contractility except that nitroglycerin and sodium nitroprusside have been shown to inhibit contractions (Diamond 1983) and it is now recognized that these two compounds are nitric oxide donors.

We recently examined the possibility that nitric oxide might be one of the factors that mediate uterine relaxation during pregnancy (Yallampalli et al. 1993). We tested the effects of L-arginine, the substrate for nitric oxide, on uterine contractility of strips of tissues from pregnant rats in vitro. L-arginine caused a rapid and substantial relaxation of spontaneous activity of the uterine strips from rats at mid to near term gestation (Fig. 6). The relaxation effects were reversed by l-nitroarginine methyl ester (L-NAME), an inhibitor of nitric oxide synthase (Fig. 6). Sodium nitroprusside, a nitric oxide donor, completely abolished spontaneous contraction (Fig. 6). Methylene blue, an inhibitor of guanylate cyclase, also prevented the inhibitory effects of L-arginine. These results strongly support the existence of an L-arginine-nitric oxide–cGMP system for regulating uterine relaxation.

When we examined the effects of L-arginine on tissues from delivering animals we were surprised to find that L-arginine had little effect. Figure 7 shows dose responses (dose of L-arginine versus duration of inhibition) of L-arginine of tissues at day 18 of pregnancy and during delivery. These studies show that the inhibitory action of the L-arginine was considerably lower during delivery and may indicate that nitric oxide may contribute to the maintenance of uterine quiescence during pregnancy but not during delivery.

Since responses to L-arginine were lower at delivery than preterm we reasoned that perhaps the generation of nitric oxide by nitric oxide synthase is lower during delivery, that guanylate cyclase produces less nitric oxide during delivery, or that cGMP is less effective during delivery. When we tested the ability of 8-bromo cGMP to inhibit uterine contractions, we found that during delivery at term or preterm the responses were greatly attenuated versus preterm nondelivery (Fig. 8). These data clearly indicate that the effector system of cGMP to produce relaxation is greatly reduced during term and preterm delivery.

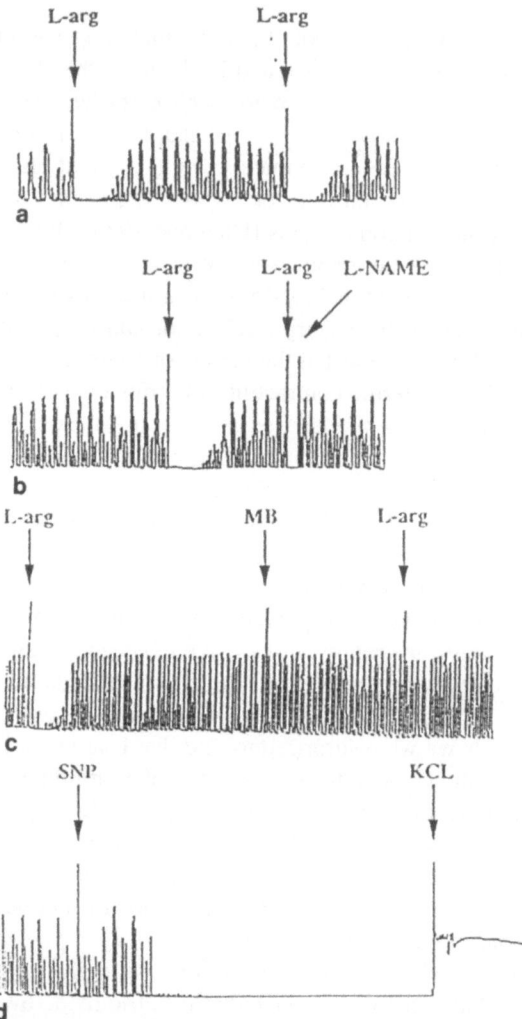

Fig. 6a–d. The effect of L-arginine, L-NAME, methylene blue and sodium nitroprusside on spontaneously contracting uterine strips from rat uterus obtained on day 18 of pregnancy. Application of L-arginine to a muscle bath caused immediate relaxation (10–15 min duration) (**a**). The effect of L-arginine was antagonized by L-NAME when added during an L-arginine-induced relaxation (**b**). Inhibition of L-arginine with methylene blue (**c**). Effects of sodium nitroprusside (**d**). These are typical recordings and each upstroke from baseline represents a contraction

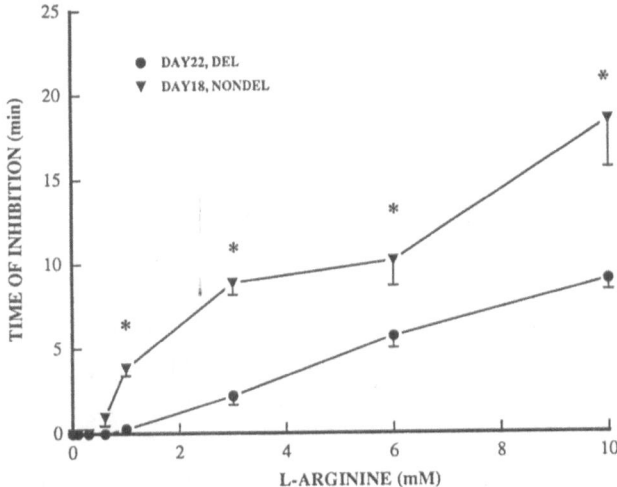

Fig. 7. Dose-dependent relaxation effects of L-arginine (0.1 mM–10 mM) on spontaneously contracting uterine strips from rats at day 18 of gestation and during delivery. The duration of inhibition of spontaneous uterine contractions are dose-dependent. The effects of L-arginine from concentrations of 1 mM are significantly ($p < 0.01$) decreased during spontaneous delivery compared to day 18. Each *data point* represents mean ± S.E.M. The total number of strips studied at each time period was 8–16 from 4–6 animals per group

The fact that the antiprogesterone compound (ZK299, Schering AG) produced similar changes to those at term show that the decrease in sensitivity to cGMP may be controlled by progesterone (Fig. 8). This finding has broader implications, in that other substances which inhibit uterine contractility may act through cGMP and be expected to decline during labor.

We treated intact pregnant rats at day 17 through to term with L-NAME and sodium nitroprusside to test the possibility that these compounds might respectively either initiate preterm labor or prevent the onset of spontaneous labor. We found that neither compound altered the timing of parturition: all animals delivered spontaneously at term along with controls. However, what we noted in the L-NAME-treated animals was really remarkable in that all animals developed severe hypertension and classic symptoms of preeclampsia (Yallampalli and

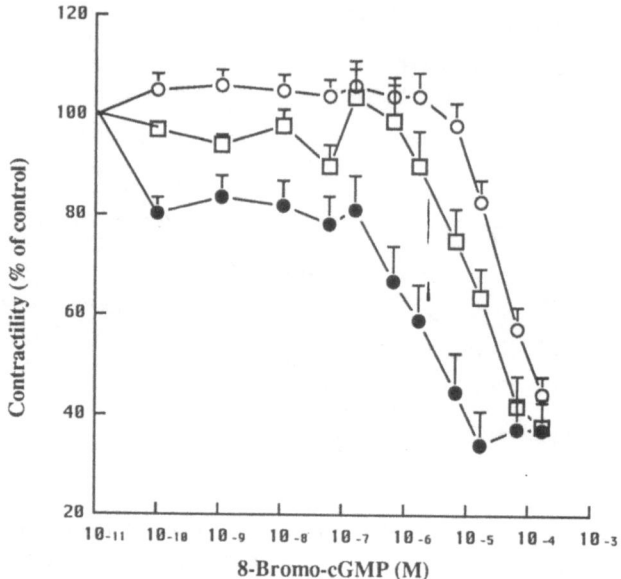

Fig. 8. Effects of 8-bromo-cGMP on uterine strips from rats at day 18 pregnancy (*closed circles*), day 18 pregnancy during preterm birth following antiprogesterone treatment (ZK299, Schering AG) (*open squares*), and during spontaneous birth at term (*open circles*). Note that during both term and preterm labor the dose–response relationship to cGMP is shifted to a less sensitive state

Garfield 1993). These included growth-retarded fetuses and proteinuria. These studies might demonstrate that nitric oxide synthesis is involved in the generation of preeclampsia, and the inhibition of nitric oxide synthesis in pregnant animals might serve as a useful model for further studies of the treatment of preeclampsia.

The fact that neither nitric oxide synthase inhibitors nor donors initiated early labor or prevented term labor suggests that nitric oxide may not be essential for the maintenance of pregnancy and its withdrawal may not initiate labor. However, nitric oxide may still play a significant role in uterine quiescence and nitric oxide donors might be effective in suppressing uterine contractility. We suggest that nitric oxide inhibits uterine contractility prior to conversion of the myometrium to an active and reactive muscle to produce effective labor contractions

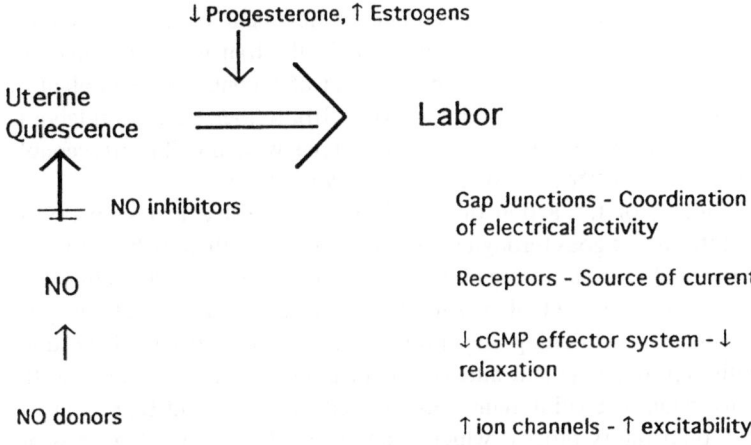

Fig. 9. Model of the inhibition of uterine contractility with nitric oxide (*NO*)

(Fig. 9). Labor is achieved when the inactive muscle is activated by changes in steroid hormones and the addition of gap junctions for electrical coordination (see above), receptors for generation of currents, and increased ion channels for excitability, and a decrease in sensitivity to cGMP which inhibits the muscle.

Nitric oxide donors may be ineffective in preventing the initiation of labor because they act at a step prior to conversion to the active stage and because of a decrease in the cGMP relaxation system (Fig. 9). On the other hand, nitric oxide synthase inhibitors may increase spontaneous contractions but they will not initiate labor because they too act prior to the conversion step. Nitric oxide would not be expected to inhibit labor because there is a decrease in the responsiveness of the cGMP effector system.

1.4 Model for Control of Labor

All previous models of events which initiate or control labor seem to be inappropriate. They neither define obligatory steps nor do they adequately explain what role contractile agonists or inhibitors play in the overall scheme of labor. Also, no model suitably describes the reversibility of steps, which is important in treatment of labor problems. What is clear is that labor is not simply the transition from inactive muscle to active muscle by the addition of a contractile stimulant or the withdrawal of a tonic inhibitor (Fig. 10). Rather, successful labor is achieved through a series of stages, some with possible irreversible steps and some stages acting dependently of others.

Studies of the action of progesterone suggested to Csapo (1981) that the uterus goes through a conversion process, thereby becoming an active and responsive organ. Similarly, investigations with antiprogesterones by Elger et al. (1990) have indicated that the uterus goes through a conditioning step prior to labor. Work that we have done with gap junctions and nitric oxide (see above) demonstrates that the conversion or conditioning step outlined by Csapo and Elger consists of a preparatory process which is an essential stage of labor, possibly irreversible and independent of other stages (Fig. 11). The preparatory process (step 1 or the primary stage) is followed by secondary stages (step 2) which are dependent upon the preparatory process, are reversible, and comprised of at least three stages. These stages include a direct route from the preparatory process to successful labor or to steps involving either stimulation or inhibition and release from inhibition to produce successful labor contractions.

We propose that the preparatory process is controlled by genomic mechanisms and involves the increase or decrease of specific proteins that control the force and frequency of phasic contractions of labor and receptors for stimulation or inhibition. The changes in proteins as part of the preparatory process consist of an increase in gap junctions (see above), ion channels (Mironneau 1990), ion pumps (see A.K. Grover this volume; Khan et al. 1991), systems for myofilament interaction, IP_3 transduction mechanisms (Harbon et al. 1990), and receptors for oxytocin (Soloff et al. 1979; Garfield and Beier 1989), stimulatory prostaglandins, and α receptors for norepinephrine (Kawarabayashi and Osa 1976). While a decrease in proteins which favor relaxation include the cGMP effector system (see above) and a decline in inhibitory receptors.

The available evidence indicates that the preparatory process is achieved in most animals with the withdrawal of progesterone and the rise in estrogens which accompanies or precedes the induction of new proteins or the decrease in others. This process is stimulated by es-

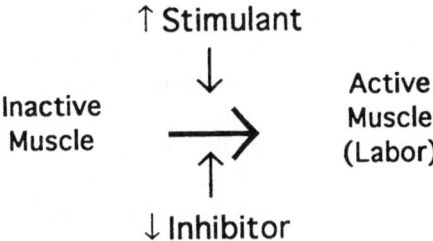

Fig. 10. Model of what labor is not!

trogens or increases in estrogen/progesterone ratios as estrogens probably stimulate the synthesis of most of the important proteins. In humans the preparatory process may occur more slowly and involve stretch, prostaglandins, IL-1, oxytocin, fetal ischemia, and other substances. In other words, there may be several pathways that lead to the preparatory process in humans or other animals.

It is possible that once the preparatory process is complete that successful labor proceeds without any additional assistance or any other

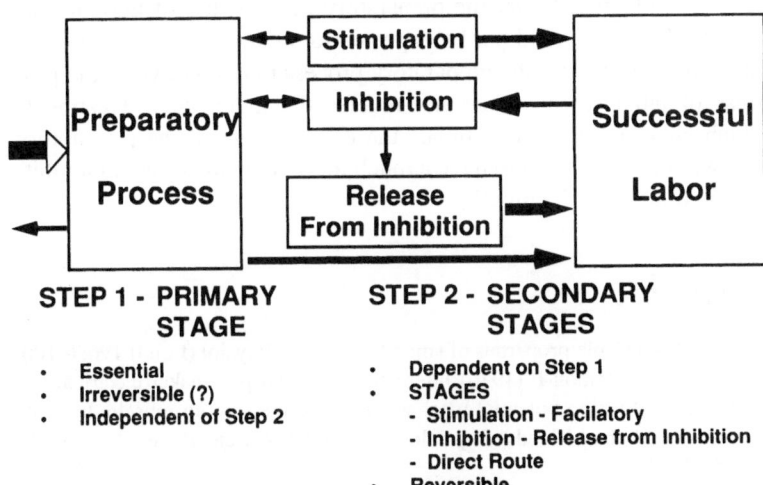

Fig. 11. Model of labor. Shown are various steps and stages in the conversion of the myometrium during labor. See text for details

inhibitory phase (the direct route, Fig. 11). However, it is at this point (stimulatory stage) that oxytocin and other agonists (prostaglandins, endothelins, etc.) stimulate contractility to facilitate labor. We believe this process is nonspecific and that successful contractions of labor can be achieved through the use of almost any agonist. This step is also reversible by the application of antagonists or the withdrawal of the stimulant. But antagonists or withdrawl of stimulates will not prevent labor because they will not reverse the steps back to conditions existing prior to the preparatory process.

The inhibitory step following the preparatory process is similar to the stimulatory step and may involve nonspecific agents that increase cAMP, close gap junctions, increase calcium transport or decrease membrane potential. In addition, this process is reversible by antagonists which may lead to withdrawal of inhibition and labor or to the stimulatory stage.

Although our model (Fig. 11) is complex, we believe that it represents the complicated steps and stages of the conversion of the uterus at the end of pregnancy and probably is representative of events which occur during preterm labor. The model has many implications for the pharmacological management of labor, either stimulation or inhibition. Since stimulants act after the preparatory process, use of these agents prior to this step would not be expected to be effective in initiation of labor. Inhibition after the preparatory process may not reverse the progress of labor since it does not reverse the major step. Treatments which either prevent or stimulate the development of the preparatory step would be expected to have a much more dramatic effect in inhibiting or stimulating the muscle.

References

Abe Y (1968) Cable properties of smooth muscle. J Physiol (Lond) 196:7–100
Anderson NC, Ramon F (1976) Interaction between pacemaker electrical behavior and action potential mechanism in uterine smooth muscle. In: Bulbring E, Shuba MF (eds) Physiology of smooth muscle. Raven, New York, pp 53–63
Bengtsson B, Chow EMH, Marshall JM (1981) Calcium dependency of pregnant rat myometrium: comparison of circular and longitudinal muscle. Biol Reprod 30:869–878

Beyer EC et al. (1989) Antisera directed against connexin 43 peptides react with a 43-dK protein localized to gap junctions in myocardium and other tissues. J Cell Biol 108:595–605

Blennerhassett MG, Garfield RE (1991) Effect of gap junction number and permeability on intercellular couplin in rat myometrium. Am J Physiol 261 (Cell Physiol):C1001–C1009

Bozler E (1938) Electrical stimulation and conduction of excitation in smooth muscle. Am J Physiol 122:616–623

Chwalisz K, Fahrenholz F, Hackenberg M, Garfield RE, Elger W (1991) The progesterone antagonist onapristone increases the effectiveness of oxytocin to produce delivery without changing the myometrial oxytocin receptor concentrations. Am J Obstet Gynecol 165:1760–1770

Cole WC, Garfield RE (1986) Evidence for physiological regulation of gap junction permeability. Am J Physiol 251:C411–C420

Cole WC, Garfield RE, Kirkaldy JS (1985) Gap junctions and direct intercellular communication between rat uterine smooth muscle cells. Am J Physiol 249:C20-C31

Csapo AI. Force of labour (1981) In: Principles and practice of obstetrics and perinatology. Iffy L, Kamientzky HA (eds) John Wiley and Sons, New York, pp 761–799

Dahl G, Azarnia R, Werner R (1980) De novo construction of cell-to-cell channels. In Vitro 16 :1068–1075

Demianczuk N, Towell ME, Garfield RE (1984) Myometrial electrophysiological activity and gap junctions in the pregnant rabbit. Am J Obstet Gynecol 149:485–491

Diamond J (1983) Lack of correlation between cyclic GMP elevation and relaxation of nonvascular smooth muscle by nitroglycerin, nitroprosside, hydroxylamine and sodium azide. J Pharm Exp Ther 225:422–426

Elger W, Chwalisz K, Fähnrich M, Hasan SH, Laurent D, Beier S, Ottow E, Neef G, Garfield RE (1990) Studies on labor conditioning and labor inducing effects of antiprogesterones in animal models. In: Garfield RE (ed) Uterine contractility: mechanisms of control. Serono Symposium, USA, pp 153–176

Finn CA, Porter DG (1975) The Uterus. Elek Science, London

Furchgott RF, Vanhoutte PM (1989) Endothelium derived relaxing and contracting factor. FASEB J 3:2007–2018

Garfield RE (1985) Cell-to-cell communication in smooth muscle. In: Grover AK, Daniel EE (eds) Calcium and contractility, chap 4. Human Press, Clifton, pp 143–173

Garfield RE (1988) Structural and functional studies of the control of myometrial contractility and labor. In: The onset of labor: Cellular and integrative mechanisms in the onset of labor, McNellis D, Challis J, MacDonald P, Nathanielsz P, Roberts J (eds), Perinatology Press, Ithaca, NY

Garfield RE, Beier S (1989) Increased myometrial responsiveness to oxytocin during term and preterm labor. Am J Obstet Gynecol 161:454–461

Garfield RE, Hayashi RH (1981). Appearance of gap junctions in the myometrium of women during labour. Am J Obstet Gynecol 140:254–260

Garfield RE, Sims S, Daniel EE (1977) Gap junctions: their presence and necessity in myometrium during gestation. Science 198:958–960

Garfield RE, Sims SM, Kannan MS, Daniel EE (1978) Possible role of gap junctions in activation of myometrium during parturition. Am J Physiol 235:C168–C179

Garfield RE, Rabideau S, Challis JRG, Daniel EE (1979) Hormonal control of gap junction formation in sheep myometrium during parturition. Biol Reprod 21:999–1007

Garfield RE, Kannan MS, Daniel EE (1980a) Gap junction formation in myometrium: control by estrogens, progesterone and prostaglandins. Am J Physiol 238:C81–C89 (Cell Physiology 7)

Garfield RE, Merrett D, Grover AK (1980b) Gap function formation and regulation in myometrium. Am J Physiol 239:C217–C228 (Cell Physiology 8)

Garfield RE, Puri CP, Csapo AI (1982) Endocrine structural and functional changes in the uterus during premature labor. Am J Obstet Gynecol 142:21–27

Garfield RE, Gasc JM, Baulieu EE (1987) Effects of the antiprogesterone RU 486 on preterm birth in the rat. Am J Obstet Gynecol 157:1281–1285

Garfield RE, Blennerhassett MG, Miller SM (1988) Control of myometrial contractility: role and regulation of gap junctions. Oxf Rev Reprod Biol 10:436–490

Grover AK (1986) Role of cellular membranes in calcium-mobilization of uterine smooth muscle. The physiology and biochemistry of the uterus in pregnancy and labor. CRC Press, Huszar G (ed): pp 93–107

Harbon S et al (1990) Multiple regulation of the generation of inositol phosphates and cAMP in myometrium. Garfield ER (ed) Serono Symposia, Uterine Contractility . pp 123–140

Ignarro LJ, Kadowitz PJ (1985) The pharmacological and physiological role of cyclic GMP in vascular smooth muscle relaxation. Ann Rev Pharm Toxicol 25:171–191

Kanda S, Kuriyama H (1980) Specific features of smooth muscle cells recorded from the placental region of the myometrium of pregnant rats. J Physiol 299:127–144

Kao CY (1989) Electrophysiological properties of uterine smooth muscle. In: Wynn RM, Jollie WP (eds) Biology of the uterus. 2nd edn. New York, Plenum Press:pp 403–54

Kawarabayashi T, Osa T(1976) Comparative investigations of alpha-and beta-effects on the longitudinal and circular muscles of the pregnant rat myometrium. Jpn J Physiol 26:403–416

Khan I, Tabb T, Garfield RE, Grover AK (1991) Changes in mRNA for Ca pumps in pregnant rat uterus. Biochem Int 27:189–196

Kuriyama H (1961) Recent studies of the electrophysiology of the uterus. In: Progesterone and the defense mechanism of pregnancy. Little Brown, Boston, Ciba Fndn Study Group Vol 9, pp 51–70.

Kuriyama H, Suzuki H (1976) Changes in electrical properties of rat myometrium during gestation and following hormonal treatments. J Physiol 260:315–333

Li CG, Rand MJ (1991) Evidence that part of the NANC relaxant response of guinea-pig trachea to electrical field stimulation is mediated by nitric oxide. Br J Pharmacol 102:91–94

MacKenzie LW, Garfield RE (1985a) Hormonal control of gap junctions in the myometrial gap junction: In vitro studies. In: Physiological Development of Fetus and Newborn,Jones C (ed), Academic Press, London, pp 411–416

MacKenzie LW, Garfield RE (1985b) Hormonal control of gap junctions in the myometrium. Am J Physiol 248:C296–C308

MacKenzie LW, Garfield RE (1986) Effects of estradiol 17β on myometrial gap junctions and pregnancy in the rat. Can J Physiol Pharmacol 64:462–466

Marshall JM (1962) Regulation of activity in uterine smooth muscle. Physiol Rev 42:213–227

Marshall JM (1974) Effects of neurohypophysial hormones on the myometrium. In: Handbook of Physiology, Section 7: Endocrinology Vol IV: The pituitary gland, Part I, Greep RO, Astwood EB (ed) pp 469–492. American Physiological Society, Washington

Miller SM et al (1989) Improved propagation in myometrium associated with gap junctions during parturition. Am J Physiol 256:C130–141 (Cell Physiology 25)

Mironneau J (1973) Excitation-contraction coupling in voltage-clamped uterine smooth muscle. J Physiol 233:127–141

Mironneau J (1976) Relationship between contraction and transmembrane ionic current in voltage-clamped uterine smooth muscle. In: Physiology of Smooth Muscle. Bulbring E, Shuba MF (ed). Raven, NY, pp 175–183

Mironneau J, Lalanne C, Mironneau C, Savineau JP, Lavie JL (1984) Comparison of pinaverium bromide manganese chloride and D600 effects on electrical and mechanical activities in rat uterine smooth muscle. Eur J Pharmacol 98:99–107

Mironneau J (1990) Electrical signals and uterine contractility: Ion channels and excitation-contraction coupling in myometrium. Garfield RE (ed) Serono Symposia, Uterine Contractility. pp 9–21

Moncado S. Palmer RMG, Higgs EA (1991) Nitric oxide: physiology, pathophysiology and pharmacology. Pharmacol Rev 43:109–142

Osa T, Fujino T (1978) Electrophysiological comparison between the longitudinal and circular muscles of the rat uterus during the estrous cycle and pregnancy. Jpn J Physiol 28:197–209

Peracchia C (1980) Structural correlates of gap junction permeation. Int Rev Cytol 66:81–146

Pickard RS, Powell PH, Zar MA (1991) The effect of inhibitors of nitric oxide biosynthesis and cyclic GMP formation on nerve-evoked relaxation of human covernosal smooth muscle. Br J Pharm 104:755–759

Puri CP, Garfield RE (1982) Changes in hormone levels and gap junctions in the rat uterus during pregnancy and parturition. Biol Reprod 27:967–975

Reynolds SRM (1949) Physiology of the Uterus, 2nd Edition, P.B. Hoeber Inc., NY

Sakai N, Blennerhassett MG, Garfield RE (1992) Effects of antiprogesterones on myometrial cell-to-cell coupling in pregnant guinea pigs. Biol Reprod 46:358–365

Sanders KM, Ward SM (1992) Nitric oxide as a mediator of nonadrenergic noncholinergic neurotransmission. Am J Physiol 262:G379-G392

Sims SM, Daniel EE, Garfield RE (1982) Improved electrical coupling in uterine smooth muscle is associated with increased numbers of gap junctions at parturition. J Gen Physiol 80:353–341

Soloff MS, Alexandrova M, Fernstrom MJ (1979) Oxytocin receptors: triggers for parturition and lactation? Science 204:1313–1315

Verhoeff A, Garfield RE, Ramondt J, Wallenburg H (1985) Myometrial activity related to gap junction area in periparturient and ovariectomized, estrogen-treated sheep. Acta Physiologica Hungarica 66:539–551

Yallampalli C, Garfield RE (1993) Inhibition of nitric oxide synthesis in rats during pregnancy produces symptoms identical to preeclampsia. Am J Obstet Gynecol (in press)

Yallampalli C, Izumi H, Garfield RE (1993) An L-arginine: nitric oxide system exists in the uterus and inhibits contractility during pregnancy. Am J Obstet Gynecol (in press)

2 Receptors and Signal Transduction in the Myometrium

Simone Harbon, Zahra Tanfin, Lien Do Khac, Olivier Goureau, and Denis Leiber

2.1 Introduction

In the myometrium, as in many other smooth muscle preparations, Ca^{2+} and cAMP, the two major intracellular second messengers, exert opposite effects at the level of contractility. The necessity of calcium for uterine contraction has long been recognized, the role of Ca^{2+} being obligatory, whether the stimulus is hormonal or voltage-induced. On the other hand, cAMP has been shown to contribute to uterine relaxion (Hardman 1981, Do Khac et al. 1986b, Diamond 1990). The increase in intracellular Ca^{2+} evoked by stimulatory agonists is considered to originate at least in part from intracellular stores (Van Breemen and Saida 1989; Somlyo and Himpens 1989; Mironneau et al. 1984). In this regard, the phosphoinositide-phospholipase C transducing mechanism that is consistently associated with Ca^{2+}-mobilizing receptors (Berridge and Irvine 1984; Berridge 1987) has been demonstrated to be activated by contracting agonists in different myometrial preparations (Marc et al. 1986, 1988; Anwer et al. 1989; Goureau et al. 1990). Additionally, a number of recently reported findings provide satisfactory correlations between the increased generation of inositol phosphates, the ability of inositol 1,4,5-trisphosphate, $InsP(1,4,5,)P_3$, to release Ca^{2+} from intracellular stores, and the accompanying Ca^{2+}-induced uterine contractions (Carsten and Miller 1985; Marc et al. 1988; Kanmura et al. 1988).

The purpose of the present review is to summarize some data from our studies which emphasize different mechanisms that trigger the modulation of the levels of both cAMP and inositol phosphates in the rat and guinea pig myometrium and their potential contribution to the regulation of uterine contractility. We concentrate most on the transduction of receptor occupancy to the activation (or inhibition) of adenylyl cyclase and phospholipase C pathways as well to the activation of putative ionic channels. The targets (receptors, guanine nucleotide regulatory "G proteins") of potential regulation during pregnancy are also considered. By way of background, brief descriptions of G proteins and G protein-coupled receptors are also included.

Uteri were obtained from immature female Wistar rats (4 weeks old) and Hartley guinea pigs (5 weeks old), pretreated with 30 and 160 mg of estradiol, respectively, for 2 days and used the following day. Myometrium was separated by stripping away the endometrium. For the pregnant rats (at day 12 and 21 of gestation), the uterine horns were excised longitudinally, and after removal of fetuses, their sites of attachment, and placenta tissues, myometrium was prepared free of endometrium essentially as reported (Marc et al. 1988; Goureau et al. 1990; Vesin and Harbon 1974). We have previously described the various methods utilized in this work, including incubation of myometrial strips for the estimation of cAMP (Vesin and Harbon 1974), as well as [^{3}H]inositol phosphate accumulation (Marc et al. 1988; Goureau et al. 1990) and the immunological characterization of the subunits of G proteins (Milligan et al. 1989; Tanfin et al. 1991).

2.2 Main Structural Features
of G Protein-Coupled Receptors

A wide variety of membrane receptors for hormones and neurotransmitters are coupled to guanine nucleotide binding regulatory "G" proteins which upon activation by ligand-occupied receptors stimulate or inhibit various effectors such as enzymes that synthesize cytoplasmic second messengers (adenylyl cyclase, phospholipase C), or ion channels. Several of the G protein-linked receptors have been purified and cloned. They include receptors for catecholamines and most other biogenic amines, muscarinic receptors, and receptors for eicosanoids and numerous peptide hormones. These receptors are integral membrane glycoproteins, similar in size, primary sequences, and overall structure. The best characterized mammalian G protein-coupled receptors are rhodopsin and the β-adrenergic receptor. Based on analogies with bacteriorhodopsin and computer modelling, all of the G protein-coupled receptors belong to the superfamily of proteins that span the plasma membrane seven times and have an extracellular amino terminus which is glycosylated and an intracellular carboxyl terminus.

The putative seven transmembrane α-helices are connected by three extra- and three intra-cellular hydrophilic loops. Potential sites of threonine and serine phosphorylation are present in the COOH terminal cytoplasmic domain. Sequence similarities among members of the receptors are greatest in the predicted membrane spanning regions. The main features for receptors that vary in both composition and length are the third cytoplasmic loop and the carboxy terminus tail; these may constitute part of the variable ligand specificity as well as variable G protein specificity (for reviews see Dohlman et al. 1987, 1991).

2.3 G Proteins – Molecular Structure and Function

In all eukaryotic organisms, a family of heterotrimeric GTP-binding and hydrolyzing proteins (G proteins) plays an essential role in linking many cell-surface receptors to effector proteins at the plasma membrane. G proteins are heterotrimers, composed of three distinct subunits: α (molecular mass = 39–46 kDa), β (37 kDa) and γ (8 kDa). The β- and γ-subunits exist as a tightly associated complex that functions as a unit. Although the same βγ-subunit complex can apparently be shared among different α-subunits to form the heterotrimer, the identity of the α-subunit is currently used to define an individual G protein oligomer. Each α-subunit contains a high-affinity binding site for guanine nucleotides and possesses GTPase activity that is crucial for the action of G proteins. In some cases α-subunits possess specific residues that can be covalently modified by bacterial toxins. Cholera toxin catalyzes the transfer of the adenosine diphosphate (ADP)-ribose moiety of nicotinamide-adenine dinucleotide (NAD) to a specific Arg residue in a certain α-subunit (e.g., α_s). Similarly, pertussis toxin ADP-ribosylates those α-subunits that possess a specific Cys residue near the carboxyl terminus. Modification of α-subunits by cholera toxin constitutively activates these proteins (by inhibiting their GTPase activity), whereas modification by pertussis toxin interferes with the interaction of appropriate receptors and the G protein. Together the application of molecular biology and peptide-directed antibodies has made it possible to rapidly and covincingly establish that G proteins are not just G_s, G_i and G_t (transducin), but rather constitute quite a large family of homologous proteins. There are at least four forms of

$G_s\alpha$, three forms of G_i ($G_i1\alpha$, $G_i2\alpha$ and $G_i3\alpha$), two forms of $G_o\alpha$ and two forms of $G_t\alpha$. Besides G_s and G_i which are involved in the stimulatory and inhibitory pathways of the adenylyl cyclase system, respectively, recent reports have identified members of the $G_q\alpha$ family that are not modified by the bacterial toxins and that appear to constitute the regulators of the phospholipase C pathway. Although none of the G protein subunits contains regions that might obviously associate with a lipid bilayer, the heterotrimer is bound to the plasma membrane. This is apparently due to the fact that γ–subunits are prenylated and at least some α-subunits (those of the G_i subfamily) are myristoylated. These lipid modifications serve to anchor the subunits to the membrane and they also increase the affinity of α for $\beta\gamma$ (for reviews see Birnbaumer et al. 1990; Hepler and Gilman 1992).

During the process of coupling of occupied receptors to specific effector systems, the G protein is required to undergo a cyclical pattern of activation followed by a subsequent deactivation. In the resting state, the G protein exists in the holomeric form with guanosine-5'-diphosphate (GDP) bound to the nucleotide binding site of the α-subunit. The interaction of the protein with the appropriately occupied receptor stimulates the dissociation of GDP, presumably as a result of a conformational change that leads to the "opening" of the guanine–nucleotide binding site. In the presence of cellular concentrations of GTP, the nucleotide binding site is rapidly filled. Then the holomeric G protein can dissociate into an active α-subunit with GTP bound and free γ–subunits. This active α-subunit is able to interact with an effector enzyme or ion channel. Some evidence now supports the hypothesis that γ, like α, can also interact with and modulate the activity of at least some effector proteins. Hydrolysis of the terminal phosphate of bound GTP by the intrinsic GTPase activity deactivates the α-subunit which is then able to reassociate with γ-subunits and to restore the inactive GDP-bound holo-G protein. The G protein is now available for an interaction with a new receptor molecule and for the reinitiation of the receptor and GTP driven turnover cycle. It is worth noting that fluoroaluminates (AlF$_4^-$) can interact with α-bound GDP to mimic GTP and thereby activates α, independently of receptor occupancy (Birnbaumer et al. 1990; Hepler and Gilman 1992).

2.4 Modulation of cAMP Generation in Rat and Guinea Pig Myometrium

A number of G protein-coupled receptors have been described to contribute to the regulation of intracellular cAMP in both the estrogen-
treated rat and guinea pig myometrium. Thus the adenylyl cyclase
cAMP forming system can be enhanced by way of receptor activation
(β-adrenergic, prostacyclin "PGI_2"), via the stimulatory regulatory protein G_s (Vesin and Harbon 1974; Harbon et al. 1984; Tanfin and Harbon 1987). cAMP accumulation could also be enhanced by cholera
toxin which activates G_s as well as by forskolin which interacts with
and directly activates the catalytic unit of adenylyl cyclase (Tanfin and
Harbon 1987; Mokhtari et al. 1988). The inhibitory arm of adenylyl
cyclase is also operating in the guinea pig myometrium where stimulation of muscarinic receptors (M2 subtype) is associated with a decrease
in cAMP, a pertussis toxin event, involving the contribution of G_i
(Leiber et al. 1990). It was further demonstrated that in the pregnant rat
myometrium (Tanfin and Harbon 1987; Goureau et al. 1992) prostaglandin (PG) receptors are also coupled, via G_i, to the inhibitory pathway of the cAMP generating system (see below). Since the first report
(Sutherland and Rall 1960) that β-adrenergic-induced relaxation of
smooth muscle might be related to an elevation of cAMP levels in the
tissue, the possible involvement of the cyclic nucleotide in the regulation of smooth muscle relaxation has been investigated widely and de-
bated (Hardman 1981; Diamond 1990). Many groups, including our
own, have attempted to provide direct and indirect evidence in support
of the general hypothesis of cAMP-mediated relaxation in the myometrium as well as in different smooth muscle preparations. Our data
(Vesin and Harbon 1974; Harbon et al. 1976; Do Khac et al. 1986a)
obtained with estrogen-treated rat myometrium did not usually show a
close correlation between rises in cAMP levels and the decrease in tension and myosin light chain dephosphorylation. In more recent experiments (Do Khac et al. 1986b), the role of cAMP in uterine relaxation
was reevaluated and it was proposed that relaxation induced by β-
adrenergic receptor stimulation was the result of the combined effects
of both a cAMP-dependent and a cAMP-independent process. The β-
adrenergic receptor-mediated event, independent of cAMP, was postulated to operate at the membrane level and would tend ultimately to de-

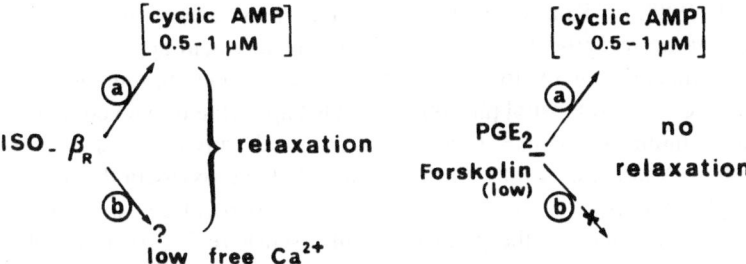

Fig. 1. Schematic representation of the postulated mechanisms involved in β-adrenergic mediated relaxation of rat myometrium. It is proposed that: (1) stimulation of β-adrenergic receptors is coupled to the activation of two distinct processes: (*a*) the adenylyl cyclase cAMP generating system and an additional process (*b*), independent of cAMP, which is presumed to operate at the membrane level and to lower intracellular Ca^{2+}. Both (*a*) and (*b*) processes are required to mediate β-adrenergic-induced relaxation. (2) PGE_2 and low doses of forskolin could not elicit relaxation: both induce rises in cAMP similar to isoproterenol (*ISO*) but are ineffective at the level of the (*b*) process. (From Do Khac et al. 1986b)

crease cytosolic Ca^{2+} (Fig. 1). The latter interpretation was substantiated by some recent observations, which will be discussed below. It is also of interest that in the guinea pig myometrium, the contractile event associated with muscarinic receptor stimulations, though triggered mainly by the M3/M5 receptor subtypes via the inositol phosphate-Ca^{2+} pathway, appeared to be modulated by a contribution of the M2 receptor subtype via its elicited decrease in cAMP (Leiber et al. 1990).

2.5 Modulation of the Generation of Inositol Phosphates in the Myometrium

2.5.1 Stimulation of the Phosphoinositide-Phospholipase C Pathway by Contractants

In the intact estrogen-treated guinea pig myometrium, prelabelled with [³H]inositol, carbachol and oxytocin were found to cause an enhancement in the generation of inositol phosphate accumulation (Marc et al.

1986, 1988; Do Khac et al. 1992). Stimulation by carbachol reflected a muscarinic receptor interaction while that of oxytocin appeared to be predominantly an oxytocin receptor-mediated event. Stimulations at the level of both inositol phosphates and tension were mimicked by the fluoroaluminate complex (AlF_4^-), suggesting the involvement of a G protein (Marc et al. 1988). For both carbachol and oxytocin the receptor-phospholipase C coupling is insensitive to pertussis toxin. It is worthnoting that α_q, the pertussis toxin insensitive G protein α-subunit, which has recently been reported to couple to phospholipase C-β (Smrcka et al. 1991), can be immunodetected in different myometrial preparations (unpublished observations). Supportive evidence was provided that phosphatidyl inositol 4,5-bisphosphate, $PtdIns(4,5)P_2$, was the target substrate for the enhanced phospholipase C activity induced by carbachol, oxytocin and fluoroaluminates. It was shown that upon addition of the stimulatory agonist to myometrial strips, there was a rapid and early production of the active $Ins(1,4,5)P_3$ isomer which preceeded the formation of $Ins(1,3,4)P_3$ and was followed by the generation of $InsP_2$ and $InsP$. Additionally, it was observed a significant decrease in the level of $PtdIns(4,5)P_2$, at a time point corresponding to the earliest rise in $Ins(1,4,5)P_3$ formation due to carbachol. Recent reports have extended these observations and clearly established that in different myometrial preparations, similar to the situation in the guinea pig tissue, phospholipase C was the target for an activation by a variety of contractants. Thus in the rat myometrium, an increased generation of inositol phosphates was induced by carbachol and oxytocin (Anwer et al. 1989; Goureau et al. 1992) and similarly by prostaglandins, via an FP receptor subtype (Goureau et al. 1992), by endothelin (ET) via an ET-A receptor subtype (Bousso-Mittler et al. 1989; our data to be submitted) and by bombesin, via a gastrin releasing peptide (GRP)-preferring receptor subtype (Amiot et al. 1993). An increased accumulation of inositol phosphates due to α_1-adrenergic receptor activation was also documented for the human myometrium (Breuiller-Fouche et al. 1991).

2.5.2 Involvement of a Dihydropyridine-Sensitive Ca^{2+} Component in the Accumulation of Inositol Phosphates Elicited by Carbachol, Oxytocin and Fluoroaluminates

Support for a Ca^{2+} entry-mediated regulation of phospholipase C in the guinea pig myometrium was provided (Harbon et al. 1990; Do Khac et al. 1992) by the ability of agents such as high K^+ and the Ca^{2+} ionophore ionomycin to cause, independently of receptor activation, an increased breakdown of PtdIns(4,5)P$_2$. Analogous to the pattern of inositol phosphate accumulation induced by a stimulatory receptor-mediated agonist, ionomycin caused an early production of Ins(1,4,5)P$_3$ followed by Ins(1,3,4)P$_3$, InsP$_2$ and InsP. The responses to both high K^+ and ionomycin were prevented when Ca^{2+} was withdrawn from the incubation medium, indicating that an increase in Ca^{2+} entry was responsible for the observed stimulations. The stimulatory effects of high K^+ and ionomycin persisted in the presence of inhibitors of the cyclooxygenase and lipoxygenase pathways (Harbon et al. 1984), ruling out the contribution of a Ca^{2+}-induced release of arachidonic acid metabolite(s) which could have activated the generation of inositol phosphates via the receptor–G protein–phospholipase C cascade. Our conclusion that an increase in Ca^{2+} influx can activate phospholipase C degrading PtIns(4,5)P$_2$ is consistent with previous observations made in many excitable cells (Eberhard and Holz 1988). It was worth noting that omission of Ca^{2+} from the incubation medium resulted in a slight but consistent attenuation of the generation of inositol phosphates induced by carbachol, oxytocin and fluoroaluminates (36%, 38%, and 25% decrease, respectively). Basal levels of inositol phosphates were also reduced under these conditions. Thus, the degree of stimulability by each agonist remained virtually unchanged, indicating that the cascade of the signaling system involving receptor–G protein–phospholipase C interaction was not regulated by an influx of Ca^{2+}. The findings nevertheless suggested that the Ca^{2+}-dependent component of phospholipase C regulation also contributed, although to a minor extent, to the overall inositol phosphate responses evoked by carbachol and oxytocin (receptor-mediated), as well as AlF_4^- (direct activation of G protein). Biochemical evidence that part of the receptor-mediated generation of inositol phosphates was dependent on an influx of Ca^{2+} through voltage-gated Ca^{2+} channels was further provided. Indeed, the production

of each inositol phosphate species triggered by carbachol, oxytocin and
AlF_4^- was similarly reduced (30%–35%) by removal of extracellular
Ca^{2+} or by the Ca^{2+} channel blocker nifedipine. Additionally, the Ca^{2+}
channel agonist Bay K 8644 was able to reverse inhibition caused by
nifedipine at the level of the inositol phosphate responses induced by
carbachol. These data imply that, in addition to the direct coupling of
cell surface receptors to phospholipase C activation, via a G protein, a
process that is insensitive to elevation in intracellular Ca^{2+} and that
contributes predominantly to the increased generation of inositol phosphates, there is also a coupling of these cell receptors to voltage-operated Ca^{2+} channels. The resulting increase in Ca^{2+} influx would then
be responsible for the additional, although modest (35%), receptor-
mediated phospholipase C activation. The mechanisms that link muscarinic and oxytocin receptors to the activation of voltage-operated
Ca^{2+} channels, whether direct or indirect, i.e., as a consequence of second messenger generation, have yet to be clarified.

2.5.3 Activation of β-Adrenergic Receptors Inhibits Ca^{2+} Entry-Mediated Generation of Inositol Phosphates in the Guinea Pig Myometrium. A Dihydropyridine-Sensitive, cAMP-Independent Inhibitory Effect

The effect of isoproterenol on the generation of inositol phosphates
was analyzed with the aim of demonstrating possible cross-talk between contractants and relaxants at the level of phospholipase C activation (Do Khac et al. 1992). Isoproterenol was found to attenuate
(35%) the generation of inositol phosphates due to both receptor- (oxytocin, carbachol) and G protein-mediated activation. Maximal inhibition was achieved at 20 nM isoproterenol with an EC_{50} of 0.5 nM . Inhibition by isoproterenol reflected a β-adrenergic-mediated event and
could be noted at the level of both $InsP_3$ isomers, $Ins(1,4,5)P_3$ and
$Ins(1,3,4)P_3$, as well as at the level of $InsP_2$ and $InsP$, with no change
in the rate of sequential appearance of $InsP_3$, $InsP_2$, and $InsP$ upon carbachol stimulation (Do Khac et al. 1992). These data, added to the observations that treatment with isoproterenol also resulted in an attenuation of the decrease in $PtdIns(4,5)P_2$ levels caused by carbachol at the
early points of stimulation, supported the interpretation that phospholi-

pase C degrading PtdIns(4,5)P_2 was the target for the inhibitory effect of isoproterenol. In a number of cell types, elevated cAMP levels were found to be associated with an attenuation of agonist-induced phosphoinositide hydrolysis (Anwer et al. 1989). Such a mechanism did not contribute to the β-adrenergic-mediated inhibition of inositol phosphate formation in the guinea pig myometrium (Do Khac et al. 1992). Indeed, no clear-cut rises in cAMP could be detected with isoproterenol at concentrations (20–100 nM) that were already supramaximal for inducing inhibition of inositol phosphate accumulation. Both forskolin and cholera toxin which caused marked elevations of cAMP failed to attenuate the generation of inositol phosphates due to oxytocin (Marc et al. 1988; Do Khac et al. 1992). Finally and more importantly, it was found that isoproterenol-mediated inhibition of inositol phosphates was prevented by pretreatment of the myometrium with pertussis toxin (Do Khac et al. 1992). The latter observation indicated that a G protein of the G_i or G_o family was involved in the coupling of the β-adrenergic receptor to the attenuated phospholipase C activity and ultimately excluded the participation of G_s and the stimulatory pathway of adenylyl cyclase in the β-adrenergic inhibition, which appeared to be a cAMP-independent, pertussis toxin-sensitive event.

It was further demonstrated (Do Khac et al. 1992) that isoproterenol was affecting the portion of the agonist-induced increase in the production of inositol phosphates that apparently depends on nifedipine-sensitive Ca^{2+} influx: (1) no additional β-adrenergic inhibitory effect could be demonstrated in a Ca^{2+}-depleted medium and, similarly, inhibitions elicited by isoproterenol and by nifedipine were not additive, suggesting a common inhibitory pathway; (2) stimulations of inositol phosphate accumulation by agents that elevate intracellular Ca^{2+}, such as ionomycin and high K^+, were insensitive to the inhibitory action of isoproterenol; (3) ionomycin as well as high K^+ were able to counteract the β-adrenergic receptor-mediated inhibition; (4) the Ca^{2+} channel agonist Bay K 8644, which antagonized nifedipine-mediated inhibition, similarly antagonized inhibition triggered by isoproterenol. It became possible to propose that β-adrenergic receptor-mediated inhibition of PtdIns(4,5)P_2 breakdown is most probably secondary to inhibition of Ca^{2+} entry through voltage-gated Ca^{2+} channels, which in turn is responsible for the attenuated phospholipase C activity.

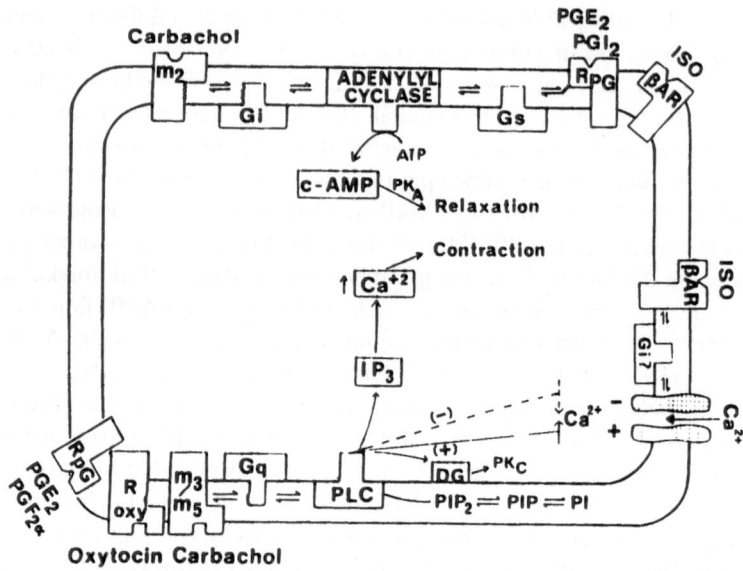

Fig. 2. Schematic representation of the stimulatory and inhibitory processes which contribute to the modulation of adenylyl cyclase and phospholipase C (*PLC*) pathways in the guinea pig myometrium. (a) Stimulatory agonists, namely, PGE_2, PGI_2, isoproterenol (*ISO*), interacting with specific receptors (R_{PG}, *bAR*) are coupled to the catalytic unit of adenylyl cyclase via the stimulatory regulatory protein G_s. Inhibitory receptors (muscarinic, m_2) are coupled to adenylyl cyclase via the inhibitory G_i. (b) Diverse stimulatory receptors to PGs (R_{PG}), to oxytocin (*R oxy*), to carbachol (Rm_3/m_5) are coupled by means of the pertussis toxin-insensitive *Gq* to phospholipase C (*PLC*) degrading PtIns(4,5)P_2, PIP2, with an increased generation of IP_3 and diacyl glycerol (*DG*). (c) Coupling of βAR by means of a pertussis toxin-sensitive G protein (*Gi* ?) to a voltage-operated ionic channel: the resulting decrease in Ca^{2+} modulated negatively (-) the PLC activity. For details see the text

   Our interpretations indirectly imply that β-adrenergic receptors in the myometrium might be linked directly, via a pertussis toxin-sensitive G protein, to an inhibiton of voltage gated Ca^{2+} channels (Do Khac et al. 1992), a situation encountered with some receptors in neuronal and endocrine cells (Schultz et al. 1990). The results could, however, also be accounted for by the coupling of β-adrenergic receptors to a K^+ channel whose activation may lead to membrane hyperpolarization, resulting in an inhibition of Ca^{2+} influx and indirectly affecting phosphoinositide hydrolysis. Figure 2 presents schematically the mechanisms, both stimulatory and inhibitory, described herein that contribute to the modulation of the phospholipase C pathway in the guinea pig myometrium. Figure 2 illustrates the dual regulation of the adenylyl cyclase transducing process, via both stimulatory and inhibitory receptors and specific G proteins.

In most myometrial preparations analyzed so far, β-adrenergic receptors have been reported to couple through G_S to the stimulatory pathway of adenylyl cyclase (Harbon et al. 1984). The present findings imply that, in the guinea pig myometrium, β-adrenergic receptors are additionally coupled to a distinct effector, namely, an ionic channel, via a pertussis toxin-sensitive G protein which is distinct from G_S. An important question that remains to be clarified is whether both pathways involve the same or distinct subpopulations of β-adrenergic receptors. In any case, the ultimate β-adrenergic-elicited reduction in Ca^{2+} influx and the associated decline in the accumulation of inositol phosphates may be of functional relevance. This may reasonably constitute the cAMP-independent pathway that we previously demonstrated (Do Khac et al. 1986b) to operate in combination with cAMP for β-adrenergic-mediated relaxation of the myometrium (Fig. 1).

2.6 Modulatory Mechanisms Regulating the Levels of cAMP in Pregnant Rat Myometrium

2.6.1 Cross-Regulation Between G Protein-Mediated Pathways, Stimulation and Inhibition of Adenylyl Cyclase

The sex steroid hormones estrogen and progesterone are known to have regulatory effects on the contractile state of the uterus and on the degree of catecholamine responsiveness. In order to assess potential permissive hormone action at the level of the different elements that compose the adenylyl cyclase cascade, we have used uteri obtained from rats at different stages of gestation. With all the stimulatory agonists (namely, β-adrenergic, prostacyclin, cholera toxin, and forskolin), the onset of gestation was accompanied by a progressive decline in cAMP responses, compared to the corresponding maximal responses observed with the estrogen-dominated myometrium (considered as day 0), with no more than 10% stimulations being expressed in the day 12 tissue (Tanfin and Harbon 1987; Harbon et al. 1990). The phase of attenuated responsiveness to the diverse stimulations was followed by a gradual and total restoration just before parturition, except for PGs whose cAMP stimulatory responses remained impaired during the whole gestation period. Of much interest was the finding that treatment of day 12 myometrium (midgestation) with pertussis toxin resulted in a reversal of the attenuated cAMP responses to both β-adrenergic and cholera toxin stimulations, suggesting that the inhibitory protein G_i was involved.

The above observation led us to characterize and tentatively evaluate the modulation of various forms of G proteins in different myometrial preparations (Milligan et al. 1989; Tanfin et al. 1991). Using varied α-subunit-specific antisera, we reported for the estrogen-treated rat myometrium (day 0), the absence of $G_i1\alpha$, the presence of both $G_i2\alpha$ and $G_i3\alpha$ as well as a novel form of G_0. An antiserum specific for $G_s\alpha$-subunits allowed the identification of both 52- (major) and 45- (minor) kDa proteins indicated the presence of a 35/36 kDa doublet, with a greater intensity of labeling for the higher molecular mass. Experiments performed under the same conditions with membranes from days 0, 12, and 21 indicated no qualitative changes in the nature of the G_i proteins. Particularly relevant were the quantitative modifications observed during gestation at the level of $G_i2\alpha$ and $G_i3\alpha$ which additionally appeared to be differentially regulated, while the level of $G_s\alpha$ (52/45 kDa) displayed no significant modifications (Fig. 3). Densitometric scanning of immunoblots showed that the amount of $G_i3\alpha$ progressively decreased with advancing gestation (day 0 > day 12 > day 21), some 75 ± 5% decrease being noted at the end of gestation. A different pattern was observed with $G_i2\alpha$ which significantly and consistently increased by 160% at midgestation compared to day 0, with a return to day 0 value at the end of gestation. It became important to note that the increased level of $G_i2\alpha$ at midgestation coincided with the absence of adenylyl cyclase stimulability. Furthermore, the decrease in $G_i2\alpha$ in day 21 myometrium to levels similar to those found in day 0 preparations coincided with recovery of full adenylyl cyclase stimulability. As G proteins are usually considered to exist as a heterotrimeric $\alpha\beta\gamma$ complex (Birnbaumer et al. 1990), it was interes-

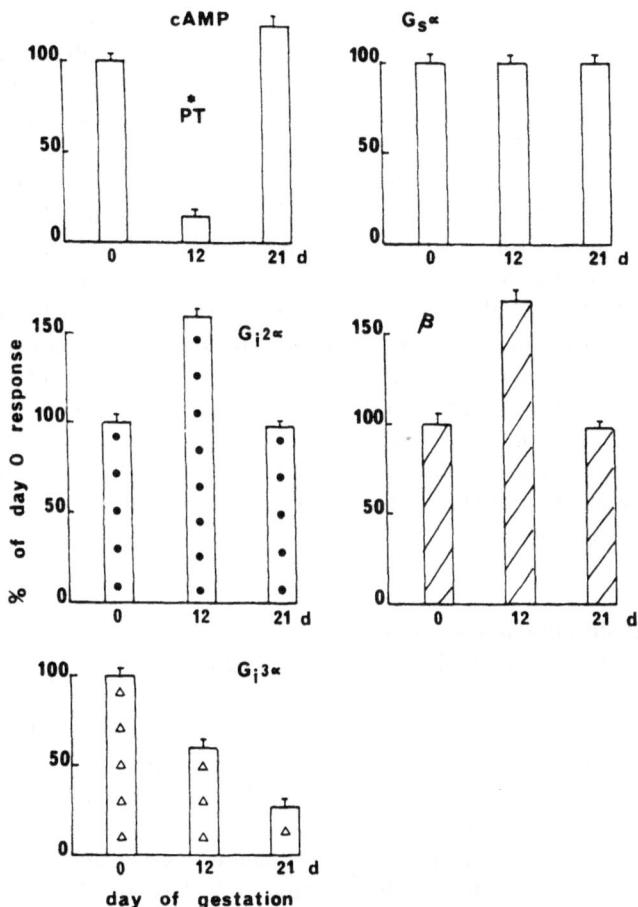

Fig. 3. Comparison between the evolution of the β-adrenergic accumulation of cAMP and the modulation of the G_i protein levels in the rat myometrium throughout gestation. For cAMP experiments myometrial strips obtained from rats at the indicated day of gestation were incubated for 7 min with 240 mM 3-isobutyl-1 methyl xanthine (IBMX) in the presence of 0.1 µM isoproterenol. Results are expressed as the percentage of control responses (day 0: estrogen-treated). *Asterisk* indicates a restored β-adrenergic-mediated cAMP response after pertussis toxin *(PT)* treatment. For quantitative determination of $G_i2\alpha$ *(bars with black circles)* and $G_i3\alpha$ *(bars with triangles)*, $G_s\alpha$ *(open bars)*, and β-subunits *(hatched bars)* myometrial membrane preparations were resolved on sodium dodecyl sulfate–polyacrylamide gel (12.5% [w/v] acrylamide) and immunoblotted as described. Data adapted from Tanfin and Harbon (1987) and Tanfin et al. (1991)

ting to note a pattern of β-subunit modulation similar to that observed with $G_i2\alpha$. The demonstrated increase in $G_i2\alpha$ and β-subunits may thus provide a reasonable interpretation for the altered adenylyl cyclase pathway revealed at midgestation. It was further conceivable to propose a potential role for G_i2 in mediating adenylyl cyclase inhibition in the myometrium, as has recently been proposed in platelets (Simonds et al. 1989). The progressive decline in $G_i3\alpha$ levels with advancing gestation may certainly underlie an important phenomenon; however, it seems difficult to implicate it in the pattern of the altered adenylyl cyclase pathway. First, the day 12 myometrium was associated with both a decrease in the G_i3 level and an attenuated adenylyl cyclase stimulability, and, second, adenylyl cyclase (see below) was normally expressed in the myometrium near term (Tanfin and Harbon 1987). The functional role of G_i3 remains to be clarified. In any case, the above observations provide a potential mechanism by which permissive regulations, linked to hormonal status during gestation, operate on G_i protein levels.

2.6.2 Evolution from a Stimulatory to an Inhibitory Prostaglandin-Mediated Effect

The latest phase of gestation (day 21) was characterized by a full restoration of adenylyl cyclase responsiveness toward isoproterenol, cholera toxin, and forskolin. In marked contrast, the inactivity of PGI_2 (and similarly of PGE_2), which emerged at the onset of gestation, was still retained and coincided with the development of a G_i-mediated inhibitory PG effect. Thus, the PG-elicited effect at the level of cAMP which was exclusively stimulatory in the day 0 myometrium appeared exclusively inhibitory at the end of gestation (Tanfin and Harbon 1987; Goureau et al. 1992).

2.7 Diverse Prostaglandin Receptors Activate Distinct Signal Transduction Pathways in Rat Myometrium

A classification of different subtypes of prostaglandin (PG) receptors has been proposed (Coleman 1987; Coleman et al. 1987), based on functional criteria using the rank order of potencies of natural PGs and a few relatively selective agonists and antagonists. The five main types of receptors are TP, DP, IP, FP and EP for thromboxane A_2, PGD_2, PGI_2, $PGF_2\alpha$ and PGE_2, respectively. The EP receptors may further be subdivided into EP_1, EP_2, and EP_3 subtypes (Coleman et al. 1987). Using such a pharmacological approach, we have attempted to characterize in rat myometrium the specificity of PG receptor interaction with different second messenger-generating systems and to associate these interactions with PG functional activity (Goureau et al. 1992).

2.7.1 Interaction of Prostaglandin Receptors with the Adenylyl Cyclase System

1. *PG receptors and the stimulatory pathway of adenylyl cyclase in the estrogen-treated (day 0) myometrium:* the increased generation of cAMP was restricted to PGI_2 = iloprost, a stable analog of PGI_2, PGE_2 with a respective EC_{50} = 0.5 and 5 μM. On the other hand, $PGF_2\alpha$ and PGD_2 were inactive, as were also inactive sulprostone and misoprostol, both considered as specific EP-receptor agonists. These findings support the involvement of an IP-receptor mediating adenylyl cyclase stimulation. The stimulatory effect of PGE_2 on cAMP response could, with good reason, be considered to result from an interaction with an IP- rather than an EP-receptor subtype.

2. *PG receptors and the inhibitory pathway of adenylyl cyclase in the pregnant (day 21) myometrium*: PGs could exert no stimulatory effect on cAMP. In contrast, diverse PGs were found to attenuate the cAMP responses, a pertussis toxin-sensitive process, reflecting an inhibition of adenylyl cyclase mediated by G_i – most probably G_i2. The close similarity revealed in the activity profile of the distinct PGs (Iloprost, PGE_2, $PGF_2\alpha$, sulprostone, misoprostol, EC_{50} = 3 nM), did not allow the proposal of a specific type of PG receptor coupled to adenylyl cyclase inhibition (Goureau et al. 1992).

2.7.2 Interaction of Prostaglandin Receptors with the Phospholipase C Transducing System

In both day 0 and day 21 myometrium, PGs were able to stimulate phospholipase C with an increased generation of inositol phosphates, $InsP_3$, $InsP_2$ and InsP. In day 0 myometrium (Fig. 4), the order of potency, $PGF_{2\alpha} > PGD_2 \gg PGE_2 \gg$ iloprost suggested the involvement of FP receptors. This clearly indicated that PG receptors (IP type) involved in adenylyl cyclase stimulation are distinct from those involved in phospholipase C activation. The inability of the specific DP receptor agonist BW245C to affect phospholipase C activation also excluded the contribution of a DP receptor. Similarly, the very weak activity displayed by PGE_2, could reflect the interaction of PGE_2 with an FP receptor rather than an EP receptor. In favor of this interpretation was the inability of misoprostol and sulprostone to increase inositol phosphate generation (Goureau et al. 1992). In day 21 myometrium, $PGF_{2\alpha}$ was still the most potent activator of phospholipase C, reflecting the coupling of an FP receptor with the metabolism of phosphoinositides. However, the relatively high affinity of PGE_2 (5 μM) and its high efficacy (Fig. 4) indirectly demonstrated the contribution of an EP receptor. This interpretation was confirmed by the high potency displayed by an EP selective agonist, namely, sulprostone. By use of the criteria defined by Coleman et al. (1987) to differentiate EP receptor subtypes, the effective response obtained with sulprostone ($EP_1 = EP_3$; $EP_2 = 0$) as opposed to the moderate activity displayed by misoprostol ($EP_3 = EP_2 \gg EP_1$), suggested the major contribution of an EP_1 receptor. Thus the pregnant rat myometrium constitutes a cellular model where two distinct PG receptors (FP and EP_1) are coupled to the same signal-transducing pathway, namely, phospholipase C. In both day 0 and day 21 myometrium, the putative G protein which couples FP and EP receptors to phospholipase C activation was insensitive to pertussis toxin (Goureau et al. 1992) and could be considered as a member of the G_q protein family (Smrcka et al. 1991).

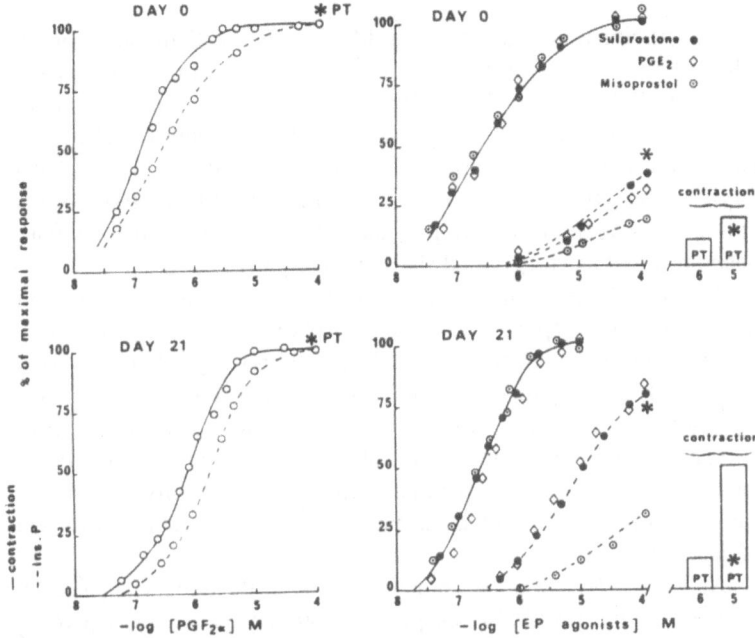

Fig. 4. Correlations between the ability of $PGF_{2\alpha}$ and EP agonists to induce contraction and to enhance the generation of inositol phosphates in the estrogen-treated (day 0) and pregnant (day 21) rat myometrium. For the estimation of total [³H]inositol phosphates (InsP), [³H]inositol prelabeled strips were incubated for 10 min with the indicated concentration of $PGF_{2\alpha}$ and different EP agonists. The contractile activity was integrated during a 2 min exposure to the indicated concentration of the prostanoid. (*) indicates the inositol phosphate and contractile responses obtained after pertussis toxin (*PT*) treatment. Data adapted from Goureau et al. (1992)

2.7.3 Prostaglandin Receptors
Contributing to Uterine Contractions

In both estrogen-treated and pregnant rat myometrium, PGs were effective contractile agonists (Coleman et al. 1987). In day 0 myometrium, $PGF_{2\alpha}$ appeared to be the most potent contractile agonist ($EC_{50} = 0.15\ \mu M$), followed by PGE_2 = sulprostone = misoprostol ($EC_{50} = 0.5\ \mu M$), and finally by iloprost, which displayed a lower activity. The situation was somewhat different in day 21 myometrium, in which the rank order of potency found was PGE_2 = misoprostol = sulprostone (EC_{50} = 0.2 μM) > $PGF_{2\alpha}$ with an EC_{50} = 0.8 μM and iloprost remaining the less active. BW245C was without effect in both day 0 and day 21 myometrium, excluding any DP receptor-mediated uterine contraction. The presence of both FP and EP receptors contributing to contraction of the pregnant rat myometrium has similarly been proposed recently (Crankshaw and Gaspar 1992).

A series of arguments has allowed us to consider a major contribution of the FP receptor coupled to phospholipase C activation in the $PGF_{2\alpha}$-induced contractions in both day 0 and day 21 myometrium (Fig. 4). These arguments include: (1) the close correlation between the dose–response curves for the contractile effect and for the generation of inositol phosphates, (2) the absence of pertussis toxin effect on the two events, and (3) the parallel decrease in the affinity of $PGF_{2\alpha}$ to cause both inositol phosphate accumulation and contraction at the end of gestation.

As far as PGE_2 was concerned, its contractile activity appeared to be mediated by specific EP receptors in both day 0 and day 21 myometrium. In view of the high efficacy displayed by sulprostone (EP_1 = EP_3; EP_2 = 0), it was reasonable to exclude any EP_2-mediated effect. The equipotency displayed by sulprostone and misoprostol (EP_3 = EP_2 >> EP_1) on uterine contraction (Fig. 4) allowed the proposal of either a major role for the EP_3 subtype and/or the copresence of both EP_1 and EP_3 subtypes contributing to uterine contraction. Of interest were the findings (Sugimoto et al. 1992) that the EP_3 receptor displayed a significant sequence similarity with other members of G protein-coupled receptors and that the EP_3 mRNA is abundantly expressed in uterus. Concerning the mechanism involved in PGE_2-induced contractions in both day 0 and 21 preparations, our data are consistent with the presence of a pertussis toxin-sensitive, EP_3 receptor-linked process, independent of the phospholipase C pathway. This interpretation was supported by two major findings illustrated in Fig. 4: (1) contractions induced by PGE_2 and also by sulprostone and misoprotol could be evidenced under conditions (micromolar concentrations) in which they caused no increment in $InsP_3$ generation and (2) pertussis toxin treatment prevented PGE_2-evoked uterine contraction, but did not modify PGE_2-stimulated inositol phosphate generation.

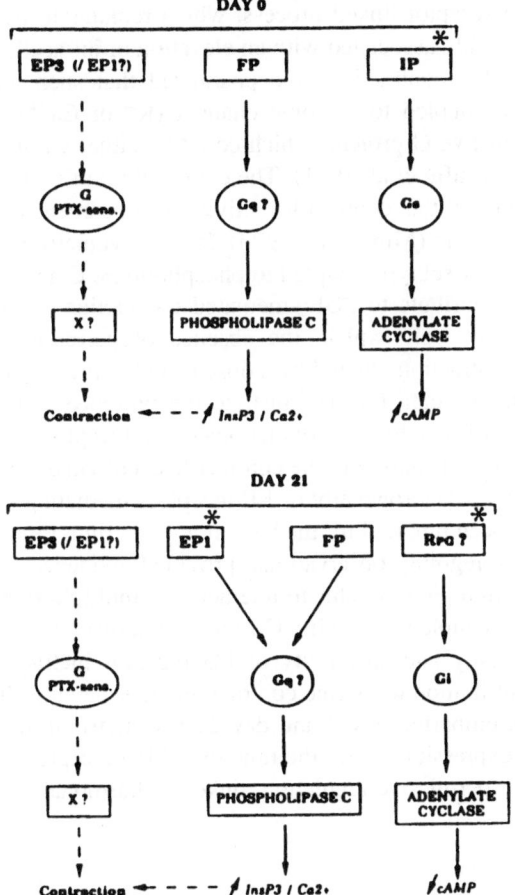

Fig. 5. Schematic representation of postulated coupling of distinct prostaglandin receptor subtypes with different transduction pathways in day 0 and 21 rat myometrium. *PTX-sens*, Pertussis toxin sensitive; *X*, transduction mechanism not yet determined; For details, see text. (From Goureau et al. 1992)

The EP_3 receptor-linked process, which remains to be determined, is most probably associated with an elevation of intracellular Ca^{2+}, a prerequisite for contraction. We postulated that such an EP_3 receptor might be coupled to an ionic channel (K^+ or Ca^{2+}), via a pertussis toxin-sensitive G protein, which could be either G_i or G_o (Milligan et al. 1989; Tanfin et al. 1991). The contribution of ionic channels to the regulation of the contractile activity is documented in the uterine smooth muscle (Mironneau 1990). In the myometrium, near term, the EP_1 receptor subtype coupled to phospholipase C activation may additionally contribute to PGE_2-mediated contraction. Such a mechanism could fairly well explain the absence of pertussis toxin effect on uterine contraction elicited by a dose of PGE_2 higher than micromollar (Fig. 4). The data further illustrate the presence in the myometrium near term of two distinct populations of EP receptors, namely, EP_1 and EP_3, each contributing to the contractile event via distinct transductory pathways. A heterogeneity of EP receptors mediating contractions has recently been reported for the human myometrium (Senior et al. 1991).

The foregoing observations provided evidence that in the rat myometrium, PGs are able to interact with multiple signal transduction pathways coupled to specific G proteins via distinct types of receptors. The diversity and specificity of PG receptor interactions as well as their contribution to uterine contraction are schematically illustrated in Fig. 5. Comparing day 0 and day 21 myometrium, the data illustrate that the expression and/or the function of PG receptors may also be the target for permissive regulation, linked to hormonal status during gestation.

2.8 Conclusions

The data illustrate that in the myometrium a diversity of G protein-coupled receptors which can be activated by neurotransmitters, neuropeptides, and eicosanoids contribute to the modulation of the two major signal transduction pathways, namely, the adenylyl cyclase (both the stimulatory and inhibitory arm) and the phospholipase C stimulatory process, through the intermediate regulation by distinct and specific G proteins. Indirect evidence was also provided for the coupling of some of these receptors to ionic channels. This is the case for the β-adrenergic and for the PG (EP$_3$ subtype) receptors whose activation might cause a decrease and increase, respectively, in intracellular Ca^{2+}, a process that in both cases is sensitive to pertussis toxin. A dialogue between different signaling pathways needs also to be considered. Thus in the rat myometrium near term, the phospholipase C–inositol phosphate pathway may contribute to the decrease in cAMP levels through a Ca^{2+}-activated phosphodiesterase (Goureau et al. 1990). Similarly, the β-adrenergic-mediated inhibiton of a voltage-operated Ca^{2+} channel contributes to the attenuation of phospholipase C activation. Finally evidence was provided that the level of G$_i$ proteins as well as the function and/or expression of PG receptors appeared to be the target for permissive regulation during pregnancy. Our findings for a differential coupling of PG receptors in the pregnant rat myometrium may be of physiological relevance. Inhibition of adenylyl cyclase, stimulation of phospholipase C and activation of the putative ionic channel pathway (resulting in an increase in cytosolic Ca^{2+}) would all lead to an enhancement of the contractile activity of the uterus, which is of crucial importance during delivery.

Acknowledgements. This work was supported by grants from the Centre National de la Recherche Scientifique (URA 1131) and by a contribution from the Institut national de la Santé et de la Recherche Médicale (CRE 900403). We are grateful to G. Thomas and G. Delarbre for technical assistance and M.H. Sarda for helping with the editing of the manuscript.
    We would like to thank Schering for the generous gift of iloprost and sulprostone, Wellcome Research Laboratories for generously providing BW245C and Searle Research Service for misoprostol.

References

Amiot F, Leiber D, Marc S, Harbon S (1993) GRP-preferring bombesin receptors increase the generation of inositol phosphates and tension in rat myometrium. Am J Physiol (in press)

Anwer K, Hovington JA, Sanborn BM (1989) Antagonism of contractants and relaxants at the level of intracellular calcium and phosphoinositide turnover in the rat uterus. Endocrinology 124:2995–3002

Berridge MJ (1987) Inositol trisphosphate and diacylglycerol: two interacting second messengers. Annu Rev Biochem 56:159–163

Berridge MJ, Irvine RF (1984) Inositol trisphosphate, a novel second messenger in cellular signal transduction, Nature 312:315–321

Birnbaumer L, Abramowitz J, Brown AM (1990) Receptor-effector coupling by G proteins. Biochim Biophys Acta 1031:163–224

Bousso-Mittler D, Kloog Y, Wollberg Z, Bdolak A, Kochva E, Sokolovsky M (1989) Functional endothelin/sarafatoxin receptors in the rat uterus. Biochem Biophys Res Comm 162:952–957

Breuiller-Fouche M, Doualla-Bell Kotto Maka F, Geny F, Ferre F (1991) Alpha-1 adrenergic receptor: binding and phosphoinositide breakdown in human myometrium. J Pharmacol Exp Ther 258:82–87

Carsten ME, Miller JD (1985) Ca^{2+} release by inositol trisphosphate from Ca^{2+}-transporting microsomes derived from uterine sarcoplasmic reticulum. Biochem Biophys Res Comm 130:1027–1031

Coleman RA (1987) Methods in prostanoid receptor classification, in prostaglandins and related substances. A practical approach, Eds, IRL Press, Oxford, UK, pp 267–303

Coleman RA, Kennedy I, Sheldrick RLG, Tolowinska IY (1987) Further evidence for the existence of three subtypes of PGE_2-sensitive (EP-) receptors. Br J Pharmacol 91:407P

Crankshaw DJ, Gaspar V (1992) Effects of prostanoids on the rat's myometrium in vitro during pregnancy. Biol Reprod 46:392–400

Diamond J (1990) β-Adrenergic receptor, cyclic AMP and cyclic GMP in control of uterine motility, in Uterine function: Molecular and Cellular Aspects, Carsten E, Miller JD, Eds Plenum, New York, pp 249–275

Dohlman HG, Caron MG, Lefkowitz RG (1987) A family of receptors, coupled to guanine nucleotide regulatory proteins. Biochemistry 26:2657–2666

Dohlman HG, Thorner J, Caron MC, Lefkowitz RJ (1991) Model systems for the study of seven-transmembranes segment-receptors. Annu Rev Biochem 60:653–688

Do Khac L, d'Albis A, Jammot C, Harbon S (1986a) Myosin light chain phosphorylation in intact rat uterine smooth muscle. Role of calcium and cyclic AMP. J Muscle Res Cell Motility 7:259–268

Do Khac L, Mokhtari A, Harbon S (1986b) A re-evaluated role for cyclic AMP in uterine relaxation. Differential effect of isoproterenol and forskolin. J Pharmacol Exp Ther 239:236–242

Do Khac L, Mokhtari A, Renner M, Harbon S (1992) Activation of β-adrenergic receptors inhibits Ca^{2+} entry-mediated generation of inositol phosphates in the guinea pig myometrium, a cyclic AMP-independent event. Mol Pharmacol 41:509–519

Eberhard DA, Holz RW (1988) Intracellular Ca^{2+} activates phospholipase C. Trends Pharmacol Sci 11:517–521

Goureau O, Tanfin Z, Harbon S (1990) Prostaglandins and muscarinic agonists induce cyclic AMP attenuation by two distinct mechanisms in the pregnant rat myometrium. Biochem J 271: 667–673

Goureau O, Tanfin Z, Marc S, Harbon S (1992) Diverse prostaglandin receptors activate distinct signal transduction pathways in rat myometrium. Am J Physiol 263:C257-C265

Harbon S, Do Khac L, Vesin MF (1976) Cyclic AMP binding to intracellular receptor proteins in rat myometrium. Effect of epinephrine and prostaglandin E1. Mol Cell Endocrinol 6:17–34

Harbon S, Tanfin Z, Leiber D, Vesin MF, Do Khac L (1984) Selective interactions of cyclooxygenase and lipoxygenase metabolites with the cyclic AMP and the cyclic GMP systems in the myometrium. Adv Cyclic Nucleotide Protein Phosphorylation Res 17:639–649

Harbon S, Marc S, Goureau O, Tanfin Z, Leiber D, Mokhtari A, Do Khac L (1990) Multiple regulations for the generation of inositol phosphates and cyclic AMP in myometrium. In Uterine Contractility, Garfield RE (ed), Plenum Publishing Co, pp 123–140

Hardman JG (1981) Cyclic nucleotides and smooth muscle contraction: some conceptual and experimental considerations, in Smooth Muscle: An assessment of Current Knowledge, Bülbring E, Brading AF, Jones AW, Tomita T (eds) Edward Arnold, London, pp 249–262

Hepler JR, Gilman AG (1992) G proteins. Trends Biochem Sci 17:383–387

Kanmura Y, Missiaen L, Casteels R (1988) Properties of intracellular calcium stores in pregnant rat myometrium. Br J Pharmacol 95:284–290

Leiber D, Marc S, Harbon S (1990) Pharmacological evidence for distinct muscarinic receptor subtypes coupled to the inhibition of adenylate cyclase and to the increased generation of inositol phosphates in the guinea pig myometrium. J Pharmacol Exp Ther 252: 800–809

Marc S, Leiber D, Harbon S (1986) Carbachol and oxytocin stimulate the generation of inositol phosphates in the guinea pig myometrium. FEBS Lett 201: 9–14

Marc S, Leiber D, Harbon S (1988) Fluoroaluminates mimic muscarinic-and oxytocin-receptor-mediated generation of inositol phosphates and contraction in the guinea pig myometrium. Biochem J 255: 705–713

Milligan G, Tanfin Z, Goureau O, Unson C, Harbon S (1989) Identification of both G_i2 and a novel immunogically distinct form of G_o in rat myometrial membranes, FEBS Lett 244: 411–416

Mironneau J (1990) Ion channels and excitation-contraction coupling in myometrium, in Uterine contractility, ed. Garfield, Plenum, New York, pp 9–20

Mironneau C, Mironneau J, Savineau JP (1984) Maintained contractions of rat uterine smooth muscle incubated in a Ca^{2+}-free solution. Br J Pharmacol 82:735–743

Mokhtari A, Do Khac L, Harbon S (1988) Forskolin alters sensitivity of the cAMP-generating system to stimulatory as well as to inhibitory agonists. Eur J Biochem 176:131–137

Schultz G, Rosenthal W, Hescheler J, Trautwein W (1990) Role of G proteins in calcium channel modulation. Annu Rev Physiol 52:275–292

Senior J, Marshall K, Sangha R, Baxter GS, Clayton JK (1991) In vitro characterization of prostanoïd EP-receptors in the non-pregnant human myometrium. Br J Pharmacol. 102:747–753

Simonds WF, Goldsmith PK, Godina J, Unson G, Spiegel AM (1989) G_i2 mediates $\alpha2$ adrenergic inhibition of adenylate cyclase in platelet membranes: In situ identification with $G\alpha$ C terminal antibodies. Proc Natl Acad Sci USA, 86:7809–7813

Smrcka AV, Hepler JR, Brown KO, Sternweis PC (1991) Regulation of phosphoinositide-specific phospholipase C activity by purified Gq. Science 251:804–807

Somlyo AP, Himpens B (1989) Cell calcium and its regulation in smooth muscle, FASEB J 3:2266–2276

Sugimoto Y, Namba T, Honda A, Hayashi Y, Negishi M, Ichikawa A, Narumiya S (1992) Cloning and expression of a cDNA for mouse prostaglandin E receptor EP_3 subtype. J Biol Chem 267:6463–6466

Sutherland EW, Rall TW (1960) The relation of adenosine cyclic 3',5'-phosphate and phosphorylase to the actions of catecholamines and other hormones. Pharmacol Rev 12:265–299

Tanfin Z, Harbon S (1987) Heterologous regulations of cAMP responses in pregnant rat myometrium. Evolution from a stimulatory to an inhibitory prostaglandin E2 and prostacyclin effect. Mol Pharmacol 32:249–257

Tanfin Z, Goureau O, Milligan G, Harbon S (1991) Characterization of G proteins in rat myometrium. A differential modulation of $G_i2\alpha$ and $G_i3\alpha$ during gestation. FEBS Lett 278:4–8

Van Breemen C, Saida K (1989) Cellular mechanisms regulating $[Ca^{2+}]_i$ in smooth muscle, Annu Rev Physiol 51:315–329

Vesin MF, Harbon S (1974) The effects of epinephrine, prostaglandins and their antagonists on adenosine cyclic 3'5'-monophosphate concentrations and motility of the rat uterus. Mol Pharmacol 10:457–473

3 Myometrial Electrophysiology

Jean Mironneau

3.1 Introduction

Ion channels involve in the generation of electrical events can be separated in at least three different groups. Voltage-gated ion channels mediate rapid, voltage-dependent changes in membrane conductance during action potentials. Calcium, sodium and potassium channels belong to this group. Ligand-gated ion channels are opened in response to activation of an associated receptor. Typical channels of this type include the nonspecific channels that are opened by activation of muscarinic receptors (Benham et al. 1985) or ATP-receptors (Benham et al. 1987; Friel 1988; Honoré et al. 1989a). Activation of ligand-gated ion channels mediates local increases in membrane conductance, producing depolarization and hyperpolarization of the membrane. Calcium-gated channels are only opened by an increase in cytoplasmic calcium and modulate the membrane potential. Potassium and chloride channels of this type have been identified in smooth muscle (Coleman and Parkington 1987; Pacaud et al. 1989).

In myometrial cells, action potentials are dependent on activation of several ion channels, including voltage-gated calcium channels. The calcium entering cells during a single action potential can be estimated from the calcium current density and the estimated cell volume. It is found that the internal calcium concentration may increase by 10 μM during an action potential. Although much of the calcium entering the cell may become buffered by uptake into intracellular stores, a sustained increase of calcium to 1–2 μM is sufficient to cause a maximal contraction in skinned strips of myometrium (Savineau et al. 1988).

In this brief review, I outline some of the main ion channels which are involved in the modulation of myometrial contractility.

3.2 Voltage-Gated Calcium Channels

3.2.1 Channel Type

Recent studies have indicated that vascular and visceral smooth muscle cells contain two or three different kinds of calcium channels (Loirand et al. 1986; Sturek and Hermsmeyer 1986; Benham et al. 1987; Yoshino et al. 1989). Two of these channels most likely correspond to the calcium channels termed T (transient) and L (long-lasting) that have been studied in cardiac and neuronal tissues (Nilius et al. 1985; Fox et al. 1987). The two types of calcium channels have different selectivity, single-channel conductance, and pharmacology. Furthermore, T-type channels are inactivated by steady-state holding potentials to about

–40 mV, while L-type channels need much more positive holding potentials for steady-state inactivation.

Components of calcium current in single rat myometrial cells might be separated by applying depolarizing test pulses from two different holding potentials (–40 and –80 mV). In 89% of cells used, neither change in kinetics of inactivation nor shift in the peak current was observed between the two families of calcium current ($n = 47$). Much evidence suggests that the calcium channel current appears to be mainly a L-type current: (a) Equimolar replacement of barium for calcium induces an increase in the peak current and a decrease in the inactivation rate. (b) Residual inward currents are recorded at the end of the pulses (400 msec in duration). (c) Membrane potential for half-inactivation was about –40 mV. However, in 11% of cells tested, two components of calcium channel currents can be separated by variations in holding potential (Honoré et al. 1989b). In human myometrium, evidence for two different calcium channels has been obtained at the single-channel level (Inoue et al. 1990).

3.2.2 Inactivation and Selectivity of L-type Calcium Channels

Calcium channels are also time-dependent such that under voltage clamp, a depolarizing step in the membrane sufficient to activate the channels results in an inward current that peaks within 10 msec and then declines (Amédée et al. 1987). Part of the decline is caused by a voltage-dependent inactivation of the calcium channels (Jmari et al. 1986). At small depolarizations only a small number of calcium channels enters the inactivated state, and it requires a positive depolarization before the maximum number of calcium channels become inactivated. There is an additional part of inactivation caused by an accumulation of calcium ions on the inside surface of the membrane. This is illustrated by the fact that conditioning depolarizations that produce maximal calcium current induce the greatest amount of inactivation. As the conditioning depolarization increases towards the expected reversal potential and the driving force for calcium influx becomes smaller, inactivation decreases but is never completely reversed. It is assumed that free calcium ions react reversibly with calcium channel binding sites. This may lead to an enzymatic dephosphorylation of a channel component, resulting in inactivation of the channels (Ekert and Chad 1984). There is no evidence that accumulation of calcium ions in a restricted extracellular or intracellular space may change the membrane potential in isolated myometrial cells (Amédée et al. 1987).

Sodium ions can permeate these calcium channels but only when the external calcium concentration is less than $10^{-6} M$ (Jmari et al. 1987). The channels are also permeable to the divalent cations barium and strontium which pass preferentially over calcium. A key element in the movement of cations through channels is the interaction between the charged ion and the electric field within the membrane. The force acting on the cation to move it into the cell is countered by reactive groups and steric configurations, i.e., energy barriers, that function to impede the flow of cations through the channel. This energy barrier or selectivity filter is structured in such a way as to permit certain types of cations to pass through the channel while excluding others (Hille and Schwarz 1978; Hess and Tsien 1984). In myometrial cells, the relative selectivity of the calcium channel for the divalent cations follows the sequence $Ca^{2+} > Sr^{2+} > Ba^{2+}$. The selectivity for calcium depends on the presence of external calcium, as shown for calcium channels of different excitable cells (Tsien et al. 1987). Other divalent cations, particularly cobalt and manganese but also nickel and cadmium, do not readily permeate the channel but bind to sites in the channel so strongly that they block the channel. Using large organic cations as probes, the pore size of the Ca^{2+} channel was estimated to be approximately 6 Å (Tsien et al. 1987). This is three times the diameter of the calcium ion and approximately the same size as a hydrated calcium ion.

Functional voltage-gated calcium channels have been recently expressed in *Xenopus* oocytes injected with poly(A^+) mRNA obtained from pregnant rat myometrium (Fournier et al. 1989). These calcium channels possess electrophysiological and pharmacological properties similar to those recorded in isolated myometrial cells and belong to the L-type calcium channel.

3.2.3 Calcium Channel Blockers

Various compounds have the capacity to alter calcium influx accross the cell membrane. These can be divided into two groups: inorganic inhibitors such as cobalt, manganese, nickel, and cadmium and organic compounds. The latter agents are generally referred to as calcium entry blockers (Fleckenstein 1977).

A number of calcium entry blockers, including nifedipine, nicardipine, diltiazem, verapamil, and gallopamil, have been shown to inhibit uterine contractions in various species, including the rat, rabbit, sheep, monkey, and human (Hollingsworth and Downing 1988). The dihydropyridines are the most potent and selective inhibitors of uterine tension and therefore are of considerable interest for both therapeutic and experimental purposes. In cardiac cells the concentrations of dihydropyridines, which give rise to half-maximal electrophysiological responses, are several orders of magnitude greater than the dissociation constants values determined from ligand experiments (Kokubun et al. 1987; Hamilton et al. 1987). Recent electrophysiological experiments in smooth muscle cells have demonstrated a voltage dependence of dihydropyridine antagonist action, and this has been postulated to account for the difference between the binding affinities and the dissociation constants determined from the pharmacological effects of these drugs (Dacquet et al. 1988; Loirand et al. 1989).

In the simplest model, calcium channels are in one of three configurations: closed, open, or inactivated. Transitions usually occur between the three configurations as random events, with each transition having a different probability. Under resting conditions, the probability of most channels being open is small as most channels are closed. The effect of the membrane depolarization is to increase the frequency of opening. The channels remain open for a brief but random period. During maintained depolarization calcium channels make the transition to the inactivated state and become nonconducting. This state, however, differs from being closed in that the channel usually requires the membrane to be repolarized before it can open again (reactivated). The probability of leaving the inactivated state is small at depolarized potentials and becomes much greater at more negative membrane potentials. Binding of calcium entry blockers to one of the calcium channel states inhibits the calcium current.

In myometrium, L-type calcium channels are largely blocked by the dihydropyridine derivatives, nifedipine at $10^{-7} M$ and (+)isradipine at $10^{-8} M$. To study the mechanism of blockade of the calcium channel current with (+)isradipine in more detail, Honoré et al. (1989b) used different protocols to assess the relative contribution of calcium channel configuration to isradipine blockade. First, when voltage clamp depolarization is applied at a frequency of 0.05 Hz, the inhibition induced by $2 \times 10^{-8} M$ (+)isradipine reaches $50 \pm 5\%$ ($n = 7$) within 5 min. Second, (+)isradipine ($2 \times 10^{-8} M$) is applied during a rest period of 5 min and blockade is assessed as the difference between peak current in the control and the first pulse after drug exposure. Under these conditions, the calcium channel current is inhibited by $48 \pm 6\%$ ($n = 4$), indicating that the blockade is not dependent on the number of voltage depolarizations applied at 0.05 Hz (absence of use-dependent inhibition). Therefore, it is suggested that (+)isradipine does not bind to the open state of the calcium channel. Third, in order to investigate whether (+)isradipine binds with a higher affinity to the inactivated state of the calcium channel, effects of isradipine have been determined on the voltage dependence of the calcium channel current (Honoré et al. 1989b). From the shift in the inactivation curve, it is possible to estimate the dissociation constant for (+)isradipine binding in the resting and inactivated state by using an approach described by Bean (1984), assuming one-to-one binding of drug to the resting and inactivated states. The dissociation constant for binding to the inactivated state (K_i) can be calculated using the equation:

$$\Delta Vh = k \ln (1 + [Isr]/K_i)/(1 + [Isr]/Kr)$$

where ΔVh is the shift of the midpoint of the steady-state inactivation curve, k is the slope factor of the inactivation curve, $[Isr]$ is the (+)isradipine concentration used, and Kr is determined as the potency of (+)isradipine for the resting calcium channel. With $\Delta Vh = 24$ mV, $k = 5.7$ mV, $Kr = 23$ nM and $[Isr] = 20$ mM, K_i is estimated to be 0.13 nM.

These findings support the idea that (+)isradipine not only binds to the resting state of calcium channels but also that it has a higher affinity for the inactivated state (voltage-dependent inhibition).

3.2.4 Characterization of (+)-[^{3}H]isradipine Binding Sites

As the potency of (+)isradipine blockade changes with the holding potential in a manner consistent with a higher affinity to the inactivated state of the calcium channel compared to the resting state, it is of interest to verify whether the isradipine concentration inhibiting 50% of calcium channels in electrophysiological data is similar to the dissociated constant of the high-affinity binding site for (+)-[^{3}H]isradipine determined with radioligand binding studies in microsomal fractions of myometrium. Scatchard analysis of the specific binding of (+)-[^{3}H]isradipine results in a linear plot, thereby indicating specific binding to a single class of sites. The dissociation constant, K_D, is 0.10 ± 0.01 nM, and the maximal binding capacity, B_{max}, is 95 ± 5 fmol/mg of protein ($n = 8$). These results indicate that the K_D value from binding experiments is similar to the apparent dissociation constant obtained from electrophysiological studies and support the existence of a single class of calcium channels.

3.2.5 Modulation of Calcium Channels by Oxytocin

It is well known that oxytocin increases the frequency and the force of spontaneous contractions of the uterus. This effect has been correlated with an increase in calcium current (Mironneau 1976) and a release of calcium from the intracellular stores. The concentration of oxytocin receptors in myometrium depends on the steroid concentration, because estrogens are capable of causing an increase in oxytocin receptor concentration and progesterone antagonizes the estrogen effect (Soloff 1984).

In the presence of estrogens in the culture medium, oxytocin ($2 \times 10^{-10} M$) produces a very small increase of the maximal calcium current. In contrast, when isolated cells from pregnant rats are cultured with $10^{-6} M$ diethylstilbestrol (DES for 12 h) the increase in calcium current induced by $2 \times 10^{-10} M$ oxytocin reaches $55 \pm 8\%$ ($n = 5$). As oxytocin does not affect the apparent reversal potential or the steady-state inactivation curve of the calcium current, it is assumed that oxytocin acts by increasing the calcium conductance of myometrial cells (Mironneau 1990). Further experiments with recordings from single channels are needed to propose a mechanism of action for oxytocin in myometrium.

3.2.6 Changes in Calcium Channels During Gestation

Calcium channel current density, normalized by cell capacitance, significantly increases during gestation (Table 1). The average density of calcium channel current is maximal at day 18 of gestation $(9.2 \pm 1.1\ \mu A/\mu F,\ n = 5)$ and remains elevated near term $(7.3 \pm 2.1\ \mu A/\mu F, n = 8)$ at day 21 of gestation. The electrophysiological properties of calcium channels are not modified during gestation. Threshold for calcium channel current (about -40 mV), the voltage at the peak inward current (about $+10$ mV), and extrapolated reversal potential (about $+65$ mV) are virtually identical in nonpregnant and pregnant rat myometrial cells between days 10 and 21 of gestation. Sensitivity of calcium current to $(+)$isradipine is not affected during gestation as the concentration producing half-inhibition at a holding potential of -80 mV is about $2 \times 10^{-8}\ M$. These results differ from those of Inoue and Sperelakis (1991) showing no change of calcium channel density during pregnancy. In order to confirm the increase in

Table 1. Effects of pregnancy on the membrane parameters of myometrial cells

	Cell surface $(10^{-4}\ cm^2)$	CM $(10^{-12}\ F)$	I_{Ba}^{2+} (pA)	I_{Ba}^{2+}/CM $(\mu A/\mu F)$
Nonpregnant				
$(n = 9)$	0.37 ± 0.04	52.5 ± 5.8	131 ± 23	2.45 ± 0.30
10 days pregnancy				
$(n = 4)$	0.45 ± 0.05	63.2 ± 7.7	208 ± 56	3.10 ± 0.70
18 days pregnancy				
$(n = 5)$	$0.73 \pm 0.11^{**}$	$102.0 \pm 15.6^{*}$	$832 \pm 86^{**}$	$9.20 \pm 1.10^{**}$
21 days pregnancy				
$(n = 8)$	$1.04 \pm 0.08^{**}$	$145.0 \pm 12.3^{**}$	$741 \pm 150^{*}$	$7.30 \pm 2.10^{*}$

CM, membrane capacitance; I_{Ba}^{2+} = maximal barium current; I_{Ba}^{2+}/Cm = barium current density.
$*p < 0.01$; $**p < 0.001$. External solution: 5 mM Ba^{2+}.

Table 2. Effects of pregnancy on the binding parameters of $(+)$-$[^3H]$isradipine to intact strips of myometrium

	K_D (nM)	B_{max} (fmol/mg wet wt)	n
Nonpregnant	0.20 ± 0.02	21.2 ± 1.1	4
10 days pregnancy	0.23 ± 0.02	$27.2 \pm 1.0*$	4
18 days pregnancy	0.23 ± 0.03	$32.5 \pm 0.5*$	4
21 days pregnancy	0.24 ± 0.03	24.0 ± 1.3	4

$*p < 0.001$. External solution: 2 mM Ca^{2+}

calcium channel density near term $(+)$-$[^3H]$isradipine specific binding is determined in intact strips of myometrium during gestation (Table 2). The dissociation constant, K_D, is unchanged while the maximal binding capacity, B_{max}, increases from 21.2 ± 1.1 fmol/mg wet wt at day 10 ($n = 4$) to 32.5 ± 0.5 fmol/mg wet wt at day 18 of gestation ($n = 4$). These results suggest that the calcium channel protein may be synthesized and incorporated in a greater amount into myometrial cells near pregnancy. The mechanisms responsible for the increase in calcium channel density are not known. However, it can be postulated that changes in hormonal status or stretching of the cells caused by fetuses may act as signals for synthesis of calcium channels proteins.

3.3 Voltage-Gated Sodium Channels

In myometrial smooth muscle cells, sodium-dependent action potentials have been first reported using the microelectrode technique (Amédée et al. 1986). These action potentials are totally blocked by micromolar concentrations of tetrodotoxin (TTX). More recently, TTX-sensitive sodium currents have been described in dissociated cells from pulmonary artery (Okabe et al. 1988), portal vein (Mironneau et al. 1990), and pregnant myometrium of rats (Ohya and Sperelakis 1989; Martin et al. 1990) and humans (Young and Herndon-Smith 1991).

3.3.1 Binding Sites for [³H]saxitoxin in Myometrial Membranes

Scatchard plot of the specific binding component for [³H]saxitoxin (STX) indicates the existence of both high- and low-affinity binding sites. The dissociation constants, K_D, and maximal binding capacities, B_{max}, are 0.53 ± 0.11 nM and 39 ± 5 fmol/mg of protein for the high-affinity site, and 27 ± 6 nM and 350 ± 45 fmol/mg of protein for the low-affinity site ($n = 6$).

Therefore, the properties of the high-affinity site are examined using 0.4 nM [³H]STX. Under these conditions about 80% of the signal originates from binding to the high-affinity sites and 20% from binding to the low-affinity sites. Increasing concentrations of unlabelled neurotoxins gradually inhibit [³H]STX binding. The inhibition constant values, K_i, for STX and TTX are 0.18 ± 0.02 and 0.94 ± 0.04 nM, respectively ($n = 4$). The Hill coefficients are close to 1, suggesting the drugs bind to a single family of binding sites (Martin et al. 1990).

The properties of the low-affinity binding site are investigated using 15 nM [³H]STX as 80% of the total binding is associated with the low-affinity binding site. Increasing concentrations of STX and TTX inhibit [³H]STX binding to low-affinity sites. The K_i values for STX and TTX are 55 ± 9 nM ($n = 3$), and 1.2 ± 0.4 μM ($n = 3$), respectively (Martin et al. 1990).

Binding of [³H]STX to dispersed myometrial cells gives similar dissociation constant values for both high- and low-affinity binding sites, indicating that both binding sites for [³H]STX are well located in the plasma membrane of myometrial cells.

3.3.2 Electrophysiological Properties of Sodium Channels

Because the fast sodium current peaks within 2–3 msec, capacitive transients have to be reduced by using small cells and electrodes with a relatively low resistance. As myometrial cells are long cylinders, ranging from 80 to 100 µm in length and from 6 to 12 µm in width (Amédée et al. 1986), the first 1–3 msec are not properly clamped and the peak current amplitude may be underestimated, especially for the high depolarizing pulses. Neurotoxins such as veratridine and toxins from sea anemone have been shown to reveal the voltage-dependent sodium current by stabilizing an open form of the sodium channel (Catterall 1980). When $10^{-4} M$ veratridine is applied in the bath solution, the inward calcium channel current is progressively inhibited within 2–3 min (Romey and Lazdunski 1982). After the pulse, there is a standing inward tail current representing a population of veratridine-modified channels that does not close at –80 mV.

The voltage gating of the sodium current is determined by studying both activation and inactivation curves. Sodium current starts to activate at about –40 mV and reaches a maximum at +20 mV. Half-maximal activation is obtained near –15 mV. The sodium current is fully available at very negative membrane potentials (–110 mV), presents a half-inactivation potential near –70 mV, and is completely inactivated at –40 mV (Martin et al. 1990).

3.3.3 Sodium Channel Blockers

Both STX and TTX applied for 5 min reduce the amplitude of the sodium current in a concentration-dependent manner. The concentrations of STX and TTX inhibiting 50% of the maximal sodium current (IC_{50}) are estimated to be 1.4 ± 0.3 and 8.8 ± 0.2 nM, respectively ($n = 5$). The slope factor of the concentration–response curves is about 0.6, suggesting that the neurotoxins have multiple sites of interaction in myometrial membranes. Subtraction of the current obtained in the presence and absence of TTX reveals that the veratridine-activated sodium current does not inactivate substantially during depolarization.

Interestingly, the dissociation constant of STX measured by electrophysiology (1.4 nM) does not correspond to the K_D of [^{3}H]STX for either high- (0.53 nM) and low-affinity sites (27 nM). These observations suggest that the high- and low-affinity receptors for STX identified by binding experiments could reflect the existence of two different subtypes of STX-sensitive sodium channels, which could correspond to different membrane proteins. Recently, Pidoplichko (1986) reported that two types of sodium currents recorded from ventricular cardiac cells can be distinguished by their sensitivity to TTX ($K_D = 80$ nM and $7 \mu M$). Unequivocal answers to these molecular questions will require more detailed electrophysiological and biochemical experiments in myometrium. However, it is now clear that mammalian sodium channels are encoded by a multigene family and that separate genes encode for high- and low-affinity STX receptors (Rogart et al. 1989).

In isolated myometrial smooth muscle cells, the dissociation constants for TTX against the sodium current range between 8 and 27 nM (Ohya and Sperelakis 1989; Martin et al. 1990), suggesting that the inhibitory effects of TTX may depend largely on an interaction with the high-affinity binding sites for neurotoxins. Thus, the low-affinity binding sites may be associated with a nonfunctional subtype of sodium channels.

3.3.4 Changes in Sodium Channels During Gestation

Changes in sodium channel density have been studied during gestation (Inoue and Sperelakis 1991). There is a quasi-linear increase in average density of sodium current which is maximal near term. This effect is not produced by an increase in sodium current density per cell, but rather from an increase in the fraction of cells which possess fast sodium channels. The role of the fast sodium channel is not known yet. However, the insertion of fast sodium channels into the cell membrane may cause an elevation of intracellular calcium concentration by the sodium–calcium exchange system functioning in the reverse mode which can, in turn, potentiate the myometrial contraction.

It has been reported, using multicellular preparations of pregnant rat myometrium, that in midpregnancy the ratio of calcium current to sodium current is 0.57, whereas at term the ratio decreases to 0.31, indicating that the importance of sodium current increases near term (Nakai and Kao 1983). However, it has to be noted that at term the resting membrane potential is rather depolarized (–45 mV) and the contribution of sodium channels to excitation–contraction is expected to be limited. In contrast, at midpregnancy, the resting membrane potential ranges between –60 and –70 mV. In this voltage range, sodium channels are largely activatable and may contribute to spike generation and cell-to-cell conduction. Further experiments are necessary to clarify the physiological role of the sodium channels in pregnant myometrial smooth muscle cells.

3.4 Ligand-Gated Nonspecific Channels

Adenosine trisphosphate (ATP) has effects similar to those of non-adrenergic, noncholinergic nerves in various smooth muscles, including myometrium (Burnstock 1972). Furthermore, evidence has accumulated indicating that ATP is involved in physiological regulation of nonvascular smooth muscle tone through liberation from "purinergic" nerve endings.

3.4.1 Electrophysiological Properties of ATP-activated Channels

Application of 1 mM ATP (for 2 s) on a single myometrial cell maintained at a holding potential of −70 mV produces an inward current that is fast in onset (less than 100 ms) and has a short time to peak (less than 1 s). Furthermore, it is possible to obtain several similar responses to ATP in the cell without any decline in the amplitude of the response. The absence of desensitization in myometrium clearly differs from the high desensitization observed in other smooth muscle. In fact, it has been reported that ATP responses are facilitated in the myometrium (Honoré et al. 1989).

The concentration–response curve is fitted well by assuming that a receptor needs to bind one ATP molecule at a single site to activate current. From the fitted curve, the interpolated ATP concentration for half-activation of the conductance is 0.66 mM. However, increasing the divalent cation concentration from 2.5 to 7.5 mM results in a shift of the concentration–response curve to higher concentrations of ATP, as expected from an ATP-chelation mechanism. Therefore, it is likely that ATP^{4-} is the only effective form of ATP acting in the myometrium, and the concentration for half-activation of the conductance is then decreased to 0.09 mM ATP^{4-}.

By changing the ionic composition of the internal and external solutions it has been established that the ATP-activated channel is cation selective but with little discrimination, at least between sodium, potassium, and caesium ions. It is interesting that there is a measurable, although very small, inward current remaining when sodium ions are substituted with choline (in the presence of atropine), suggesting that the channel could be permeable to divalent cations. However, increasing the external barium concentration decreases and suppresses the ATP-activated current. This observation does not imply that the channel activated by ATP in myometrial cells is not permeable to calcium ions, since, owing to calcium chelation of ATP, the free-acid form is reduced to subthreshold levels (with $10^{-3}\,M$ ATP in the standard conditions, ATP free-acid form is reduced to $1.2 \times 10^{-6}\,M$). Further work is needed to resolve the question of whether calcium is permeant through the ATP-activated channel in myometrium.

3.4.2 Pharmacology of ATP-activated Channels

The ligand specificity and pharmacology of the ATP-activated channel in myometrium is most similar to the P_2 type of receptors. Like other responses mediated by P_2 receptors in smooth muscle, the ATP-activated response in myometrium is absolutely specific for ATP over adenosine, adenosine monophosphate (AMP), and adenosine diphosphate (ADP). The finding that ATP is by far the most effective phosphorylated adenosine derivative in activating the channels raises the possibility that ATP might actually be used as a substrate in some enzymatic reactions. ATP-γ-S and α,β-methylene ATP are found to be capable of mimicking the action of ATP, although with a lower potency. These results are consistent with the idea that the channel pathway is governed by a ligand receptor with a very high specificity for ATP but not necessarily involving hydrolysis of ATP.

The cellular mechanism underlying ATP-activated channels is calcium-independent, as ATP-activated current is obtained in the presence of both external barium (replacing calcium) and internal ethyleneglycol-bis-β-aminoethylether-N,N,N',N'-tetraacetic acid (EGTA). The two other possibilities would be either a direct control of a channel closely linked to a ligand receptor or the generation of a chemical second messenger. It is quite possible that activation of the ATP-activated channel may involve a chemical second messenger as the onset kinetics of ATP-activated are relatively fast (< 100 msec) and, furthermore, ATP-activated current does not undergo any desensitization. These possibilities remain, however, to be explored in more detail.

3.4.3 Changes in ATP-activated Channels During Gestation

The response of myometrial cells to ATP is dependent on the stage of pregnancy. In rat myometrial cells isolated at day 21 of gestation, the amplitude of the ATP-activated current is tenfold smaller than the response elicited in cells isolated at day 10 and 19 of gestation. Furthermore, pretreatment of cultured cells isolated at day 19 of gestation with $10^{-5}\,M$ DES for 15 h depresses the amplitude of the ATP response, which becomes similar to that obtained at day 21 of gestation. Therefore, the ATP-induced response appears to be strongly reduced near term in response to estrogen synthesis.

3.5 Relationship Between Ion Channels and Contraction

Control of myometrial contractility can be understood in terms of ion channel functioning. Many of these channels are primarily voltage dependent and the probability of their being in the open, conducting state is determined by membrane potential. Although some channel types can be directly activated by ligands, it is accepted that membrane potential during pregnancy is important in determining myometrial contractility.

In the myometrium, as in other smooth muscles, contraction is initiated when the concentration of free calcium in the cytoplasm increases from $10^{-7}\,M$ to $10^{-6}\,M$. This calcium may be released from stores within the cell, but the most important source of calcium for durable and repeated contractions is that which enters the cell from the extracellular space (Mironneau 1973). Therefore, calcium channels are considered to be the fundamental units regulating myometrial contractility.

Other channels provide specific conductance pathways for sodium, potassium, and chloride or are permeable for cations without strong discrimination between mono- and divalent cations. Although it is not well established that the nonspecific cation channels are permeable to calcium ions, all these channels control the membrane potential and thus regulate activity of the calcium channels because of their voltage dependence. Therefore, it appears that it is the total of the channel activity in the myometrial cell membrane which governs action potential discharge and contraction.

Acknowledgements. This work was supported by grants from Centre National pour la Recherche Scientifique, Institut National de la Santé et de la Recherche Médicale, Centre National des Etudes Spatiales, Association Française contre les Myopathies, and Région Aquitaine. I thank Ms. Biendon for her excellent assistance.

References

Amédée T, Renaud JF, Jmari K, Lombet A, Mironneau J, Lazdunski M (1986) The presence of Na channels in myometrial smooth muscle cells is revealed by specific neurotoxins. Biochem Biophys Res Comm 137:675–681

Amédée T, Mironneau C, Mironneau J (1987) The calcium channel current of pregnant rat single myometrial cells in short-term primary culture. J Physiol 392:253–272

Bean B (1984) Nitrendipine block of cardiac calcium channels: high affinity binding to the inactivated state. Proc Natl Acad Sci USA 81:1–30

Benham CD, Bolton TB, Lang RJ (1985) Acetylcholine activates an inward current in single mammalian smooth muscle cells. Nature 316:345–347

Benham CD, Hess T, Tsien R (1987) Two types of calcium channels in single smooth muscle cells from rabbit ear artery studied with whole-cell and single-channel recordings. Circ Res 61:I10-I16

Burnstock G (1972) Purinergic nerves. Pharmacol Rev 24:509–581

Catterall WA (1980) Neurotoxins that act on voltage-sensitive sodium channels in excitable membranes. Ann Rev Pharmacol Toxicol 20:15–43

Coleman HA, Parkington HC (1987) Single channel Cl$^-$ and K$^+$ currents from cells of uterus not treated with enzymes. Pflügers Arch 410:560–562

Dacquet C, Pacaud P, Loirand G, Mironneau C, Mironneau J (1988) Comparison of binding affinities and calcium current inhibitory effects of a 1,4-dihydropyridine derivative (PN 200-110) in vascular smooth muscle. Biochem Biophys Res Comm 152: 1165–1172

Eckert R, Chad JD (1984) Inactivation of Ca channels. Prog Biophys Mol Biol 44:215–267

Fleckenstein A (1977) Specific pharmacology of calcium ion in myocardium, cardiac pacemakers, and vascular smooth muscle. Annu Rev Pharmacol Toxicol 17:149–166

Fournier F, Honoré E, Brûlé G, Mironneau J, Guilbault P (1989) Expression of Ba currents in Xenopus oocyte injected with pregnant rat myometrium mRNA. Pflügers Arch 413:682–684

Fox A, Nowycky M, Tsien R (1987) Kinetic and pharmacological properties distinguish three types of calcium currents in chick sensory neurones. J Physiol 394:149–172

Friel D (1988) An ATP-sensitive conductance in single smooth muscle cells from the rat vas deferens. J Physiol 401:361–380

Hamilton S, Yatani A, Brush K, Schwartz A, Brown A (1987) A comparison between the binding and electrophysiological effects of dihydropyridines on cardiac membranes. Mol Pharmacol 31:221–226

Hess P, Tsien RW (1984) Mechanism of ion permeation through calcium channels. Nature 309:453–455

Hille B, Schwarz W (1978) Potassium channels as multi-ion single-file pores. J Gen Physiol 72:409–442

Hollingsworth M, Downing S (1988) Calcium entry blockers and the uterus. Med Sci Res 16:1–16

Honoré E, Martin C, Mironneau C, Mironneau J (1989a) An ATP-sensitive conductance in cultured smooth muscle cells from pregnant rat myometrium. Am J Physiol 257:C297-C309

Honoré E, Amédée T, Martin C, Dacquet C, Mironneau J (1989b) Calcium channel current and its sensitivity to (+)isradipine in cultured pregnant rat myometrial cells. Pflügers Arch 414:477–483

Inoue Y, Sperelakis N (1991) Gestational change in Na^+ and Ca^{2+} channel current densities in rat myometrial smooth muscle cells. Am J Physiol 260:C658-C663

Inoue Y, Nakao K, Okabe K, Izumi H, Kanda S, Kitamura K, Kuriyama H (1990) Some electrical properties of human pregnant myometrium. Am J Obstet Gynecol 162:1090–1098

Jmari K, Mironneau C, Mironneau J (1986) Inactivation of calcium channel current in rat uterine smooth muscle: evidence for calcium and voltage-mediated mechanisms. J Physiol 380:111–126

Jmari K, Mironneau C, Mironneau J (1987) Selectivity of calcium channels in rat uterine smooth muscle: interactions between sodium, calcium and barium ions. J Physiol 384:247–261

Kokubun S, Prod'Hom B, Becker C, Porzig H, Reuter H (1987) Studies on Ca channels in intact cardiac cells: voltage-dependent effects and cooperative interactions of dihydropyridine enantiomers. Mol Pharmacol 30:571–584

Loirand G, Pacaud P, Mironneau C, Mironneau J (1986) Evidence for two distinct calcium channels in rat vascular smooth muscle cells in short-term primary culture. Pflügers Arch 407:566–568

Loirand G, Mironneau C, Mironneau J, Pacaud P (1989) Two types of calcium currents in single smooth muscle cells from rat portal vein. J Physiol 412:333–349

Martin C, Arnaudeau S, Jmari K, Rakotoarisoa L, Sayet I, Dacquet C, Mironneau C, Mironneau J (1990) Identification and properties of voltage-sensitive sodium channels in smooth muscle cells from pregnant rat myometrium. Mol Pharmacol 38:667–673

Mironneau J (1973) Excitation-contraction coupling in voltage-clamped uterine smooth muscle. J Physiol 233:127–141

Mironneau J (1976) Effects of oxytocin on ionic currents underlying rhythmic activity and contraction in uterine smooth muscle. Pflügers Arch 363:113–116

Mironneau J (1990) Ion channels and excitation-contraction coupling in myometrium. In: Garfield R (ed) Uterine contractility, Serono Symposia, Norwell, pp 9–19

Mironneau J, Martin C, Arnaudeau S, Jmari K, Rakotoarisoa L, Sayet I, Mironneau C (1990) High affinity binding sites for [^{3}H]saxitoxin are associated with voltage-dependent sodium channels in portal vein smooth muscle. Eur J Pharmacol 184:315–319

Nakai Y, Kao CY (1983) Changing properties of Na^+ and Ca^{2+} components of the early inward current in the rat myometrium during pregnancy. Fed Proc 42:313

Nilius B, Hess P, Lansman JB, Tsien RW (1985) A novel type of cardiac calcium channels in ventricular cells. Nature 316:443–446

Ohya Y, Sperelakis N (1989) Fast Na^+ and slow Ca^{2+} channels in single uterine muscle cells from pregnant rats. Am J Physiol 257:C408-C412

Okabe K, Kitamura K, Kuriyama H (1988) The existence of a highly tetrodotoxin sensitive Na channel in freshly dispersed smooth muscle cells of the rabbit main pulmonary artery. Pflügers Arch 411:423–428

Pacaud P, Loirand G, Mironneau C, Mironneau J (1989) Calcium-activated chloride current in rat vascular smooth muscle cells in short-term primary culture. Pflügers Arch 413:629–636

Pidoplichko VI (1986) Two different tetrodotoxin-separable inward sodium currents in the membrane of isolated cardiomyocytes. Gen Physiol Biophys 6:593–604

Rogart RB, Cribbs LL, Muglia LK, Kephart DD, Kaiser MW (1989) Molecular cloning of a putative tetrodotoxin-resistant rat heart Na^+ channel isoform. Proc Natl Acad Sci USA 86:8170–8174

Romey G, Lazdunski M (1982) Lipid-soluble toxins thought to be specific for Na^+ channels block Ca^{2+} channels in neuronal cells. Nature 297:79–80

Savineau JP, Mironneau C, Mironneau J (1988) Contractile properties of chemically skinned fibers from pregnant rat myometrium: existence of an internal Ca-store. Pflügers Arch 411:296–303

Soloff MS (1984) Regulation of oxytocin action at the receptor level. In: Bottari S, Thomas JP, Vokaer A, Vokaear R (eds) Uterine contractility. Masson, Paris, pp 261–264

Sturek M, Hermsmeyer K (1986) Calcium and sodium channels in spontaneously contracting vascular muscle cells. Science 233:475–478

Tsien RW, Hess P, McCleskey EW, Rosenberg RL (1987) Calcium channels: mechanisms of selectivity, permeation and block. Annu Rev Biophys Chem 16:265–290

Yoshino M, Someya T, Nishio A, Yazawa K, Usuki T, Yabu H (1989) Multiple types of voltage-dependent Ca channels in mammalian intestinal smooth muscle cells. Pflügers Arch 414:401–409

Young RC, Herndon-Smith L (1991) Characterization of sodium channels in cultured human uterine smooth muscle cells. Am J Obstet Gynecol 164:175–181

4 Role of Uterine Ca^{2+} Pumps and Na^+ Pumps in Labor: A Molecular Biology Approach

Ashok K. Grover, Islam Khan, Thomas Tabb,
and Robert E. Garfield

4.1 Introduction

Premature delivery and postterm pregnancy are not just medical problems facing health care, but the accompanying complications and problems also lead to morbidity and mortality. One of the hurdles in management and prevention of such problems is a lack of understanding of the underlying physiological processes which may lend themselves to better pharmacological or other interventions. This has led to studies on understanding processes involved in the onset of labor.

4.2 Ion Movements in Contractility

Ca^{2+} ions are key messengers in the onset of labor because they play a pivotal role in the contractility of uterine smooth muscle (Carafoli 1987; Grover 1986, 1987; Grover and Daniel 1985; Savineau et al. 1988). The contractility machinery for uterine smooth muscle consists of actomyosin molecules. In general, less than 0.1 μM cytosolic $[Ca^{2+}]$ leads to relaxation and an increase in this concentration results in contraction. In addition, the activation of actomyosin may also be modulated by other second messengers.

Cytosolic concentrations of Na^+ and K^+ are 15 and 130 mM, respectively, while extracellular fluid contains 130 mM Na^+ and 5 mM K^+. There is also a membrane potential of –40 to –60 mV. Cytosolic $[Ca^{2+}]$ in resting cells is approximately 0.1 μM (Anwer et al. 1990), while it is 1 mM or higher outside the cells and in the sarcoplasmic reticulum (SR). The cells may be depolarized in response to electrical or chemical stimulation and the depolarization results in opening of voltage sensitive calcium channels (Batra 1990; Bengtsson et al. 1984; Boyle et al. 1987; Breuiller et al. 1990; Enyedi et al. 1989; Grover et al. 1983b,1984; Grover 1987; Koch et al. 1990; Magocsi and Penniston 1991) and thus, allows cytosolic $[Ca^{2+}]$ to rise. Cytosolic $[Ca^{2+}]$ may also be increased by release of Ca^{2+} sequestered in the SR. The relaxation process therefore has to involve the restoration of membrane potential and removal of cytosolic Ca^{2+} (Grover 1985).

The Na^+ pump hydrolyzes $MgATp^{2-}$ and utilizes the energy of this hydrolysis to remove 3 Na^+ from cytosol and introduce 2 $K^{+\cdot}$ Thus, the pump is responsible not only for maintenance of Na^+ and K^+ gradients but is also electrogenic. The Na^+ pump is pivotal in membrane repolarization after a membrane excitation.

Once a contraction has occurred, the cells must relax before contracting again. The removal of Ca^{2+} from the cytosol is thus equally important for the contraction–relaxation cycle. Obviously, the removal of Ca^{2+} from cytosol is not thermodynamically favorable and hence it must be coupled to some other energetically favored reactions. Although other mechanisms may be important in this process (Grover et al. 1981,1983a; Grover 1987), the major mechanisms for Ca^{2+} removal are Ca^{2+} pumps (Carafoli 1987; Grover 1987; Inesi et al. 1990; Strehler 1990). These pumps can utilize the energy of hydrolysis of adenosine triphosphate (ATP) transport Ca^{2+} (Inesi et al. 1990). There are two types of Ca^{2+} pumps – one type located in the plasma membrane (PM) and another in the SR (Grover 1985; Grover and Khan 1991). The PM Ca^{2+} pump removes Ca^{2+} from the cell to the outside, and the SR pump can sequester Ca^{2+} into the SR. The two pumps are distinct in their size, structure, immunoreactivity and regulation.

Physiologically, the existence of two types of Ca^{2+} pumps can be understood as follows: The excitation involving electromechanical coupling in several tissues has an obligatory dependence on extracellular Ca^{2+} which may be supplemented by the Ca^{2+} released from SR. The contraction of skeletal and cardiac muscle depends heavily on the release of Ca^{2+} from SR. The excitation of uterine smooth muscle by agents such as oxytocin can also occur by utilizing the Ca^{2+} released from SR even when there is no extracellular Ca^{2+} (Anwer et al. 1990). When the cell excitation follows Ca^{2+} entry from the extracellular milieu, repeated excitation would result in an accumulation of this Ca^{2+} in the cell and this Ca^{2+} accumulation would eventually lead to cell death. Therefore, the cell needs the Ca^{2+} pumps in the PM to remove the Ca^{2+} acquired after such excitation. Similarly, when Ca^{2+} released from the SR is used for the excitation, a failure to refill the SR Ca^{2+} pool would result in an inability to respond to repeated challenges. Therefore, Ca^{2+} pumps are needed for transporting Ca^{2+} back to SR, but alternative models of agonist inducing SR-refilling have also been proposed. Thus, it is obligatory to have Ca^{2+} pumps to remove Ca^{2+} from the cell into the extracellular space as well as to the SR. For finer control, the two types of pumps must also dimer in their regulatory properties.

Rat myometrium membrane is more depolarized in pregnant than in nonpregnant animals (Bengtsson et al. 1984; Table 1). This change accompanies an increase in the current density of fast Na$^+$ channels from 0 at day 5 to 0.9 pA/pF at day 21 (Inoue and Sperelakis 1991). These channels carry a fast inward Na$^+$ current which increases cytosolic Na$^+$ and results in membrane depolarization. There is also an increase in the slow activating K$^+$ channels during pregnancy (Boyle et al. 1987). Compared to the uterus in the earlier stages of gestation, in late pregnancy the myometrium is more excitable electrically, its threshold for K$^+$-induced contraction is decreased, and its excitability to a number of other agents is increased (Bengtsson et al. 1984). The number of Ca^{2+} channels does not change (Batra 1990). However, delivering animals

Table 1. Membrane properties of pregnant rat uterus

Membrane potential	Decreased during pregnancy (Bengtsson et al. 1984)
Electrical excitability	In late pregnancy (Bengtsson et al. 1984)
Excitability to high K^+	In late pregnancy (Bengtsson et al. 1984)
Oxytocin receptors	During delivery (Soloff et al. 1979; Soloff and Sweet 1982)
Na^+ channels (fast)	Increased during pregnancy (Inoue and Sperelakis 1991)
K^+ channels (slow-activating)	Increased during pregnancy (Boyle et al. 1987)
Ca^{2+} channels (L-type)	No change in number (Batra 1990)
PM Ca^{2+} pump	No change in mRNA or protein from day 15 to day 2 (detailed studies needed; Khan et al. 1992b; Magocsi and Penniston 1991)
SR Ca^{2+} pump	Increase in mRNA, protein and acylphosphates from day 15 to day 22 (detailed studies needed; Khan et al. 1992a,b)
Na^+ pump	Decrease in K^+-PNPPase from day 15 to day 22 (Fig. 1)

PNPP, p-nitrophenylphosphate

have more oxytocin receptors, which have been reported to excite the myometrial cells using an intracelluar Ca^{2+} pool (Anwer et al. 1990; Soloff et al. 1979; Soloff and Sweet 1982). At delivery, there is also an increase in the number of gap junctions which allow the uterus to act as a syncytium (see Chap. 1, this volume; Lang et al. 1991). The remainder of the chapter will be focused on possible roles of Na^+ and Ca^{2+} pumps in relation to these changes.

4.3 Ca^{2+} Pumps

4.3.1 Biochemical and Physiological Studies

Contractility, ion flux, and biochemical studies have contributed significantly to establishing the existence of Ca^{2+} pumps as Ca^{2+}–Mg^{2+}–ATPases which utilize the energy of hydrolysis of ATP to transport Ca^{2+} (Carafoli 1987; Grover 1986, Grover and Kwan 1983; Grover et al. 1980,1982; Soloff and Sweet 1982; Spencer et al. 1990,1991; Spencer and Grover 1990; Ver et al. 1989). It has also been reported that the PM Ca^{2+} pumps were activated by calmodulin and the SR pumps by phospholamban. However, for studying the Ca^{2+} pumps in the uterus, there are experimental problems due to lack of availability of pure SR fractions, degradation of Ca^{2+} pumps during membrane isolation by proteases or by reactive oxygen-induced pathways (Grover and Kwan 1983; Magocsi and Penniston 1991), and the suggestion that there may also be subtle differences between various tissues (Eggermont et al. 1990). These have made the interpretation of some of the data and further progress difficult.

4.3.2 Molecular Biology of Ca^{2+} Pumps

4.3.2.1 Molecular Biology of PM Calcium Pumps

Based on cloning information and the subsequent northern blot studies, there are at least four genes which encode for the PM Ca^{2+}pump, namely, PMCA1, PMCA2, PMCA3, and PMCA4 (De Jaegere et al. 1990; Gonzalez et al. 1992; Greeb and Shull 1989; Grover and Khan 1991; James et al. 1989; Khan and Grover 1991, Khan et al. 1992b; Shull and Greeb 1988; Verma et al. 1988). The mRNA of all the four genes encode 130- to 140-kDa proteins containing sites for binding to ATP, calmodulin, fluorescein isothiocyanate (FITC), calcium (putative), high energy acylphosphate formation and the domain for translocation. Only the product of PMCA1 is distributed widely (Greeb and Shull 1989). The PMCA4 gene is expressed at much lower levels in several tissues and its expression is the highest in kidney and brain (Gonzalez et al. 1992; Greeb and Shull 1989).

The PMCA1 gene contains a 154bp exon in its 3'-region of the protein coding sequence. This exon can be spliced so as to retain 0, 87, 114 or 154 bp in the mRNA (Khan and Grover 1991). The downstream sequence in the splices containing 0, 87 or 114 bp encodes for a peptide which can be phosphorylated by cyclic nucleotide-dependent protein kinase. In erythrocyte ghosts, the cAMP protein kinase-dependent phosphorylation of this peptide leads to an increase in velocity and affinity of the pump for Ca^{2+} (James et al. 1989). This exon was excluded in the cDNA clones obtained from human teratoma, rabbit and pig stomach smooth muscle and included in a clone reported from rat brain (De Jaegere et al. 1990; Greeb and Shull 1989; Grover and Khan 1991; James et al. 1989; Khan and Grover 1991; Khan et al. 1992a; Shull and Greeb 1988; Verma et al. 1988). Based on studies using S1-nuclease or reverse transcription followed by the polymerase chain reaction (PCR), the existence of various transcripts was established (Khan and Grover 1991). The abundant mRNA expressed in most tissues is PMCA1b, which encodes for the cyclic nucleotide sensitive isoform, but brain mRNA contains equal proportions of mRNA encoding for the sensitive and the insensitive isoforms (Khan and Grover 1991).

4.3.2.1.1 Expression of PM Ca^{2+}Pumps in Uterus

The uterus expresses predominantly PMCA1b, although it contains some transcripts for PMCA1a and PMCA2 (Khan et al. 1992a; Shull and Greeb 1988). Based on two limited preliminary studies, there is no change in the level of expression of PM Ca^{2+} during pregnancy (Khan et al. 1992a; Ver et al. 1989). The mRNA for PMCA1a or PMCA1b mRNA does not change from day 15 to 22 of gestation in rats. Furthermore, PM Ca^{2+} pump content measured using Ca^{2+} transport or using antibodies does not change during this period (Magocsi and Penniston 1991). The delivering uterus, however, may express PM Ca^{2+} pumps which are partially inhibited by oxytocin (Enyedi et al. 1989, Magocsi and Penniston 1991; Soloff and Sweet 1982). Whether this property represents a change in another PMCA gene expression or a difference in the microenvironment of the PM Ca^{2+} pump remains to be determined.

4.3.2.2 Molecular Biology of SR Ca^{2+} Pumps

The cDNA for the internal Ca^{2+} pump has been cloned from several tissues and three genes encoding it have been identified, namely, SERCA1, SERCA2 and SERCA3 (Burk et al. 1989; de la Bastle et al. 1988; Eggermont et al. 1989; Gunteski-Hamblin et al. 1988; Khan and Grover 1990; Khan et al. 1990; Lytton and MacLennan 1988; Mac-Lennan et al. 1985). SERCA1 gene encodes the SR Ca^{2+} pump isoforms expressed mainly in the fast twitch skeletal muscle. SERCA2 gene encodes the isoforms expressed in the slow twitch skeletal, cardiac or smooth muscle, and several other tissues. SERCA3 is expressed mainly in the large intestine and spleen. All three pumps contain the protein sequences necessary for binding to ATP and for the acylphosphate formation. SERCA1 and SERCA2 Ca^{2+} pump proteins can bind phospholamban. It has been reported that SERCA3 protein has sequences which are potential sites for the cAMP-dependent protein kinase phosphorylation but it remains to be shown that the protein is phosphorylated and/or regulated in this manner (Burk et al. 1989).

The cardiac muscle RNA is also spliced alternatively so as to produce the isoforms SERCA2a or SERCA2b. SERCA2a and SERCA2b differ in their 3'-sequences such that the last four amino acids of SERCA2a are replaced by 49 different ones in SERCA2b, and the 3'-untranslated regions are also different. Further splicing in the 3'-untranslated region has also been reported. From the cDNA data, it is predicted that SERCA2a encodes a 110-kDa protein and SERCA2b encodes a protein of 115-kDa. The heart and the slow twitch skeletal muscle express mainly the splice SERCA2a, where as smooth muscle, brain, and most other tissues express the splice SERCA2b. The presence of these splices has also been confirmed using antibodies against splice-specific segments and by the size of the acylphosphate intermediates formed. However, it is not known whether these proteins differ in any function due to this difference in the C-terminal region. There is no more identity between the PM and the internal Ca^{2+} pumps than between these and the other ion pumps.

The SR Ca^{2+} pump isoforms SERCA1 and SERCA2 can be regulated by phospholamban. Phospholamban is a small protein which can be phosphorylated by cyclic AMP-dependent protein kinase. The dephosphorylated but not the phosphorylated phospholamban inhibits the SR Ca^{2+} pump (Inesi et al. 1990).

4.3.2.2.1 Expression of SR Ca^{2+} Pumps in Uterus

The uterus expresses mainly the SR Ca^{2+} pump isoform SERCA2b, but it also contains some transcripts for SERCA3 (Greeb and Shull 1989; Khan et al. 1992c; Lytton and MacLennan 1988). Since SERCA2b is the major SR Ca^{2+} pump expressed in the uterus, there is a potential for regulation by phospholamban. However, phospholamban cannot be detected in the rat uterus (Khan et al. 1992b). In a preliminary study, the expression of SERCA2 mRNA was examined

Ashok K. Grover et al.

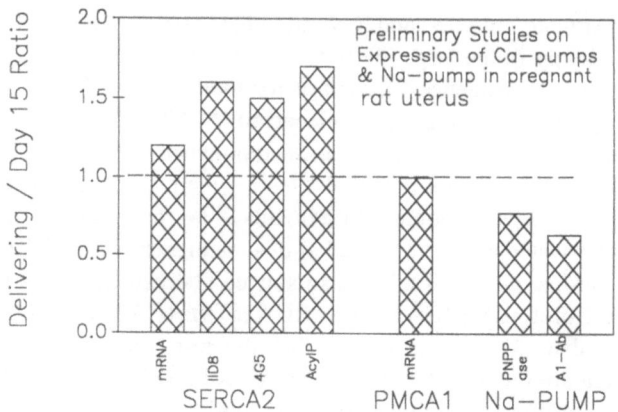

Fig. 1. All values are ratios of mean values of six to nine delivering rat uteri to those of six to nine day 15 rats. The value of the ratios was significantly higher than 1 ($p < 0.05$) for SERCA2 mRNA, protein, and acylphosphates, but not for the PMCA1 mRNA. SERCA2 mRNA was estimated as SERCA2 mRNA to 28S rRNA ratio by PARIS. Protein was estimated using a monoclonal (4G5) and a polyclonal (anti-87) antibody against SERCA2. Ca^{2+}-dependent acid stable alkali labile formation of 115-kDa acylphosphates was examined as described elsewhere (Khan et al. 1992a,b). K^+-activated ouabain-sensitive p-nitrophenylphosphatase (PNPPase) and reactivity to anti-α-1 Na^+ pump subunit antibody reactivity were significantly ($p < 0.05$) lower in day 22 pregnant rats than in day 15 rats

using PARIS (PCR assay of RNA with internal standard) and the mRNA/28S ribsomal RNA (rRNA) ratio was higher in delivering rats than at day 15 of gestation (Fig. 1) (Khan et al. 1992a,b). The expression was also examined using two antibodies, and a higher level of expression was observed in tissues from delivering rats over day 15 uteri. The activity of the SR Ca^{2+} pump was examined as Ca^{2+}-dependent acylphosphate formation of 115-kDa band and the activity was higher in uterine tissues of the delivering animal than in tissues from day 15 animals (Fig. 1). In another study, it was reported that there was a partial oxytocin inhibition of the SR calcium pump in delivering animals (Magocsi and Penniston 1991). It is not clear if a different isoform is expressed in delivering animals, whether there is a change in the microenvironment of the pump, or whether the reported

calcium uptake was really due to SR. There is need for a more comprehensive study using the various tools and uterine tissues at various
stages of gestation.

4.4 Na$^+$ Pump

4.4.1 Biochemical and Physiological Studies

Na$^+$ pump in rat uterus has been examined using contractility assays
and isotopic fluxes and biochemically. However, quantitative information on the Na$^+$ pump levels during pregnancy is scarce. Biochemical
studies are complicated by the existence of a very active MgATPase
which contributes to extremely high background activity when activation by Na$^+$ and K$^+$ or inhibition by ouabain or vanadate are studied
(Grover 1986, 1987; Grover et al. 1980). K$^+$-activated ouabain-sensitive hydrolysis of p-nitrophenylphosphate (PNPP), which is a partial
reaction of Na–K–ATPase, has been used extensively. However, even
with this assay there are no comprehensive studies during pregnancy.

4.4.2 Molecular Biology

Na$^+$ pump consists of two types of subunits – α and β (Herrera et al.
1987). The α-subunit is responsible for most functions and the β-sub-
unit is used for assembly of the Na$^+$ pump into the membrane.

   The α-subunit of Na$^+$ pump is encoded by a family of genes desig-
nated as α-1, α-2 and α-3 (Herrera et al. 1987). All the isoforms are
90–100 kDa proteins and contain domains for binding to Na$^+$, K$^+$,
ATP, FITC and the inhibitor ouabain. The different isoforms dimer in
their affinities for Na$^+$, K$^+$ and ATP (Jewell and Lingrell 1991). Particularly interesting is the α-3 isoform which has a lower affinity for
Na$^+$. β-subunits are much smaller proteins (30 kDa) than the α-sub-
units and are encoded by two genes: β-1 and β-2 (Herrera et al. 1987).
Na$^+$ pumps have been induced or repressed in an isoform-specific
manner in response to changes in ionic status and hormones and in
some forms of hypertension.

4.4.3 Expression in Uterus

Of the various isoforms of the α-subunit, the uterus contains transcripts mainly for α-1, but may also express α-2 and α-3 at very low
levels (Herrera et al. 1987). Of the β-isoforms, it expresses mainly β-2
isoforms but may also contain some mRNA for β-1. Isoform-selective
antibodies are commercially available and can be used to confirm the
expression. Using these antibodies and using ouabain-sensitive K^+-
PNPPase, it has been shown that there is a decrease in Na^+ pump expression from day 15 to day 21 of pregnancy in rat uterus (Fig. 1).
Comprehensive studies to explore this further and to understand its implications are needed.

4.5 Significance of Further Molecular Biology Studies

Difficulties in Ca^{2+} transport studies and ATPase assays to monitor
PM and SR calcium pumps have been described. There are further
problems in using these methods for studying expression since many
of these activities may be inactivated by free radicals whose production may vary with the isolation methods, the gestational stage, the
membrane permeability, and the vesicle orientation or size may also
differ. Today the availability of appropriate probes and more sensitive
PCR-based methods for estimating mRNA would permit the examination of changes in mRNA. Similarly, the availability of sequence information has allowed the preparation of sequence-specific antibodies
for such studies. These studies do not account for the amount or for the
isoforms expressed. The knowledge of the isoforms expressed is important because different isoforms may be regulated differently. It is
anticipated that molecular biology studies on the Ca^{2+} and Na^+ pumps
will provide a clearer understanding of the relaxation processes involved in uterine contractions and perhaps contribute to their rhythmicity.

These studies should lead to homologous expression, resulting in
enrichment of desired messages by introducing specified DNA (Nabel
et al. 1990) into the cells of the myometrium and examining the effects
on consequent labor.

Acknowledgments. The authors gratefully acknowledge the secretarial assistance of T. Roketta. AKG is a recipient of a Career Investigator Award from the Heart and Stroke Foundation of Ontario.

References

Anwer K, Hovington JA, Sanborn BM (1990) Involvement of protein kinase A in the regulation of intracellular free calcium and phosphinositide turnover in rat myometrium. Biol Reprod 43:851–859

Batra S (1990) Influence of chronic oestrogen treatment on the density of muscarinic cholinergic receptors and calcium channels in the rabbit uterus. J Endocrinol 125:185–189

Bengtsson B, Chow EMH, Marshall JM (1984) Activity of circular muscle of rat uterus at different times in pregnancy. Am J Physiol 246:C216-C223

Boyle MB, MacLuskky NJ, Naftolin F, Kaczmarek LK (1987) Hormonal regulation of K$^+$-channel messenger RNA in rat myometrium during oestrus cycle and in pregnancy. Nature 330:373–375

Breuiller M, Doualla-Bell F, Litime MH, Leroy MJ, Ferrd F (1990) Disappearance of human myometrial adenylate cyclase activation by prostaglandins at the end of pregnancy. Comparison with β-adrenergic response. In: Sameulsson B et al. (eds) Advances in prostaglandin, thromboxane, and leukotriene research. Raven, New York, Vol 21:811–814

Burk S, Lytton J, MacLennan DH, Shull G (1989) cDNA cloning, functional expression, and mRNA distribution of a third organellar Ca-pump. J Biol Chem 264:18561

Carafoli E (1987) Intracellular calcium homeostasis. Ann Review Biochem 56:395

De Jaegere S, Wuytack F, Eggermont JA, Verboomen H, Casteels R (1990) Molecular cloning and sequencing of the plasma-membrane Ca^{2+} pump of pig smooth muscle. Biochem J 271:655

de la Bastie D, Wisnewsky C, Schwartz K, Lompre A (1988) (Ca^{2+} + Mg^{2+}) dependent ATPase mRNA from smooth muscle SR differs from that in cardiac and fast skeletal muscles. FEBS LETT 229:45

Eggermont J, Wuytack F, De Jaegere S, Nelles L, Casteels R (1989) Evidence for two isoforms of the ER Ca-pump in pig smooth muscle. Biochem J 260:757

Eggermont JA, Wuytack F, Verbist J, Casteels R (1990) Expression of endoplasmicreticulum Ca^{2+}-pump isoforms and of phospholamban in pig smooth-muscle tissues. Biochem J 271:649

Enyedi A, Vorherr T, James P, McCormick D, Filoteo A, Carafoll E, Pennis-
    ton J (1 989) The calmodulin binding domain of the PM Ca-pump interacts
    both with calmodulin and with another part of the pump. J Biol Chem
    264:12313
Gonzalez JM, Dalmeida W, Abramowitz J, Suki WN (1992) Evidence for a
    fourth rat isoform of the plasma membrane calcium pump in the kidney.
    Biochem Biophys Res Comm 184:387
Greeb J, Shull G (1989) Molecular cloning of a third isoform of the calmo-
    dulin-sensitive PM Ca^{2+}-transporting ATPase that is expressed predomi-
    nantly in brain and skeletal muscle. J Biol Chem 264:18569
Grover AK (1985) Ca-pumps in smooth muscle: one in plasma membrane and
    another in endoplasmic reticulum. Cell Calcium 6:227
Grover AK (1986) Role of subcellular membranes in Ca^{2+}-mobilization in
    uterine smooth muscle. In: Huszar G (ed) The Physiology and Biochem-
    istry of the Uterus in Pregnancy and Labour. CRC Press Inc, Boca Raton,
    Florida, p 93
Grover AK (1987) Calcium transport and ATPases in smooth muscle. In: Kidwai
    AM (ed) Muscle sarcolemmal biochemistry, CRC Press, Boca Raton, FL, p 99
Grover AK, Daniel EE (1985) Calcium handling studies using isolated smooth
    muscle membranes. In: Grover AK, Daniel EE (eds) Calcium and Smooth
    Muscle Contractility. Humana Press, Clifton, NJ, p 245
Grover AK, Khan I (1991) Calcium pump isoforms: diversity, selectivity and
    plasticity. Cell Calcium 13:9
Grover AK, Crankshaw J, Garfield RE, Daniel, EE (1980) Smooth muscle
    vesicle orientation: a study of intactness and sidedness of rat myometrium
    plasma membrane vesicles. Can J Physiol Pharmacol 58:1202
Grover AK, Kwan CY, Daniel EE (1981) Na-Ca interactions in smooth
    muscle: evidence for Na-Ca exchange by rat myometrium plasma mem-
    brane vesicles. Am J Physiol 240:CI75
Grover AK, Kwan CY (1983) Oxalate stimulation ATP-dependent Ca uptake
    is diminished during smooth muscle subcellular fractionation. Life Sci
    32:2655
Grover AK, Kwan CY, Daniel EE (1982) Ca^{2+} dependence of Ca-uptake by
    rat myometrium plasma membrane enriched fraction. Am J Physiol
    242:C278
Grover AK, Kwan CY, Daniel EE (1983a) Na-Ca exchange in a rat myome-
    trium plasma membrane enriched fraction: further evidence and properties.
    Am J Physiol 244:C158
Grover AK, Kwan CY, Daniel EE (1983b) High affinity pH dependent passive
    calcium binding by rat myometrium plasma membrane enriched fraction.
    Am J Physiol 244:C61

Grover AK, Kwan CY, Luchowski E, Daniel EE, Triggle DJ (1984) Subcellular distribution of [^{3}H]-nitrendipine binding in smooth muscle. J Biol Chem 259:2223

Guntski-Hamblin A, Greeb J, Shull G (1988) A novel Ca-pump expressed in brain, kidney, and stomach is encoded by an alternative transcript of the slow-twitch muscle SR Ca-ATPase gene. J Biol Chem 263:15032

Herrera VLM, Emanuel JR, Ruiz-Opazo N, Levenson R, Nadal-Ginard N (1987) Three differentially expressed Na, K-ATPase subunit isoforms: structural and functional implications. J Cell Biol 105:1855–1865

Inesi G, Sumbilla C, Kirtley ME (1990) Relationships of molecular structure and function in Ca^{2+}-transport ATPase. Physiol Rev 70:749

Inoue Y, Sperelakis N (1991) Gestational change in Na^{+} and Ca^{2+} channel current densities in rat myometrial smooth muscle cells. Am J Physiol 260:C658-C663

James PH, Pruschy M, Vorherr TE, Jenniston JT, Carafoli E (1989) Primary structure of the cAMP-dependent phosphorylation site of the plasma membrane calcium pump. Biochem 28:4253

Jewell EA, Lingrell JB (1991) Comparisons of the substrate dependance properties of the rat Na, K-ATPase α1, α2 and α3 isoforms expressed in Hela cells. J Biol Chem 266:16925–16930

Khan I, Grover AK (1990) Cloning and sequence of CDNA for rabbit stomach smooth muscle internal calcium pump. Nucl Acids Res 18:4026

Khan I, Grover AK (1991) Expression of cyclic nucleotide sensitive and insensitive isoforms of plasma membrane Ca pump in smooth muscle and other tissues. Biochem J 277:345

Khan I, Spencer GG, Samson SE, Boileau G, Crine P, Grover AK (1990) Abundance of sarcoplasmic reticulum calcium pump isoforms in stomach and cardiac muscles. Biochem J 268:415

Khan I, Tabb T, Garfield RE, Jones LW, Fomin VP, Samson SE, Grover AK (1992a) Expression of the internal calcium pump in pregnant rat uterus. Cell Calcium 14:111-117

Khan I, Tabb T, Garfield RE, Grover AK (1992b) Changes in mRNA for Ca pumps in pregnant rat uterus. Biochem Int 27:189

Khan I, Tabb T, Garfield RE, Grover AK (1992c) Polymerase chain reaction assay of mRNA using 28S RNA as internal standard. J Biochem Biophys Methods 147:114–117

Koch WJ, Ellinor PT, Schwartz A (1990) cDNA cloning of a dihydropyridine-sensitive calcium channel from rat aorta. Evidence for the existence of alternatively spliced forms. J Biol Chem 265:17786–17791

Lang LM, Beyer EC, Schwartz AL, Gitlin JD (1991) Molecular cloning of rat
    uterine gap junction protein and analysis of gene expression during gesta-
    tion. Am J Physiol 260:E787-E793
Lytton J, MacLennan DH (1988) Molecular cloning of cDNAs from human
    kidney coding for two alternatively spliced products of the cardiac Ca-AT-
    Pase gene. J Biol Chem 263:15024
MacLennan DH, Brandl C, Korczak B, Green N (1 985) Amino-acid sequence
    of a Ca^{2+} Mg^{2+}-dependent ATPase from rabbit muscle SR, deduced from
    its complementary DNA sequence. Nature 316:696
Magocsi M, Penniston JT (1991) Oxytocin pretreatment of pregnant rat uterus
    inhibits Ca^{2+} uptake in plasma membrane and sarcoplasmic reticulum. Bio-
    chim Biophys Acta 1063:7–14
Nabel EG, Plautz G, Nabel GJ (1990) Site specific gene expression in vivo by
    direct gene transfer into the anterial wall. Science 249:1285–1288
Savineau JP, Mironneau J, Mironneau C (1988) Contractile properties of
    chemically skinned fibers from pregnant rat myometrium: existence of an
    internal Ca-store. Pflügers Arch 411:296–303
Shull G, Greeb J (1988) Molecular cloning of two isoforms of the PM Ca^{2+}-
    transporting ATPase from rat brain. J Biol Chem 263:8646
Soloff M, Sweet P (,1982) Oxytocin inhibition of $(Ca^{2+} + Mg^{2+})$-ATPase ac-
    tivity in rat myometrial plasma membranes. J Biol Chem 257:10687–
    10693
Soloff M, Alexandrova M, Fermstrom MJ (1979) Oxytocin receptors: triggers
    for parturition and lactation? Science 204:1313–1315
Spencer G, Yu X, Khan I, Grover AK (1 99 1) Expression of isoforms of inter-
    nal calcium pump in rabbit stomach smooth muscle and heart. Biochim
    Biophys Acta 15:1063
Spencer GG, Grover AK (1990) Subcellular distribution of Ca pump in
    smooth muscle: studies using a monoclonal antibody against erythrocytes.
    Biochem Arch 6:201
Spencer G, Khan I, Grover AK (1990) Regulation of calcium in smooth
    muscle in uterine contractility mechanisms. Serono Symposium USA, p 53
Strehler EE (1990) Advances in the molecular characterization of the plasma
    membrane Ca^{2+} pumps. J Memb Biol 120:1
Ver A, Mullner N, Somogyi J (1989) Comparison of Ca^{2+}-extrusion systems
    in the myometriual plasma membrane of pregnant and non-pregnant rats.
    Biomed Biochim Acta 48:S393-S398
Verma A, Filoteo A, Stanford D, Wieben E, Penniston J, Strehler E, Fischer R,
    Heim R, Vogel G, Mathews S, Strehler-Page M, James P, Vorherr T,
    Krebs J, Carafoli E (1988) Complete primary structure of a human plasma
    membrane Ca-pump. J Biol Chem 263:14152

5 Mechanism of Action of Steroid Hormones and Antihormones: A Mini-overview

Etienne-Emile Baulieu

Intracellular receptors of steroid hormones are ligand-regulated, DNA-binding transcription factors (TF). They form a subgroup in a super-family of receptors which includes receptors for the five classes of steroid hormones known thus far (estrogens, progestins, androgens, glucocorticosteroids, mineralocorticosteroids) and receptors for calcitriol (vitamin D_3 active metabolite), thyroid hormones, and retinoic acid; there are also "orphan" receptors whose ligands have not yet been detected [1].

The structure of steroid receptors, of molecular weights between ~ 60 000 and 120 000 daltons, comprises a ligand-binding domain (LBD) of ~ 250 amino acids (aa), insuring specific steroid binding for each class. The LBD is at the C-terminal end of the receptor and is preceeded by the DNA-binding domain (DBD) with two zinc fingers decisively involved in both nonspecific DNA binding and precise recognition of the corresponding hormone response element (HRE) of the promoter region of hormone-regulated genes. Between the C-terminal second zinc finger and LBD, the C-terminal region of DBD includes a stretch of positively charged aa. The N-terminal region, particularly immunogenic, is of variable length (~ 200–600 aa according to the individual case) and involved in cell-specific modulation of transcriptional activity. Besides binding hormones, LBD is involved in transactivation of transcription and receptor homodimerization, the latter

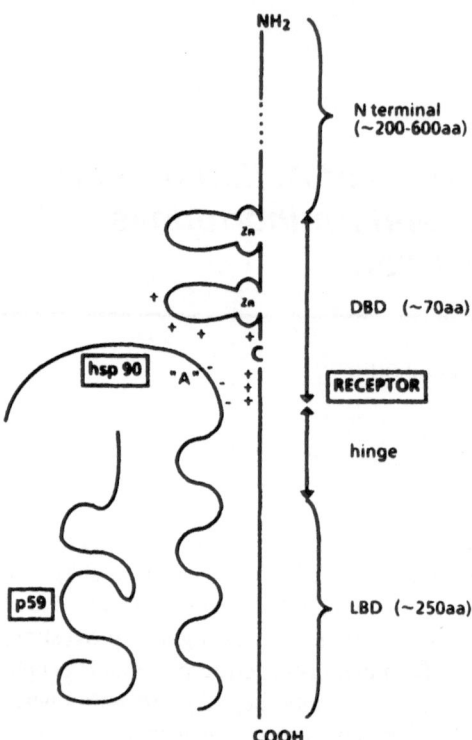

Fig. 1. Steroid receptor bound to hsp90 and p59/HBI bound to hsp90. *DBD*, DNA-binding domain; *LBD*, ligand-binding domain; *hsp90*, heat shock protein of molecular weight ~90 000 daltons (see text and references [4–9]); *p59/HBI*, hsp90 binding immunophilin: *C*, cysteine; *Zn*, zinc; *A*,A region

being implicated in the binding to the palindromic or almost palindromic structure of HRE [1–3].

We have shown that LBD also binds hsp90 (Fig. 1), and recent studies from our and other laboratories suggest multipoint interaction between hsp90 and LBD (review in [4–7]). In addition, we have observed a region of interaction [8] between a charged sequence "A" of hsp90 [7] and the positively charged C-terminal sector of DBD cited above. Such a hsp90–receptor complex does not bind DNA and is included in the so-called nonactivated, native form of the steroid receptor (of "8S" sedimentation coefficient), which may also involve other heat shock proteins, including a p59 entity which we recently cloned in the rabbit [9–11].

It is our hypothesis that hsp binding maintains the hormone binding site of LBD in a conformation appropriate to interact with the cognate agonist and antagonist ligands. In case of binding of an agonist, the change in conformation of LBD provokes, inter alia, the release of hsp90, thus giving to DBD the possibility of binding DNA. This is the first and necessary ligand-induced step of hormone action, allowing DNA binding and thus response. However, the same binding to HRE may be obtained artificially in vitro with receptor in absence of hormone or with pharmacological antihormones, thus demonstrating that although indispensable, this initial hormone-dependent release of hsp cannot be the only step involved in hormone action. Indeed, with thyroid, retinoic acid, and probably calcitriol receptors, the LBD does not bind hsp90, even though there is a positively charged sector in the DBD of these receptors; thus the difference with steroid receptors may be within the LBDs.

In other words, the thyroid and retinoic receptors are already beyond the step of hormone action in terms of hsp release. It is interesting to note that in the absence of hormone, thyroid receptor interaction with its HRE blocks transcription of the corresponding genes; binding of thyroid hormones activates the transcriptional response, likely via transconformation of the receptor permitting the activation of TFs. The binding of the steroid receptor to HRE can provoke change(s) of chromatin structure. HRE is precisely positioned on a nucleosomal structure [12,13] which is modified upon binding of the receptor (as attested by the DNase test). This may make possible interaction with and function of TFs, whether these TFs themselves bind to DNA, RNA polymerase, and/or other intermediary protein(s) involved in the complex modifying transcription.

Therefore the second molecular step in the function of the hormone involves DNA binding with consequences on chromatin structure and TF function. Whether the hormone favors receptor dimerization is still unclear, since experiments are contradictory: We found for the estrogen receptor that in its nonactivated 8S form, two molecules of ER are already present (associated to a dimer of hsp90) [14], and thus the dimeric receptor would be prepared to bind HRE (here, an estrogen response element). It has been also proposed that estrogen binding provokes receptor dimerization [15] and this may also be the case for the progesterone receptor [16], for which there is only one molecule of progesterone receptor (PR) in the 8S structure [17]. The potential importance of this dimerization in the molecular mechanism of action relates to the possibility that if there is a dimeric form before a hormone enters the cell, there is less possibility of heterodimer formation upon the introduction of hormone: for example, a dimer composed of one molecule of the steroid receptor and one of another DNA-binding receptor forming a combination of possible functional importance as is the case with thyroid and retinoic acid receptors.

The details of the mechanism of action of antisteroid is unknown. What is clear, however, is that there is no demonstrated change of chromatin structure upon binding of antisteroid to the receptor, again observed in analyses with the DNase test. Receptor and hsp90 may remain bound to each other after binding of the antihormone; this is the case for RU486 with progesterone and glucocorticosteroid receptors, as observed in in vitro experimental conditions [18]. However, it is probable that kinetically the dissociation of hsp90 from RU486-binding receptor occurs secondarily, and that further TF function is made impossible through transconformation provoked by RU486 binding. In any case, there is no difference between release of hsp90 from the estrogen receptor bound to agonist (estrogen) from that bound to antagonist (hydroxytamoxifen), and the released antagonist–estrogen receptor complexes bind to DNA [19]. The antiestrogen-induced transconformation of the estrogen receptor, however, does not stimulate TF function (as when the thyroid receptor binds to its HRE in the absence of hormone). Again differently, the action of antimineralocorticosteroids such as spironolactone suggests that upon the association of such an antagonist to the hsp90–receptor complex, there is rapid dissociation of the antagonist coupled with rapid dissociation of hsp90 from receptor and no detected change of chromatin structure either. We have proposed that the receptor bound to spironolactone is in a conformational state unable to act at the chromatin level [20].

Recently, we have cloned a so-called "p59" protein that we previously demonstrated as binding to hsp90 in nontransformed steroid receptor forms [21]. It is a "new" immunophilin, binding the immunosuppressants FK506 and rapamycin [22]. Its N-terminal region is 55%

identical to FBBP12, a FK506 binding protein involved in the immune response (review in [23]). We called it p59/hsp-binding immunophilin, "HBI". It also binds calmodulin [24] and it is associated to (auto) protein kinase activity (Le Bihan, Renoir et al., in preparation). The effects of immunosuppressants on steroid functions have now been studied [25], this protein being at the cross-road of immunological and endocrinological events.

In summary, we have detected two steps in hormone action: (1) release of the receptor from its p59/HBI-hsp90-bound inactive state to a conformation newly able to bind to DNA, and (2) transconformation of the receptor, involving chromatin change and activation of the TF function. Antisteroid compounds necessarily act at this second level, forming abortive DNA-binding antihormone–receptor complexes. The first step, release of associated proteins from the receptor, could be either kinetically retarded by antihormone binding thereby precluding DNA binding, or alternatively, be hastened, with the antisteroid dissociating rapidly and leaving the unliganded receptor unable to promote hormone action.

It is clear that much work remains to be done in order to understand all parameters of steroid action and antagonist activity at the cellular and molecular level. For example, the role of phosphorylation at the receptor, hsp90, and p59/HBI levels is still not understood. Pharmacologically, the activities of antihormones which behave as pure agonists or mixed agonist/antagonists must be studied to elaborate their dependence on animal species, cell differentiation, or even functional states of the cell. These studies are critical to the understanding of the therapeutic use of agents as antiestrogens, RU486, and spironolactone.

References

1. Evans RM (1988) The steroid and thyroid hormone receptor family. Science 240:889–895
2. Beato M (1989) Gene regulation by steroid hormones. Cell 56:335–344
3. Green S, Chambon P (1988) Nuclear-receptor enhance our understanding of transcription regulation. Trends Genet 4:309–314
4. Baulieu EE (1987) Steroid hormone antagonists at the receptor level. A role for the heat-shock protein MW 90,000 (hsp 90). J Cell Biochem 35:161–174

5. Pratt WB, Jolly DJ, Pratt DV, Hollenberg SM, Giguere V, Cadepond F, Schweizer-Groyer G, Catelli MG, Evans RM, Baulieu EE (1988) A region in the steroid binding domain determines formation of the non-DNA binding, 9S glucocorticoid receptor complex. J Biol Chem 263:267–273

6. Cadepond F, Schweizer-Groyer G, Segard-Maurel I, Jibard N, Hollenberg SM, Giguere V, Evans RM, Baulieu EE (1991) Heat shock protein 90 as a critical factor in maintaining glucocorticosteroid receptor in a nonfunctional state. J Biol Chem 266:5834–5841

7. Binart N, Chambraud B, Dumas B, Rowlands DA, Bigogne C, Levin JM, Garnier J, Baulieu EE, Catelli MG (1989) The cDNA-derived amino acid sequence of chick heat shock protein Mr 90,000 (hsp 90) reveals a "DNA like" structure, potential site of interaction with steroid receptors. Biochem Biophys Res Commun 159:140–147

8. Chambraud B, Berry M, Redeuilh G, Chambon P, Baulieu EE (1990) Several regions of human estrogen receptor are involved in the formation of receptor hsp90 complexes. J Biol Chem 265:20686–2069

9. Baulieu EE, Binart N, Cadepond F, Catelli MG, Chambraud B, Garnier J, Gasc JM, Groyer-Schweizer G, Oblin ME, Radanyi C, Redeuilh G, Renoir JM, Sabbah M (1989) Do receptor-associated nuclear proteins explain earliest steps of steroid hormone function? In: Carlstedt-Duke J, Eriksson H, Gustafsson JA (eds) The Steroid/Thyroid Hormone Receptor Family and Gene Regulation. J , pp 301–318, Birkhäuser Verlag, Basel

10. Renoir JM, Radanyi C, Faber LE, Baulieu EE (1990) The non-DNA binding heterooligomeric form of mammalian steroid hormone receptors contains a hsp90-bound 59 kDa protein. J Biol Chem 265:10740–10745

11. Lebeau MC, Massol N, Herrick J, Faber LE, Renoir JM, Radanyi C, Baulieu EE (1992) P59, an hsp90 binding protein; cloning and sequencing of its cDNA. Preparation of a peptide-directed polyclonal antibody. J Biol Chem 267:4281–4284

12. Richard-Foy H, Hager GL (1987) Sequence-specific positioning of nucleosomes over the steroid-inducible MTV promoter. EMBO J 6:2321–2328

13. Carr KD, Richard-Foy H (1990) Glucocorticoids locally disrupt an array of positioned nucleosomes on the rat tyrosine aminotransferase promoter in hepatoma cells. Proc Natl Acad Sci USA 87:9300–9304

14. Sabbah M, Redeuilh G, Baulieu EE (1989) Subunit composition of the estrogen receptor. Involvement of the hormone-binding domain in the dimeric state. J Biol Chem 264:2397–2400

15. Kumar V, Chambon P (1988) The estrogen receptor binds tightly to its responsive element as a ligand-induced homodimer. Cell 55:145–156

16. Guiochon-Mantel A, Lossfelt H, Lescop P, Sar S, Atger M, Perrot-Applanat M, Milgrom E (1989) Mechanisms of nuclear localization of the pro-

gesterone receptor: evidence for interaction between monomers. Cell 57:1147–1154

17. Renoir JM, Radanyi C, Jung-Testas I, Faber LE, Baulieu EE (1990) The non-activated progesterone receptor is a nuclear heterooligomer. J Biol Chem 265:14402–14406

18. Groyer A, Schweizer-Groyer G, Cadepond F, Mariller M, Baulieu EE (1987) Antiglucocorticosteroid effects suggest why steroid hormone is required for receptors to bind DNA in vivo but not in vitro. Nature 328:624–626

19. Sabbah M, Gouilleux F, Sola B, Redeuilh G, Baulieu EE (1991) Structural differences between the hormone and antihormone estrogen receptor complexes bound to the hormone response element. Proc Natl Acad Sci USA 88:390–394

20. Couette B, Lombes M, Baulieu EE, Rafestin-Oblin ME (1992) Aldosterone antagonists destabilize the heterooligomericstructure of the mineralocorticosteroid receptor. Biochem J 282:697–702

21. Renoir JM, Radanyi C, Jung-Testas I, Faber LE, Baulieu EE (1990) The non-activated progesterone receptor is a nuclear heterooligomer. J Biol Chem 265:14402–14406

22. Callebaut I, Renoir JM, Lebeau MC, Massol N, Burny A, Baulieu EE, Mornon JP (1992) An immunophilin that binds Mr 90,000 heat shock protein: main structural features of a mammalian p59 protein. Proc Natl Acad Sci USA 89:6270–6274

23. Schreiber SL (1991) Chemistry and biology of the immunophilins and their immunosuppressive ligands. Science 251:283–287

24. Massol N, Lebeau MC, Renoir JM, Faber LE, Baulieu EE (1992) Rabbit FKBP59-heat shock protein binding immunophilin (HBI) is a calmodulin binding protein. Biochem Biophys Res Commun 187:1330–1335

25. Renoir JM, Radanyi C, Baulieu EE (1992) Un effet des immunosuppresseurs FK506 et rapamycine sur le fonctionnement du récepteur de la progestérone: la protéine "p59-HBI", un carrefour de l'immunologie et de l'endocrinologie? CR Acad Sci Paris 315:421–428

6 Role of Progesterone in the Control of Labor

Kristof Chwalisz

6.1 Introduction

In mammals, labor can be defined as the onset of regular uterine contractions of high amplitude and short duration which are associated with
the dilatation of the uterine cervix and which bring about delivery of the
fetus and placenta. Prior to the initiation of labor the uterine motor function is suppressed and the uterine cervix which is "unripe," firm, and
rigid remains closed. Uterine quiescence during pregnancy is essential
for the normal development of the fetus. The onset of normal labor
leading to parturition should be considered as a precisely timed biological event among all mammals and every aberration in the timing of parturition, i.e., premature or post-term parturition, is associated with high
fetal morbidity and mortality. By definition, a term gestation in humans
is one that is completed between 38 and 42 weeks. However, in approximately 5–15% of women birth occurs prematurely i.e., before 37
completed weeks. Preterm labor and subsequent preterm birth is a major
problem of perinatology in general. In humans, premature birth is responsible for 75% of infant mortality and 50% of long-term neurologi-
cal handicaps (Creasy 1989). To date, there are no effective methods of
treatment and prevention for either preterm labor or preterm birth. The
current use of β-mimetics in the treatment of preterm labor is only of
limited value. The meta-analyses of the results of controlled trials show
little or no effect of β-mimetics on the frequency of the respiratory distress syndrome or on perinatal mortality (King et al. 1988).

Despite extensive research efforts over the past several decades, the
basic mechanism of the initiation of human labor remains uncertain,
although our knowledge about changes in the peripheral hormone levels as well as molecular processes taking place in the myometrial

smooth muscle cells (for review see Chap. 1, this volume) and in the uterine cervix around the time of parturition has increased considerably. However, the precise trigger for the initiation of labor in humans and nonhuman primates is still unknown. In addition, the mechanisms responsible for the initiation of term and preterm labor may vary. Preterm labor has to be considered a syndrome in which intrauterine or systemic infections play a key role (for review see Chap. 8, this volume). Furthermore, our understanding of the initiation of labor is complicated by the fact that different mechanisms operate in different species (see below). However, the results of experiments with various antiprogestins performed in different animal species and human studies with RU 486 suggest that there may be a greater similarity than difference in the basic physiological mechanism for labor initiation in related species. These studies clearly demonstrate that progesterone plays an essential role in the control of labor in all species investigated to date.

On the other hand, there is a need for improved labor induction methods due to fetal and/or maternal indications in about 10–15% of pregnancies. It has been known for at least three decades that successful labor induction with oxytocin or prostaglandins is largely dependent on the condition of the cervix. If labor induction is performed in the presence of an "unripe" cervix, the duration of labor is prolonged, resulting in a high failure rate of induction. As a consequence, there is a significant increase in the overall incidence of instrumental deliveries and cesarean sections as well as a variety of other complications. The introduction of prostaglandins, in particular locally administered PGE_2, into the clinical routine to soften the cervix before labor induction with oxytocin or intravenous PGE_2, has to be considered a major advance in obstetrics during the last decade. The two main approaches currently used in clinical practice involve the vaginal and endocervical administration of PGE_2. Controlled evaluation of preinduction cervical ripening with prostaglandins and either placebo or no prostaglandin treatment have clearly revealed that prostaglandins are effective agents in the ripening of the cervix. However, in about 20–30% of women, depending on preparation and cervical status, there is a need for repeated PGE_2 instillation due to the low ripening effect. Some studies have also reported an increased risk of hyperstimulation of the uterus and higher fetal heart rate abnormalities after local application of PGE_2 (for review see Keirse 1992) and a number of authors even employ the concomitant administration of β-mimetic agents during preinduction cervical ripening with prostaglandins (Egarter et al. 1990). Therefore, further refinements and improvements in the techniques of cervical ripening seem to be necessary, especially in terms of increased efficacy and a reduced risk in uterine hyperstimulation.

This chapter briefly reviews the animal and clinical studies with progesterone receptor antagonists (antiprogestins), focusing on their mechanism of action during pregnancy and on differences between various antiprogestagenic compounds. The results of these studies have changed our view of the mechanism of labor initiation and suggest that the uterus (myometrium and cervix) goes through a conditioning step controlled by progesterone (withdrawal). This step or process may be irreversible prior to the onset of labor and may be common to all species investigated to date. There is no doubt that antiprogestins will find broad application in obstetrics. The potential obstetrical indications for antiprogestins are tremendous and are the topic of this review.

6.2 Progesterone Functions During the Ovarian Cycle and Pregnancy

In women, progesterone is secreted during both the luteal phase of the ovarian cycle and pregnancy in a quantity which is far higher than that of any other steroid hormone. The daily production of progesterone from the corpus luteum is estimated at about 40–50 mg/day in women during the luteal phase compared with little or no secretion during the follicular phase (Lin et al. 1972a,b). The placental progesterone production during the advanced pregnancy exceeds 250 mg/24 h (Lin et al. 1972b).

6.2.1 Ovarian Cycle and Early Pregnancy

Progesterone plays a crucial role in female reproduction in mammals during almost all stages of the ovarian cycle and pregnancy. It is involved in the control of ovulation, prepares the endometrium for implantation, and in later stages of pregnancy is responsible for its maintenance. Progesterone controls growth and differentiation of target cells and regulates a variety of cell functions directly by stimulating or inhibiting structural and functional proteins, but also indirectly by functionally opposing the estradiol action. Progesterone plays most likely an important role during folliculogenesis and ovulation, acting directly on the ovary (Croxatto and Salvatierra 1990). Furthermore, it acts synergistically with estradiol in inducing the luteinizing hormone surge prior to ovulation at the hypothalamus/hypophysis level (Hoff et al. 1983). There are different progesterone effects on uterine cell proliferation, in particular on the estrogen-mediated cell proliferation, which vary among the different species (for review see Clarke and Sutherland 1990). For example, progesterone inhibits the estrogen-induced mitotic activity in the functional zones of the endometrium in primates and humans but shows a stimulatory effect on the endometrial epithelium in the rabbit. In contrast, progesterone seems to stimulate the proliferation of endometrial stem cells, which are responsible for endometrial regeneration after menstruation, during the luteal phase of the nonfertile cycle (Padykula 1991). In addition, progesterone stimulates epithelial proliferation in the mammary gland and controls lactogenesis (for review see Clarke and Sutherland 1990).

The sudden withdrawal of progesterone action at the end of the luteal phase induces the constriction of spiral arteries and in turn menstruation in humans and nonhuman primates. Recently, it has been suggested that the endometrial endothelin may be the major mediator in the vasoconstriction of the spiral arteries (MacDonald et al. 1991). It has been found that the enkephalinase which catalyzes the degradation of endothelin is present in the endometrial stroma in relatively high concentrations. Moreover, a high correlation has been found between the enkephalinase activity in the endometrium and the plasma progesterone levels (MacDonald et al. 1991; Casey et al. 1992). These findings may be important for our understanding of the regulation in the onset of labor since endothelin receptors have also been described in the myometrium (Maggi et al. 1991) and since endothelin is a very potent agent in stimulating myometrial contractions in vitro (Kozuka et al. 1989). Progesterone may also control nitric oxide synthesis or action in spiral arteries and thereby produce constriction and coagulation during menstruation (see Chap. 1, this volume, for effects of progesterone on nitric oxide). Therefore, it is logical to assume that the mechanisms controlling menstruation and labor may be similar.

In the fertile cycle, progesterone facilitates the transport of the fertilized egg through the oviduct and induces secretory changes required for implantation in the endometrium. Implantation is a precisely timed event in mammals. In women, it has recently been shown that successful implantation may only take place between days 15–20 ("implantation window") of a histologically defined 28-day cycle, i.e., during the period of highest progesterone levels (Navot et al. 1991). The secretory endometrial proteins (Beier et al. 1989) and probably other intracellular and cell surface proteins, such as integrins (Lessey et al. 1992) produced by endometrial glands as a result of progesterone stimulation, are necessary for implantation to take place. Finally, progesterone may control the expression of endometrial cytokines and growth factors which are believed to play an important role in implantation and early embryo development (see Chap. 9, this volume).

6.2.2 Role of Progesterone During Advanced Pregnancy: Uterine Quiescence

The contractility of the uterus at the end of pregnancy depends on the release of endogenous uterotonic agents such as prostaglandins and oxytocin and an increase in responsiveness of the myometrium to these stimuli. Progesterone has been proposed by A. Csapo (1975, 1981) as the hormone responsible for the quiescent state of the uterine musculature. Accordingly, the level of uterine activity depends on the balance between those factors promoting uterine contractions and progesterone. Csapo believed that progesterone withdrawal changes the uterus from the inert state to a highly responsive organ by both increasing the uterine responsiveness due to the lowering of the threshold for excitation by oxytocic agents and by inducing the release of "intrinsic uterine stimulants" (Fig. 1). Csapo also suggested that progesterone withdrawal results in the release of endogenous stimulants and eventually in the onset of labor. This concept has highly influenced the research on prostaglandins and antiprogestins. However, Csapo's theories never really defined the mechanism of progesterone action. Unfortunately, Csapo also never had the advantage of working with antiprogestins because his life ended prior to this exciting development.

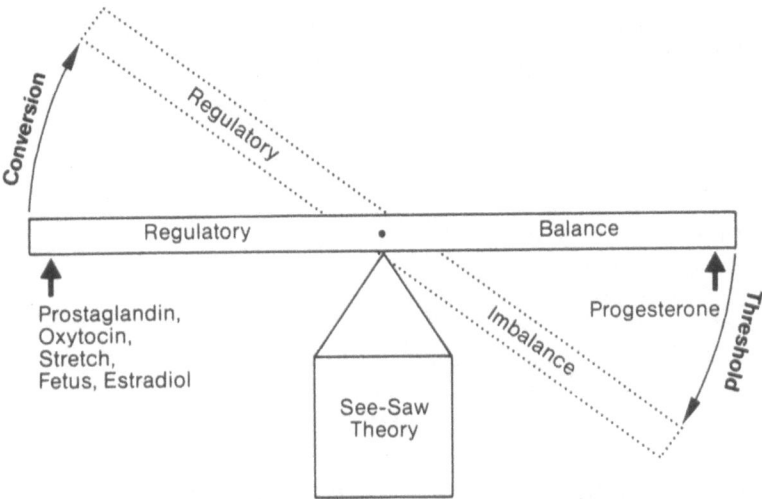

Fig. 1. "Progesterone-block" theory of A. Csapo (1975)

6.2.3 Comparative Aspects of Parturition: Problems in Understanding the Basic Mechanisms of Parturition Control

Csapo developed his "progesterone block theory" predominantly on the basis of the results of experiments performed in rabbits and rats. In these two species, as well as in goats and pigs, the corpus luteum is the major source of peripheral progesterone with a sudden progesterone withdrawal and concomitant increase in estradiol preceding parturition (Thorburn and Challis 1979). In rats and rabbits ovariectomy performed during advanced pregnancy results in premature parturition (Csapo and Wiest 1969). The progesterone withdrawal in corpus luteum-dependent species may be the result of increased luteolytic activity and/or decreased luteotropic support. There is considerable evidence indicating that $PGF_{2\alpha}$, of uterine origin, is responsible for the luteolysis in pregnant rats. Furthermore, progesterone prolongs pregnancy in rats and rabbits (Soloff 1989) and prevents the ability of the luteolytic $PGF_{2\alpha}$ to induce preterm birth in rats (Fuchs et al. 1974). Pregnant rats are also very sensitive to exogenous estrogens, which effectively induce abortion or preterm parturition (Elger 1979). Proges-

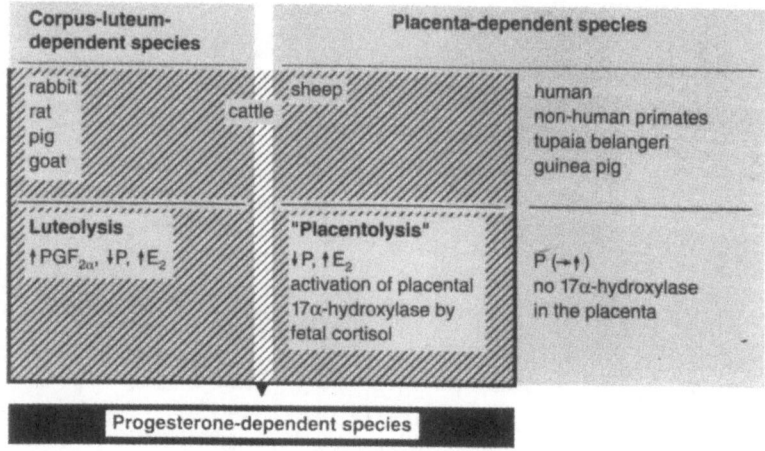

Fig. 2. Comparative aspects of parturition. *PGF2α*, prostaglandin F2α; *P*, progesterone; *E2*, estradiol

terone withdrawal at term is not limited to the corpus luteum-dependent species. During late pregnancy in sheep, the placenta is the major source of progesterone. In this species, the increase in fetal cortisol seems to be the signal for parturition. Fetal cortisol increased the activity of 17α-hydroxylase in the placenta which is responsible for the metabolism of progesterone in C_{19} steroids, e.g., androstenedione, substrates for the placental aromatase. As a result, there is a prepartum decrease in progesterone ("placentolysis") and an increase in unconjugated estrogen levels which are associated with the increase in prostaglandins the utero-ovarian venous blood (for review see Thorburn and Challis 1979 and Soloff 1989). Elevation of serum cortisol levels, either by direct administration or by adrenocorticotropic hormone (ACTH) treatment of the fetus, induces premature parturition and hormonal changes which are similar to those observed at term. These observations suggest that the initiation of labor in sheep is linked to the maturation of the fetal hypothalamo–pituitary–adrenal axis and that the fetus determines the time of parturition (Fig. 2).

However, in humans, nonhuman primates, and guinea pigs there is no progesterone decrease prior to the onset of labor. On the contrary, in humans there is a continuous increase in peripheral progesterone concentrations until birth. In humans and guinea pigs, circulating levels of progesterone reach relatively high concentrations at term (ca.$10^{-6}\,M$) (Pasqualini and Kincl 1985). In these species there is no 17α-hydroxylase activity in the placenta and fetal cortisol does not seem to play a role in triggering parturition. Csapo (1975) believed that in species in which the placenta is the major source of progesterone during advanced pregnancy, the measurement of peripheral plasma levels of progesterone do not accurately reflect progesterone concentrations in uterine target cells. The results of experiments with antiprogestins presented below indicate, however, that progesterone plays an important role in controlling the onset and progression of labor also in "progesterone-independent species."

6.2.4 The Guinea Pig and *Tupaja belangeri* as the Animal Models of Term and Preterm Birth

We selected the guinea pig and the tree shrew *Tupaja belangeri* as our primary models of parturition. There are many similarities between guinea pigs and humans in the endocrinology of pregnancy and parturition. Neither species shows any decline in progesterone before parturition (see Figs. 3,4). In guinea pigs, progesterone is secreted predominantly from the placenta during advanced pregnancy, and ovariectomy performed after the luteoplacental shift does not interrupt pregnancy (Elger 1979). Exogeneous progesterone also does not prolong pregnancy in guinea pigs (Porter 1970). In contrast to rats, treatment with estrogen after day 20 of pregnancy does not induce abortion or premature parturition in guinea pigs (Heap et al. 1973, Elger 1979). Similarly, dexamethasone treatment does not induce preterm parturition but has a rather inhibitory effect on parturition (K. Chwalisz, unpublished data). There are also similarities with regard to myometrial responses to prostaglandins. In guinea pigs PGE_2 is much more effective than $PGF_{2\alpha}$ in terminating pregnancy. The opposite situation is observed in pregnant rats ($PGF_{2\alpha}$ terminates pregnancy in rats predominantly by inducing luteolysis). The duration of pregnancy in guinea pigs is relatively long (spontaneous parturition in our strain of guinea pigs, Pirbright White, takes place overnight between days 67 and $68 + 3$ of gestation).

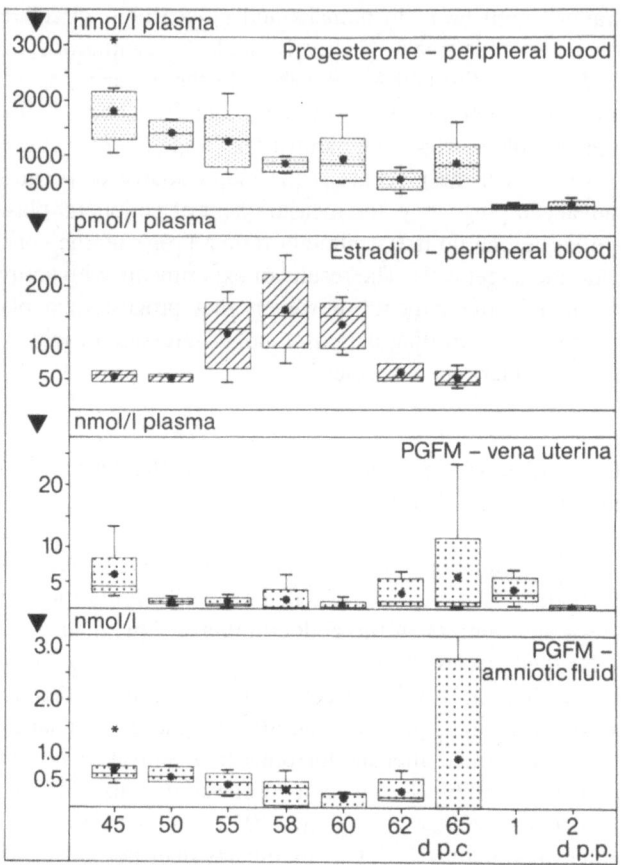

Fig. 3. Concentrations of progesterone, estradiol, and prostaglandin $F_{2\alpha}$ metabolite *(PGFM)* in guinea pigs in the course of pregnancy ($n = 5$–8/group)

In contrast to rats and rabbits showing a rapid onset of labor, in guinea pigs as in humans and nonhuman primates, labor develops slowly and is protracted. Furthermore, in guinea pigs the myometrial responsiveness to oxytocin and prostaglandins develops continuously during the course of pregnancy in the presence of high progesterone levels. However, in contrast to other mammalian species, the circulating progesterone is almost totally bound to the high-affinity binding protein

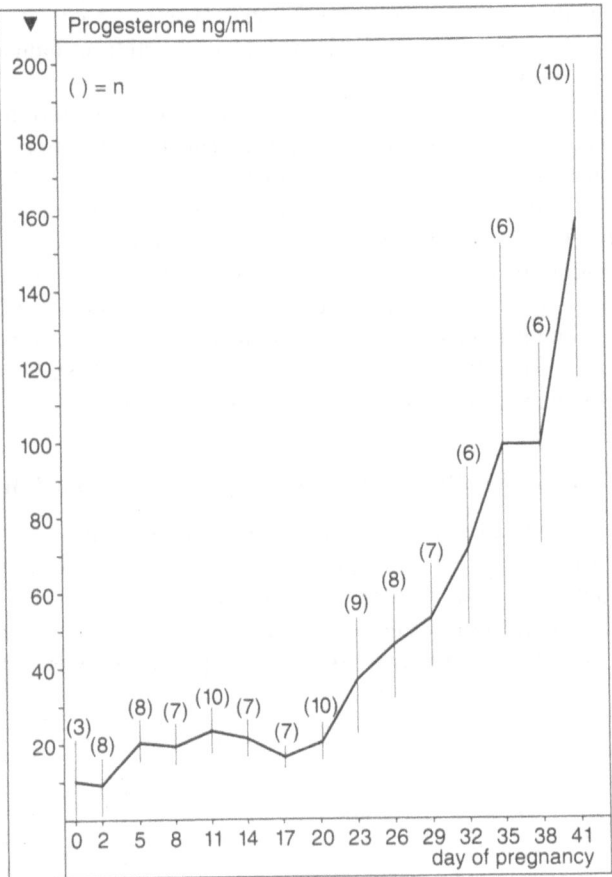

Fig. 4. Peripheral blood progesterone concentrations during pregnancy in *Tupaja belangeri*

(PBG) in pregnant guinea pigs (Westphal 1986). In pregnant women, there is relatively little specific binding of circulating progesterone to plasma proteins (i.e., CBG). Another important difference between guinea pigs and nonhuman primates and humans is the uterine control over the corpus luteum function. In guinea pigs the uterus secretes $PGF_{2\alpha}$ in pulsatile fashion which in turn induces luteolysis. Hysterectomy (elimination of uterine $PGF_{2\alpha}$) performed in cyclic guinea pigs after ovulation prolongs the luteal life span. There is no evidence that uterine prostaglandins control corpus luteum function in humans and nonhuman primates.

Tupaja belangeri, a small (160–200 g) nocturnal animal from Indochina, belongs to a family of tree shrews (Tupaide). The tree shrews, once considered to be primitive primates, do not seem to be very closely related to primates due to their long history of independent evolution (Luckett 1980). However, with regard to the endocrionology of parturition, this species shows similarities with humans. Peripheral levels of progesterone increases continuously in the course of pregnancy, reaching high levels at term with no detectable progesterone drop before parturition (see Fig. 4) (Schweier et al. 1975). Our strain gives birth between day 44 and 46 post conception (p.c.).

6.2.4.1 The Guinea Pig Model of Ascending Intrauterine Infection

Recently, we demonstrated that the pregnant guinea pig may also be used to study the mechanisms of preterm labor. Lipopolysaccharide (LPS), tumor necrosis factor-α (TNF-α), and interleukin-1β (IL-1β) after local (intrauterine) injections effectively induced labor and deliveries in midpregnant (day 42–43 p.c.) and preterm (day 60 p.c.) gui-

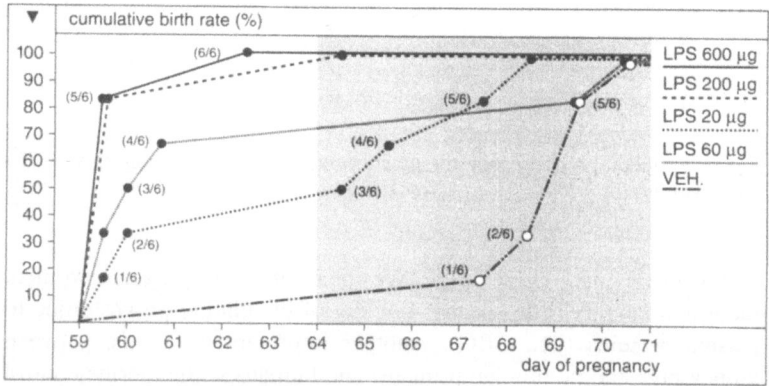

Fig. 5. Dose-dependent induction of preterm parturition with lipopolysaccharide *(LPS)* in guinea pigs. LPS was injected transcervically on day 61 p.c. (Bukowski 1993)

nea pigs. Intrauterine application of LPS consistently induced preterm parturition within 24–48 h in both periods of pregnancy in a dose-dependent manner (Fig. 5). It should be mentioned that in pregnant rats, LPS induces resorptions but does not induce labor (R.E Garfield, personal communication).

6.3 Progesterone Antagonists

Progesterone withdrawal can be achieved by (1) inhibiting ovarian and/or placental progesterone biosynthesis, by (2) neutralizing circulating progesterone by passive immunization with monoclonal antibodies, and by (3) blocking the progesterone action at the receptor levels. The most potent 3β-hydroxysteroid dehydrogenase inhibitor, epostane, inhibits the conversion of pregnenolone to progesterone and has been shown to lower plasma progesterone levels and to interrupt pregnancy in various animal species and in women (see Van Look and Bygdeman 1989 for review). However, this compound needs to be used at high doses for several consecutive days to achieve a drop in serum progesterone and termination of pregnancy in women (Birgerson et al. 1987). This treatment has been associated with frequent side effects. Moreover, the effect of epostane is not specific solely to progesterone biosynthesis, since this compound lowers plasma levels of other steroids such as estrogens and corticosteroids (Creange et al. 1981). Progesterone antibodies have only been used as pharmacological tools to study progesterone action in lower animals (Ellis et al. 1988). The discovery of the specific progesterone receptor antagonist mifepristone (RU 486) by scientists from Roussel Uclaf in 1981 (Phillibert et al. 1985; Baulieu 1985) has to be considered as a major milestone in steroid chemistry and reproductive physiology and medicine. The 11β-aryl-substituted steroidal progesterone receptor antagonists (antiprogestins, antigestagens) mifepristone (RU 486), onapristone (ZK 98 299), lilopristone (ZK 98 734), and ZK 112 993, whose chemical structures are shown in Fig. 6, bind with high affinity to the progesterone (PR) and glucocorticoid (GR) receptors (Phillibert et al. 1985, Neef et al. 1984, Elger et al. 1986). In comparison with the structurally related compound RU 486, onapristone and lilopristone have been shown to have similar antiprogestagenic but reduced antiglucocorticoid

Antiprogestational compounds

RU-38.486 (Mifepristone)

11β-(4-N,N-Dimethylamino--phenyl)-17β-hydroxy-17α--propinyl-4,9(10)-estradien--3-one

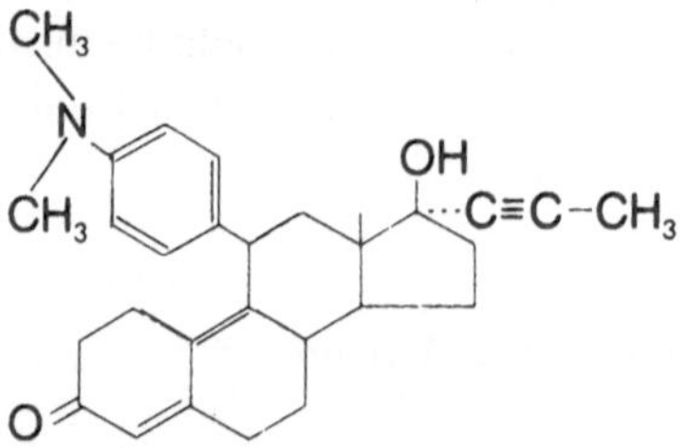

ZK 98.734 (Lilopristone)

11β-(4-N,N-Dimethylamino--phenyl)-17β-hydroxy-17α--(3-hydroxy-prop-1(Z)-enyl)--4,9(10)-estradien-3-one

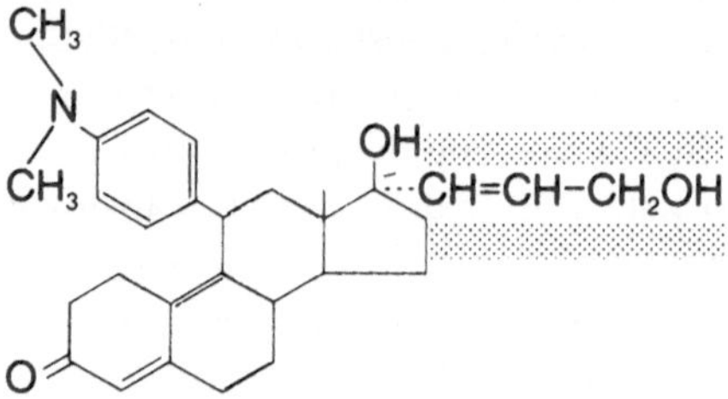

ZK 98.299 (Onapristone)

11β-(4-N,N-Dimethylamino--phenyl)-17α-hydroxy-17β--(3-hydroxy-propyl)-13α--methyl-4,9-gonadien-3-one

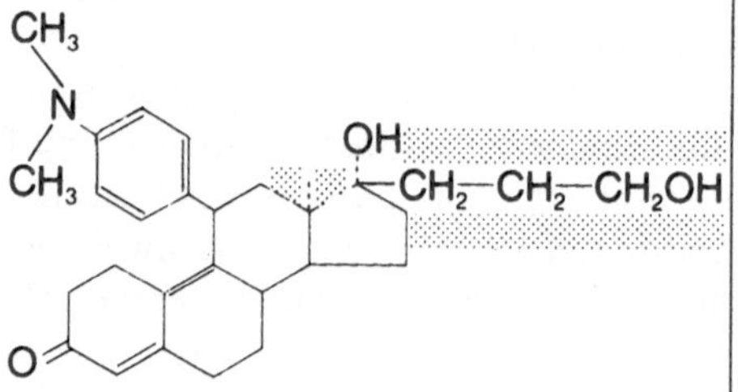

ZK 112.993

11β-(4-Acetylphenyl)-17β--hydroxy-17α-(1-propinyl)--4,9-estradien-3-one

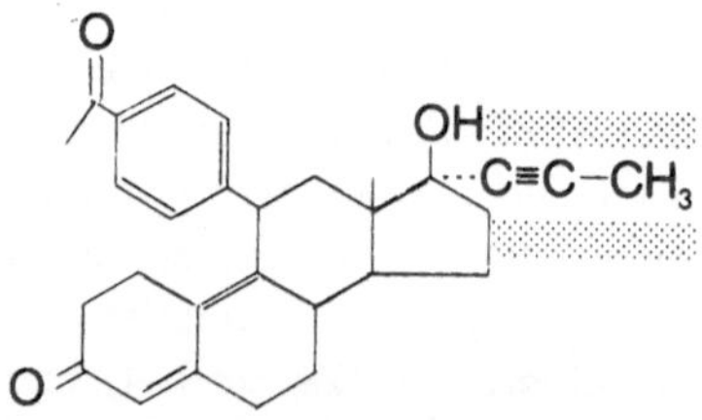

Fig. 6. Chemical structures of the antiprogestins RU 486, onapristone, lilopristone, and ZK 112

activities (Neef et al. 1984; Elger et al. 1986). RU 486 and onapristone also bind to the androgen receptor and show slight antiandrogenic activity in vivo. Lilopristone has some androgenic and glucocorticoid-agonistic activities at high doses (Elger et al. 1986; Elger et al. 1990).

6.3.1 Different Molecular Mechanism of Onapristone Action

Onapristone, which is a 13α-configurated (retro) steroid, shows a different stereochemical structure compared to the 13β-configurated anti-progestins such as RU 486 (Moguilewski and Philibert 1985), lilopri-stone, and ZK 112 993 (Neef et al. 1984). The molecular mechanisms of action of onapristone and RU 486 are also different. RU 486 promotes dimerization of the PR and in its binding to DNA (Meyer et al. 1990; Horwitz 1992; Gronemeyer et al. 1992). Therefore, RU 486 may act as a mixed agonist/antagonist. The agonistic activity of RU 486 in the endometrium has been reported in ovariectomized monkeys (Koering et al. 1986) and postmenopausal women (Gravanis et al. 1985). In contrast, onapristone and some other 13α-configurated progesterone antagonists fail to promote the formation of stable receptor dimers and prevent the binding of PR-complexes to the progesterone responsive elements of DNA (Klein Hitpass et al. 1991; Horwitz 1992). As a result, onapristone (and other 13α-antiprogestins) does not express agonistic activities and may be regarded as a "pure" PR antagonist (Klein Hitpass et al. 1991; Horwitz 1992). The absence of agonistic activity of onapristone may explain some of the special properties of this compound seen in late pregnant animals, especially its high labor-inducing activity (see below).

6.3.2 Interaction of Antiprogestins with Estrogens

None of the investigated antiprogestagenic compounds bind to the estrogen receptors (ER). This excludes estrogenic or antiestrogenic activity at the receptor level. There were also no inhibitory effects of anti-progestins on estrogen secretion in various animal experiments. However, there is evidence from previous studies that antiprogestins inhibit some of the estrogen effects in both the pregnant and nonpreg-

nant uterus. The interaction of antiprogestins with estrogens and antiestrogens has been extensively studied in pregnant guinea pigs during
late pregnancy (Elger et al. 1990). In this species antiprogestin-induced
labor was enhanced with antiestrogen tamoxifen and was completely
prevented by 17β-estradiol (E2) and the estrogen precursor androstenedione (Elger et al. 1990). Furthermore, it has been demonstrated that
both RU 486 and onapristone have a functional (noncompetitive) antiestrogenic effect at the endometrial level in both intact and ovariectomized monkeys. The comparative studies of the antiestrogenic
properties of various antiprogestins performed in ovariectomized rats
(Chwalisz et al. 1993a), rabbits (Chwalisz et al. 1991a), and monkeys
(Van Uem et al. 1989; Wolf et al. 1989; Chwalisz et al. 1992; Slayden
et al. 1993) suggest that this action is tissue, species, and compound
dependent. From these data no assumption can be made concerning a
general inhibition of estrogen action as the result of antiprogestin treatment. The mechanism of the "antiestrogenic" activity of antiprogestins
is not clear. Progesterone also shows functional antiestrogenic action
by either reducing the levels of ER or stimulating local estradiol metabolism by inducing 17β-hydroxysteroid dehydrogenase (17β-HSD)
(Whithead et al. 1981; King and Whitehead 1986). On the contrary, the
functional antiestrogenic activity of antiprogestins is associated with a
marked increase in ER and ERmRNA levels (Neulen et al. 1990;
Chwalisz et al. 1991c). The upregulation of ERs in the endometrium
and other uterine tissues following antiprogestin treatment is the most
consistent finding in all species investigated to date. The functional
antiestrogenic activity of antiprogestins may explain some of the effects of antiprogestins seen in pregnant animals, at least in the guinea
pig model (Elger et al. 1990).

6.4 Overview of the Effects of Antiprogestins
During the Ovarian Cycle and Early Stages of Pregnancy

Treatment with progesterone antagonists effectively inhibits progesterone action in target tissues. In nonhuman primates and in humans,
treatment with antiprogestins during the ovarian cycle results in the inhibition of ovulation, in the inhibition of endometrial transformation,
and in the induction of premature menstruation, depending on the stage

of the cycle (for review see Puri and Van Look 1991; Van Look and Bygdeman 1989). The extensive experimental studies performed during the last decade with various antiprogestagenic compounds in different animal species such as rats and guinea pigs and in humans indicate that all stages of pregnancy are susceptible to antiprogestins. Depending on the stage of pregnancy, antiprogestins effectively inhibit nidation, induce abortions, and induce preterm labor in all species including humans studied to date. However, the effects of antiprogestins during more advanced periods of pregnancy vary in different species (see below). During early pregnancy the endometrium (inhibition of receptivity before implantation) and decidua seem to be the major targets of antiprogestin treatment. However, there is evidence from animal studies that antiprogestins may also accelerate tubal egg transport and indirectly delay egg development by acting on the oviducts (Lejeune et al. 1986; Psychoyos 1986; Roblero et al. 1987). The results of clinical studies with RU 486 during early pregnancy are consistent with the animal data. It has been well established that RU 486 can be effectively and safely used to prevent implantation and in combination with a prostaglandin to induce early abortion in humans (for review see Van Look and Bygdeman 1989; Heally 1990; Puri and Van Look 1991; Spitz and Bardin 1993).

6.5 Labor-Inducing Activity of Antiprogestins During Late Pregnancy

During the more advanced stages of pregnancy the myometrium and the uterine cervix become the major targets for antiprogestin treatment. Labor and delivery depend (a) on the release of endogenous uterotonic agents such as prostaglandins and oxytocin, (b) on the high responsiveness of the myometrium to these stimuli, and finally (c) on the state of the cervix.

The assessment of the occurrence of preterm birth after treatment with antiprogestins alone during advanced pregnancy predominantly describes their labor-inducing activity i.e., successful labor contractions may only develop if the uterus (myometrium and cervix) is prepared for labor and delivery. In some species such as rats, particularly at preterm, it is difficult to distinguish between the labor-inducing and labor-conditioning effects of antiprogestins on the uterus because both effects are observed almost at the same time. However, in guinea pigs, nonhuman primates, and in humans, antiprogestins show predominantly conditioning effects in the uterus.

6.5.1 Induction of Parturition Studies
in Species with Progesterone Withdrawal at Term

In "progesterone-dependent" species such as rats, rabbit, sheep, pigs, and cows treatment with antiprogestins near term effectively induces preterm parturition which is very similar to the normal parturition at term. In these species there is a prepartal fall in serum progesterone prior to spontaneous birth.

Studies in Rats. Preterm parturition of living fetuses can be effectively induced after a single treatment with onapristone, RU 486, or lilopristone on day 20 or day 21 of pregnancy in rats (Chwalisz 1986;Chwalisz and Elger 1986; Bosc et al. 1987) (normal spontaneous delivery occurs during the evening of day 22 and morning of day 23) (Fig. 7). This effect is dose dependent and deliveries were observed approximately 24 h after treatment at effective doses. Lilopristone was the most effective compound in inducing preterm parturition in rats (Chwalisz and Elger 1986). Treatment with onapristone on day 18 did not effectively induce parturition even at high doses but resulted in bleeding and very prolonged deliveries. Similar effects were observed with RU 486 after treatment of pregnant rats on day 16 p.c. (Garfield et al. 1987). This finding is surprising because pregnant rats are very sensitive to progesterone withdrawal. It seems that the release of intrinsic uterine stimulants does not solely occur under progesterone control at earlier stages of pregnancy even in a "progesterone-dependent" species like the rat. However, infusion of oxytocin after onapristone treatment induced delivery, suggesting that antiprogestin treatment caused cervical ripening and enhanced myometrial responsiveness to oxytocin. Normally, this process occurs sharply about day 22 p.c., i.e., after the prepartal fall in serum progesterone (Fuchs et al. 1983).

Studies in Pigs, Sheep, and Cattle. In pigs, preterm parturition was successfully induced with ZK 112 993 at dose levels of 200 and 400 mg/sow. This compound was given on day 109 of pregnancy, i.e.,

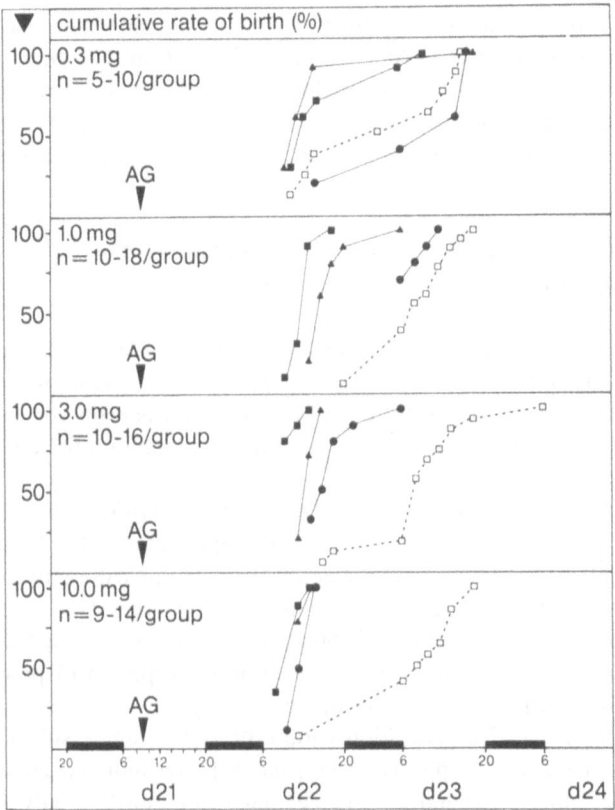

Fig. 7. Induction of parturition with antigestagens *(AG)* in rats (white squares, vehicle control; black squares, ZK 98.734, lilopristone; black circles, RU 486, nifepristone; black triangles, ZK 98.299, onapristone). The animals were treated with 0.3–10 mg s.c. on day 21 p.c (9–10 a.m.)

or the mother animals (Höfig et al. 1988). RU 486 given to late pregnant sheep also induced preterm parturition, with healthy newborn(s) and normal delivery of placenta(s) (Gazol et al. 1993). Similarly, RU 486 (2 mg/kg body weight i.m. administered on days 277 and 278) was very effective in inducing preterm parturition in cattle (Li et al. 1991). Parturition of living vigorous calves occurred 55 h after treatment. Acute luteolysis was evident in the significant decline in serum

progesterone concentrations after RU 486 treatment. There was no incidence of difficult birth (distocia) after RU 486 in this study. On the
contrary, dystocia, neonatal calf losses, and post partum retention of
placenta were frequently observed complications after birth induction
with prostaglandins or glucocorticoids in cows (Paislay et al. 1986).

6.5.2 Induction of Parturition Studies in Guinea Pigs
and *Tupaja belangeri*

In guinea pigs and Tupaja belangeri in which the placenta is the major
source of circulating progesterone (Figs. 3,4), there is no progesterone
withdrawal prior to parturition. Effects of antiprogestins on labor induction is dependent on the stage of pregnancy and on the compound
used in both species.

Labor-inducing Activity of Antiprogestins in Guinea Pigs. Experiments performed in guinea pigs during midpregnancy (day 42–43 p.c.)
using different compounds revealed that treatment with antiprogestins
alone was not very effective in inducing expulsions. Deliveries were
induced in only approximately 50% of the animals after a latency of
several days (Elger et al. 1986). In the preterm period of pregnancy,
onapristone and other 13α-configurated antiprogestins were effective
in inducing preterm parturition within 48 h after single subcutaneous
treatment on day 61 p.c. However, the 13β-configurated compounds
RU 486, lilopristone and ZK 112 993 induced parturition only in about
50% of animals and after longer latency (Fig. 8). The progestins R5020
and gestodene totally inhibited the labor-inducing activity of onapristone which clearly indicates that this action is mediated by the progesterone receptor (Fig. 9). It should be noted that onapristone-induced
deliveries on day 61 p.c. were not always normal. In some animals protracted deliveries in spite of an open cervix were observed, indicating
perhaps that the uterine contractions were not fully effective. Interestingly, epostane was ineffective in inducing preterm parturition in spite
of lowering the progesterone plasma levels in this particular species
(Fig. 10).

Mechanism of Action: Interaction with Estrogens. In midpregnant
guinea pigs (day 42–43 p.c.) the labor-inducing activity of antiprogestins (RU 486, onapristone, lilopristone, ZK 112 993) were dramatically

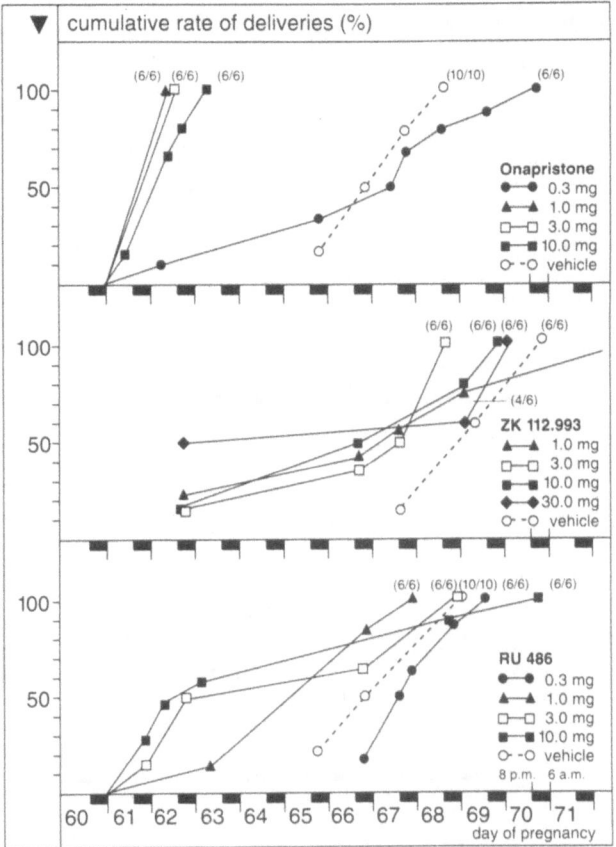

Fig. 8. Induction of parturition with onapristone, RU 486, and ZK 112 993 in guinea pigs. Animals were treated s.c. on day 61 p.c. (9–10 a.m.)

enhanced by adding epostane or the antiestrogen tamoxifen (Elger et al. 1990). Since the synergistic effect of antiprogestins with epostane was reversible with estrogen or estrogen precursors such as andros-tenedione, but not with a progesterone substitution, it was concluded that the increase of labor-inducing activity of this combination was due to a reduction in estrogen levels by epostane (Elger et al. 1990). Fur-thermore, and quite surprisingly, treatment with estradiol almost totally

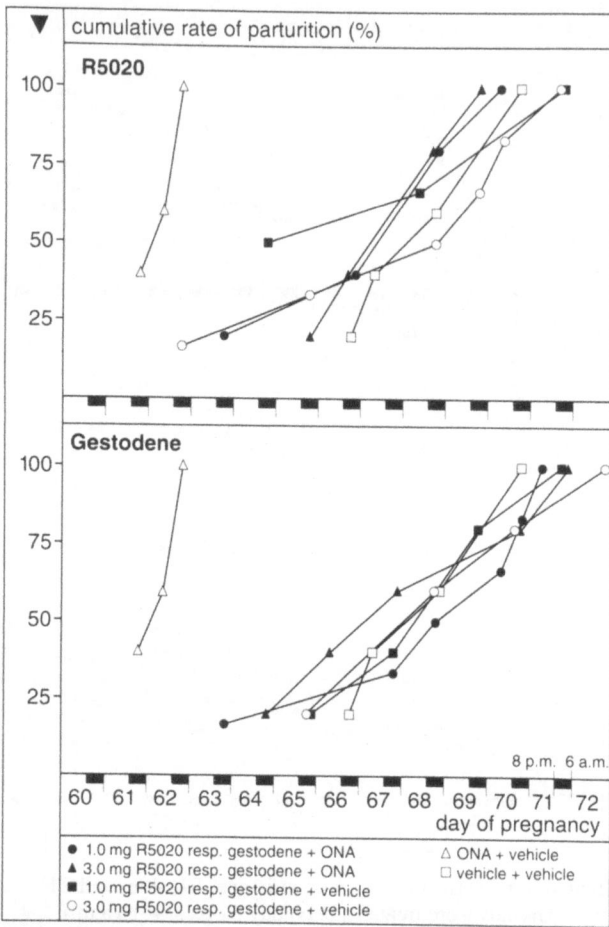

Fig. 9. Inhibition of onapristone-induced preterm parturition with the progestins R 5020 and gestodene in guinea pigs. Onapristone (*ONA*; ZK 98 299, 3 mg) was given s.c. on day 61 p.c. R 5020 and gestodene (1 and 3 mg/animal per day) were administered s.c on days 60–63 p.c. (*n* = 5–6/group)

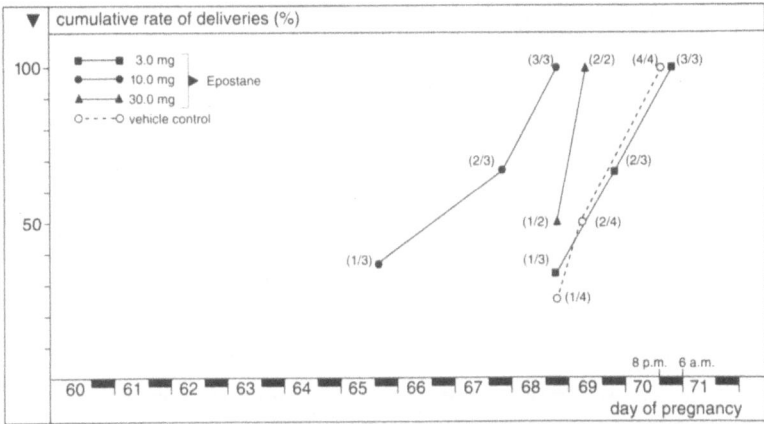

Fig. 10. Failure of epostane to induce preterm parturition in guinea pigs. Single subcutaneous injections on day 61 p.c.

inhibited the labor-inducing activity of onapristone and other antiprogestins at this stage of pregnancy (Elger et al. 1990).

On the contrary, we were unable to demonstrate any synergistic effect of antiprogestins with tamoxifen and the aromatase inhibitor CGS 16 949 (Ciba-Geigy) in the prepartal phase (day 61 p.c.) of guinea pig pregnancy (K. Chwalisz, unpublished data). However, estradiol (0.01 mg/animal.day) (Fig. 11) and the estradiol precursor androstenedione (Fig. 12) clearly inhibited the labor-inducing activity of onapristone. Surprisingly, estradiol treatment alone had an inhibitory effect on spontaneous parturition in guinea pigs. These data clearly demonstrate that estrogen has an inhibitory effect on the spontaneous and antiprogestin-induced parturition by probably inhibiting the release of endogenous uterotonins in guinea pigs. This inhibitory effect of estradiol seems to be unique to this species. These data also demonstrate that the synergism of antiprogestins with antiestrogens is dependent on the stage of pregnancy and probably due to the plasma levels of estradiol. As demonstrated in Fig. 3, estradiol levels starts to decline around day 60 p.c. This is consistent with the measurements of the aromatase activity in the placenta and ovary which reveal that there is a continuous decrease in the activity of this enzyme in both organs in

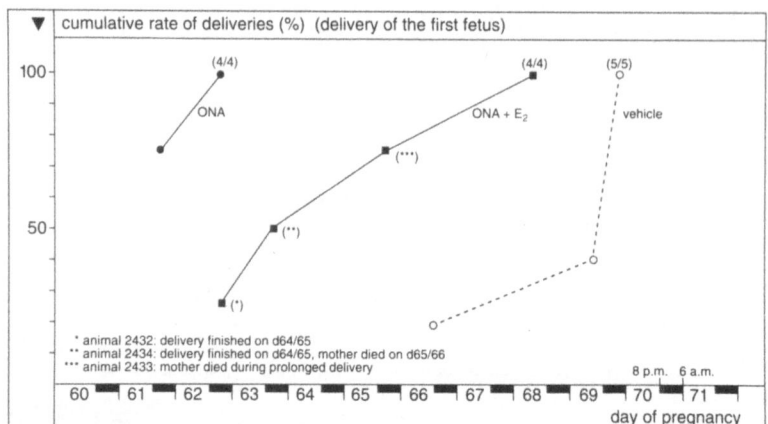

Fig. 11. Inhibitory effect of estradiol (E_2) on onapristone-induced preterm parturition in guinea pigs. Onapristone (*ONA*; 3 mg/animal s.c) was given on day 61 p.c. Estradiol (0.01 mg/animal.day) was given on days 60–62 p.c.

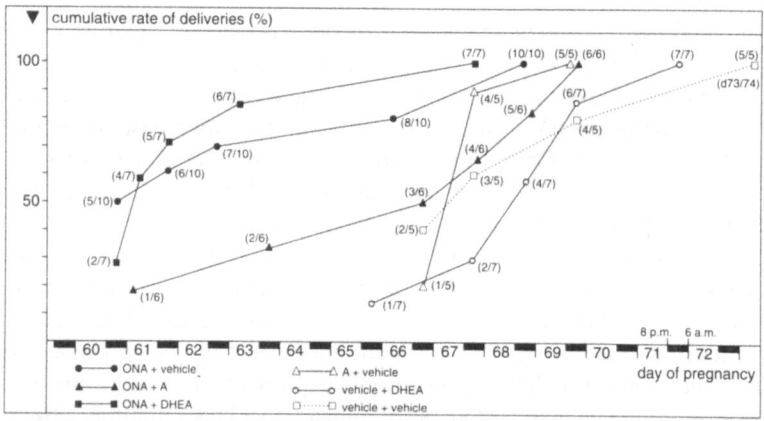

Fig. 12. Effect of androstenedione *(A)* and dehydroepiandrosterone *(DHEA)* on onapristone-induced preterm parturition in guinea pigs. Onapristone (*ONA*; 3 mg/animal s.c) was given on day 61 p.c. A (10 mg/animal per day, s.c.) and DHEA (10 mg/animal per day, s.c.) were given s.c. on days 59–61 p.c. with and without (controls) onapristone. Note the inhibitory effect of androstenedione

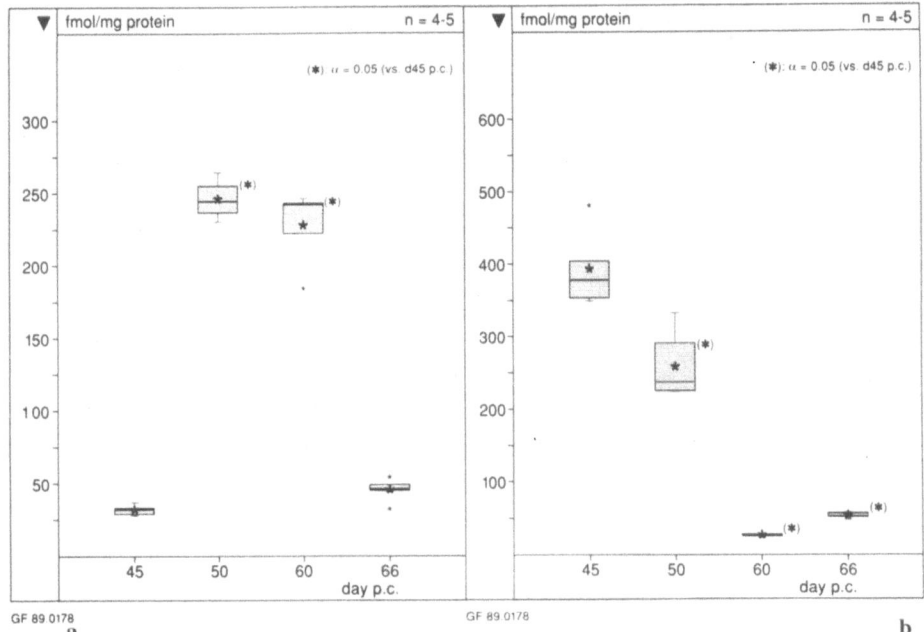

Fig. 13a,b. Changes in the ovarian (**a**) and placental (**b**) aromatase activity in the course of pregnancy in guinea pigs (S. Eggert, unpublished data)

the course of pregnancy with very little activity at term (Fig. 13). This observation may explain why the inhibition of estrogen action or synthesis did not increase the labor-inducing activity of antiprogestins at term.

The mechanism of the inhibitory action of estradiol on the antiprogestin-induced labor in guinea pigs is not fully apparent. Clearly, this is an important observation that could lead to a better understanding of the steps involved in the induction of labor. Studies performed in midpregnant guinea pigs (day 42–43 p.c.) indicate that estradiol has no inhibitory effect on the onapristone-induced myometrial responsiveness to sulprostone (Elger et al. 1990). This observation might suggest that estrogen inhibits the release of uterine stimulants during both antiprogestin-induced and spontaneous labor in guinea pigs. A similar effect of estradiol has been observed in pregnant rats with combinations of antiprogestins and estradiol: A combined treatment of RU 486 with estradiol on day 16 p.c. inhibited expulsions of the fetuses (R.E. Gar-

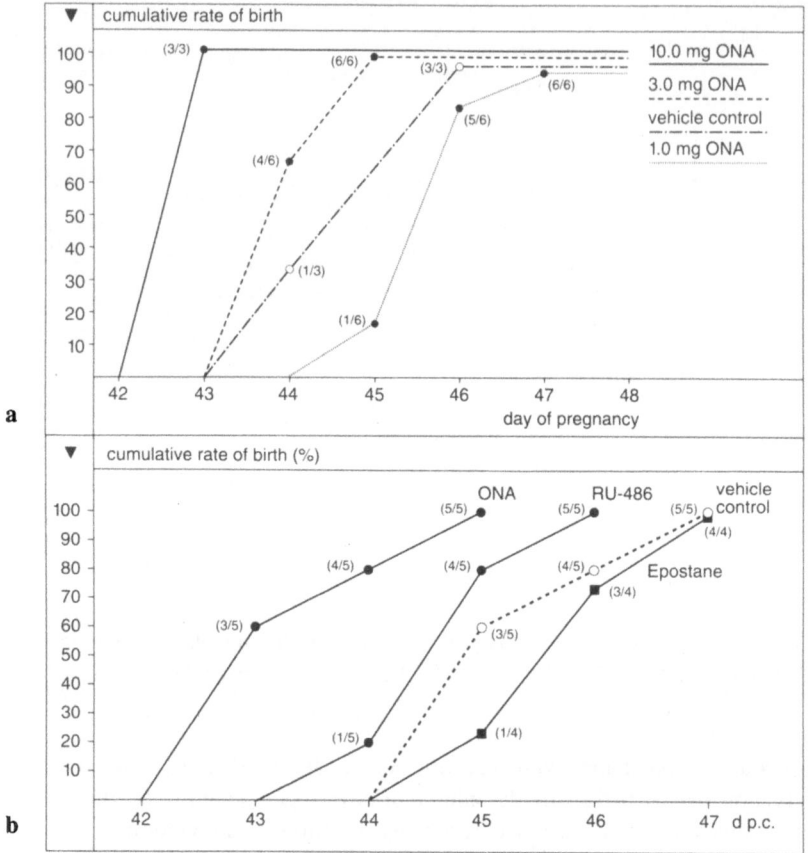

Fig. 14. a Induction of parturition with onapristone *(ONA)* in *Tupaja belangeri*. The animals were injected daily s.c. starting on day 42 p.c. until delivery. **b** Effects of RU 486, epostane and onapristone on parturition in Tupaja belangeri. 10 mg/animal of each compound was given subcutaneously on day 42 p.c.

field, personal communication). The macroscopic evaluation during autopsy suggested that the uteri of rats treated with a combination of RU 486 and estradiol were in a state of tonic contraction. Therefore, the inhibitory effect of estradiol on RU 486-induced labor may be due to the hyperstimulation of the uterus in rats (R.E. Garfield, personal communication). Further studies using measurements of the intrauterine pressure are needed to clarify the inhibitory effect of estradiol on antiprogestin-induced labor in guinea pigs and rats.

Induction of Parturition Studies in Tupaja belangeri. The effects of both onapristone and RU 486 were very similar to those seen in the guinea pig. In this species, onapristone effectively induced preterm parturition at doses of 3 and 10 mg/animal administered daily s.c., starting on day 42 p.c. (until parturition). This effect was dose dependent (Fig. 14a). The effect of RU 486 (10 mg/animal s.c.) was only marginal. Interestingly, epostane (10 mg/animal.day s.c) did not induce preterm birth in spite of lowering plasma progesterone levels (Fig. 14b).

6.5.3 Studies in Nonhuman Primates

RU 486 alone and in combination with oxytocin has been evaluated in monkeys for labor induction (Germain et al. 1985; Wolf et al. 1989; Haluska et al. 1987,1990). To date, there is only one preliminary study of labor induction with onapristone (Haluska et al. 1993). All studies discussed below demonstrate that RU 486 alone does not successfully induce preterm parturition in monkeys. However, it is very effective in inducing deliveries in combination with oxytocin.

In the first study in late pregnant cynomolgus monkeys (*Macaca fascicularis*) reported by Germain et al. (1985), RU 486 was given orally (20 mg/kg body weight, $n = 6$) for 3 days between days 130 and 155 (term = day 160 of pregnancy). The outcome of this treatment was very poor: Two monkeys were unaffected by the treatment, one animal delivered a dead fetus on day 171, and the remaining three animals delivered dead fetuses between 1 and 7 days after the onset of treatment. The labor-inducing activity of RU 486 and onapristone has also been studied by Novy's group in rhesus monkeys (*Macaca mullata*) (Haluska et al. 1987, 1990, 1993). Chronically instrumented animals were treated with RU 486 (20 mg/kg per day for 3 days orally, $n = 5$) approximately on day 130 of pregnancy (term = 168 days). Due to a lack in the progress of labor, all RU 486-treated animals were delivered by cesarean section between 48–72 h after the first treatment. There was one intrauterine fetal death in the RU 486-treated group. A continuous increase in the total uterine activity (area under the curve, AUC), assessed by intrauterine pressure measurement, was observed from the onset of treatment. In two of four animals, the cervices changed from firm to soft within the first 24 h with no changes thereafter. There was no apparent change in amniotic fluid prostaglandins (PGFM, PGEM-II, $PGF_{2\alpha}$, 6-keto $PGF_{1\alpha}$) within 48 h of treatment with RU 486. The authors concluded that RU 486 stimulated intense uterine activity but not the orderly sequence of changes in prostaglandins and cervical status observed during normal parturition in monkeys. The same group came to similar conclusions in another study recently performed using onapristone (Haluska et al. 1993). However, in this study the onapristone dose was three times lower (10 mg/kg orally for 3 days starting on day 135) and, therefore, it is difficult to interpret the results. On the other hand, treatment of late pregnant rhesus monkeys with epostane was effective in inducing parturition without complications (Haluska et al. 1991). These authors concluded that inhibition of progesterone secretion by epostane, but not the inhibition of progesterone action at the receptor level induces physiological changes in the uterus leading to normal parturition. However, epostane cannot be considered as a specific inhibitor of progesterone secretion, since estradiol levels were also decreased after epostane in this study. Therefore, the high labor-inducing activity of epostane may be due to the inhibition of secretion of other steroids in monkeys.

A study performed in Hodgen's laboratory demonstrates that sequential treatment with RU 486 and oxytocin is a very effective strategy in inducing preterm parturition in monkeys (Wolf et al. 1989). A single oral dose of 25 mg RU 486 alone, administered on day 160 of pregnancy (term = 167+ 3.3 days) was not effective in inducing preterm parturition (within 48 h) in cynomolgus monkeys (*Macaca fascicularis*). A manual examination of the cervix revealed the ripening effect of RU 486 in terms of an increased dilatation and a shortening and softening of the cervix. Additional treatment with oxytocin (20 U i.v.) 12 h after treatment with RU 486 effectively induced vaginal deliveries in 9 of 14 monkeys within 4 h after the start of oxytocin treatment. In 12 of 14 cases, vaginal delivery was achieved within 36 h after oxytocin treatment. Oxytocin administered alone was ineffective in achieving delivery. No overt toxicity on fetuses, newborns, or mothers was detected after RU 486 treatment. RU 486 had a stimulatory effect on colostrum and transiently enhanced weight gain in infants. This study demonstrates that RU 486 was effective in inducing cervical ripening and in increasing myomerial responsiveness to oxytocin in cynomolgus monkeys at the investigated dose, but when RU 486 was given alone it did not effectively induce uterine contractions to achieve deliveries. This study also shows that progesterone withdrawal is a trigger for lactogenesis in primates.

The final results of studies performed by Haluska and coauthors are not very different from the study performed in Hodgen's laboratory (Wolf et al. 1989). However, the interpretation of the results was quite different. Both studies demonstrated that RU 486 does not effectively induce parturition when given alone. Moreover, the increased effectiveness of oxytocin in inducing preterm parturition in RU 486-treated animals in Hodgen's study indicates that the cervix was prepared for delivery and that the poor progress in labor RU 486-treated animals was due to insufficient uterine contractions. The quality of uterine contraction induced by RU 486 differs from that normally occurring during spontaneous parturition (Haluska et al. 1987). RU 486 induced irregular long-lasting contractions of low amplitude in late pregnant rhesus monkeys in this study. We made similar observations after both onapristone and RU 486 treatment in the guinea pig model (Chwalisz et al. 1991b). This type of contraction clearly differs from contractions seen during spontaneous or oxytocin-induced labor, which are phasic and of short duration and long amplitude. A comparison of the total uterine activity (AUC) of spontaneous and RU 486-induced labor (as was done in Haluska's study), may provide similar results, but the RU 486-induced contractions were clearly not sufficient to induce deliveries in monkeys. This observation suggests that the assessment of the total uterine activity alone is not a proper parameter of uterine contractility and may lead to false conclusions. Furthermore, all other animal studies conducted with different antiprogestins (see above), and all human studies with RU 486 performed at different stages of pregnancy (see below), have demonstrated (directly or indirectly) a softening effect of antiprogestins on the cervix.

6.5.4 Human Studies with RU 486
During Advanced Pregnancy and at Term

It is remarkable that the administration of RU 486 alone is only effective in preventing or terminating pregnancy during the perinidatory stage when the endometrium is the major target for antiprogestins. The efficacy of RU 486 alone in inducing expulsions decreases significantly during the progress of pregnancy. As a postcoital agent, RU 486 is as effective as any other currently available method, having fewer side effects, however (Glasier et al. 1992; Webb et al. 1992). When given alone during the late luteal phase as a once-a-month contraceptive, RU 486 does not effectively terminate pregnancy (Couzinet et al. 1990). The efficacy of RU 486 alone in inducing complete abortions, irrespective of the dose, is about 80–85% within 10 days after missed menses, about 65% within 56 days of amenorrhea and less than 40% in later stages of pregnancy (Van Look and Bygdeman 1989). The administration of RU 486 to pregnant women with amenorrhea of 7 weeks or less induced complete abortion in 65–85% of the subjects. However, the addition of a prostaglandin increases the efficacy to 95% or more (Van Look and Bygdeman 1989; Aubeny and Baulieu 1991). To date, worldwide over 100 000 early pregnancy terminations have been performed with RU 486 in combination with a prostaglandin. The results have confirmed the initial studies with regard to efficacy and safety aspects (Spitz and Bardin 1993). These observations strongly indicate that myometrial contractions are not effectively activated by RU 486 (without a prostaglandin) in women during more advanced stages of pregnancy.

Second-Trimester Pregnancy Termination. For this reason, the clinical studies with RU 486 during the second trimester of gestation were performed in combination with prostaglandins. All these studies have demonstrated that pretreatment with RU 486 (200–600 mg orally) has significantly reduced the induction/abortion interval compared to the prostaglandins treatment alone (Table 1). Pretreatment with RU 486 reduced the required prostaglandin dose and consequently the prostaglandin-related side effects. Furthermore, women primed with RU 486 experienced considerably less pain during abortion. The shortening of the induction–abortion interval indirectly indicates the softening effect of RU 486 on the cervix (see also Table 1). However,

Table 1. Cervical ripening and labor induction with RU 486: Clinical studies

Study	Measurement of the cervical status	Ripening effect	Additional treatment	Reference
Nonpregnant women				
1.	Hegar dilators + force transducer	+	–	Gupta and Johnson (1990)
First trimester				
1.	Hegar dilators	+	Vacuum aspiration	Durlot et al. (1988)
2.	Hegar dilators	+	Vacuum aspiration	Rådestadt et al. (1988)
3.	Hegar dilators	+	Vacuum aspiration	WHO (1990)
4.	Hegar dilators force transducer	+	Vacuum aspiration	Urquhart and Templeton (1990a)
5.	Cervical gouge	+	Cervagem	Johnson and Bryce (1990)
6.	Hanks probe	+	Vacuum aspiration	Lefebvre et al. (1990)
Second trimester				
1.		(+)	Extra-amniotic PGE_2	Urquhart and Templeton (1990b)
2.	Hegar dilators	+	Vaginal PGE_2	Frydman et al. (1988)
3.		(+)	Extra amniotic PGE_2	Hill et al. (1990)
4.		(+)	Vaginal gemeprost	Rodger and Baird (1990)
5.		(+)	Vaginal 9-meth. PGE_2	Gottlieb and Bygdeman (1991)
Intrauterine fetal death (second and third trimester				
1.	–	(+)	PG or OT[a]	Cabrol et al. (1985)
2.	–	–	–	Cabrol et al. (1990)
Term pregnancy				
1.		(+)	OT[a]	Frydman et al. (1991)
2.		(+)	OT[a]	Frydman et al. (1992)

Numbers in parentheses indicate indirect evidence.
PG, prostaglandin; OT, oxytocin.
[a]In nonresponders to RU 486 alone.

RU 486 is more effective in shortening the induction–abortion interval than dilapan (which is an intracervical synthetic tent effectively inducing cervical dilatation), when used in combination with gemeprost for the termination during the second trimester (Thong and Baird 1992). The advantages of RU 486 may be due to its additional effect on myometrial responsiveness.

Intrauterine Fetal Death. RU 486 has also been shown to be effective in inducing labor in women with an intrauterine fetal death (Cabrol et al. 1985,1990). In a double-blind, controlled multicentric study involving 94 patients, RU 486 (600 mg/day for 2 days) effectively induced labor within 72 h in 63% of patients compared to 17.4% in the placebo group. It has to be expected that further studies with a combination of antiprogestins and a prostaglandin will improve the outcome of labor induction in intrauterine fetal death. The success of RU 486 administered alone in intrauterine fetal death is much higher than that achieved in second-trimester abortion in the presence of living fetus (Cabrol et al.1990). This observation might indicate that the labor-inducing activity of antiprogestins is dependent on the presence of living fetus and/or normal placenta and suggests that the conceptus has some inhibitory effect on the uterus in humans.

Induction of Labor at Term. The use of antiprogestins to prepare the uterus and the cervix for labor induction near term seems to be one of the most exciting indication for antiprogestins. In France, RU 486 has been used for labor induction in women near term requiring termination of pregnancy because of medical indications (Frydman et al. 1991, 1992). In the randomized, placebo-controlled study reported recently (Frydman et al. 1991), 200 mg RU 486 or placebo were given for 2 days in 120 women at term (after 37.5 weeks of amenorrhea). Four days after the first treatment, labor was induced with oxytocin in those women who had not delivered before. This study was designed to mainly evaluate the labor-inducing activity of RU 486, since the interval between priming and labor induction with RU 486 was relatively long. In the RU 486 group 54% of the women underwent spontaneous labor within 4 days of RU 486 administration, compared with 18% in the placebo group. RU 486-treated women needed much less oxytocin than the placebo-treated women, which indicates that both the uterine cervix and myometrium were prepared for labor and delivery. The rates of cesarean sections and operative vaginal deliveries were comparable in both groups. Neither infants nor women experienced any side effects in either the preliminary (Frydman et al. 1991) or the main study (Frydman et al. 1992).

   Transplacental Passage of RU 486. Only preliminary animal and human studies evaluating safety aspects of the use of RU 486 at term pregnancy have been published. Transplacental passage of RU 486 was demonstrated in monkeys during the second and third trimester of gestation (Wolf et al. 1988) and in women undergoing second-trimester pregnancy termination (Frydman et al. 1985; Hill et al. 1991). In monkeys, the fetal–placental index (AUC) decreased from 31.2% to 17.8% between the second and third trimester of pregnancy. In women, maximum fetal plasma concentrations were measured as early as 4 h after treatment, indicating rapid placental transfer of the drug. The levels of RU 486 and its N-monodemethylated metabolite (RU 42 633) in fetal blood were approximately 6–12% of the corresponding maternal plasma levels at all time intervals (4, 24 and 48 h after 600 mg RU 486). In the human study, no significant changes occurred in fetal cortisol, but there was a significant increase in fetal aldosterone. The importance of this finding is uncertain.

   In summary, the results of human studies with RU 486 alone performed during advanced pregnancy indicate that this antiprogestin does not effectively induce labor and delivery. Therefore, the addition of an oxytocic agent such as a prostaglandin or oxytocin is necessary to achieve delivery in women.

6.5.5 Mechanism of Action of Antiprogestins on Uterine Contractility: Effects on Uterine Prostaglandins During Pregnancy

The low effectiveness of antiprogestins to produce labor and delivery may be considered as indirect evidence of depressed liberation of uterine prostaglandins. In late pregnant guinea pigs, there was a rapid elevation of myometrial sensitivity to exogenous prostaglandins following antiprogestin treatment in pregnant guinea pigs. However, there was no parallel increase in a stable prostaglandin $F_{2\alpha}$-metabolite (PGFM), which one should expect assuming that there are inhibitory effects of progesterone on uterine prostaglandin secretion in this species (Elger et al. 1989). Treatment of cyclic guinea pigs with onapristone during the luteal phase resulted in the dose-dependent inhibition of luteolysis and in the decrease in PGFM concentrations in peripheral blood, indicating that the release of uterine prostaglandins responsible for luteolysis was suppressed (Elger et al. 1989, 1990). Studies with RU 486 performed in late-pregnant monkeys also demonstrate that the increase in amniotic fluid prostaglandins took place 40 h after the onset of uterine activity or 48 h after treatment began (Haluska et al. 1987). In contrast, amniotic fluid prostaglandins increased 1–2 days before spontaneous parturition in control monkeys. There are few data on the effect of RU 486 on prostaglandin levels in women during late pregnancy. In a second trimester abortion study, there was no significant increase in PGEM or PGFM concentrations in peripheral maternal plasma within 24 h after the administration of 600 mg RU 486, i.e., prior to PGE_2 treatment. However, there was a significant increase in both metabolites after the therapeutic injection of PGE_2.

On the basis of experiments in guinea pigs, we have proposed that the downregulation of uterine prostaglandins during pregnancy may not be the function of progesterone but rather an embryonic function (Elger et al. 1989,1990). The inhibitory effect of antiprogestins on the uterine prostaglandins, which we called "antiprogestin syndrome" may explain their low effectiveness in activating myometrial contractions and the occurrence of prolonged labor in the presence of a highly responsive uterus. Moreover, this may also be the reason for the poor effectiveness of mifepristone in inducing first and second-trimester abortions in humans when given without a prostaglandin (see above).

6.6 Effects of Antiprogestins on Myometrial Responsiveness to Oxytocic Agents During Advanced Pregnancy

A number in vitro (Garfield and Beier 1989) and in vivo studies in laboratory animals, monkeys, and humans (see below) have clearly shown that antiprogestins increase the myometrial responsiveness to both prostaglandins and oxytocin. These studies indicate that the antiprogestins may sensitize the uterus to uterotonic agents by increasing the receptor concentrations of oxytocin and prostaglandins, or by postreceptor events including an elevation in gap junctions, ion channels or a decrease in sensitivity to cGMP which inhibits contractions (see below).

6.6.1 Animal Studies

In late-pregnant guinea pigs there was a substantial increase in the effectiveness of prostaglandins and oxytocin to induce abortion and parturition after antiprogestins. A 30-fold increase in the uterine responsiveness to sulprostone (PGE_2-analog) was found in midpregnancy (day 42–43 p.c.) guinea pigs (Elger et al. 1987, 1990). Evaluation of the active threshold oxytocin doses in midpregnancy in the presence or absence of onapristone priming indicates that there is an increase in uterine responsiveness to oxytocin by about a factor of 30 (Chwalisz et al. 1991b). Normally, at this stage of pregnancy the uterus is insensitive to oxytocin and even very high oxytocin doses do not effectively induce abortion without priming with antiprogestins. In the preterm period of pregnancy (day 61 p.c.), onapristone and RU 486 increased the myometrial responsiveness approximately seven- to tenfold. It was possible to define doses of onapristone (0.3 mg and 0.1 mg/animal, s.c.), RU 486 (1.0 mg and 3.0 mg/animal s.c.) and oxytocin (25 mU/h, s.c.) which alone did not induce preterm parturition but given sequentially effectively induced preterm birth within a few hours (Fig. 15). At this stage of pregnancy, the uterus is responsive to oxytocin, but doses higher than 167 mU/h (1000 mU/day) are needed to induce preterm parturition alone (Elger and Hasan 1985). The efficacy of onapristone in elevating myometrial responsiveness to oxytocin and sulprostone increased in the course of pregnancy. Comparable effects can be obtained with approximately 30 times lower doses of this antiprogestin at preterm (day 61 p.c.) as compared to late pregnancy (day 42-43 p.c.). This study also shows that onapristone is 10–30 times more effective than RU 486 in increasing the myometrial responsiveness to oxytocin in preterm guinea pigs. Intrauterine pressure recording performed in midpregnant guinea pigs (around day 43 p.c.) revealed phasic, labor-like contractions in response to oxytocin in onapristone-primed animals in contrast to tonic reactions in controls (Chwalisz et al. 1991b).

In summary, the results of the studies performed in midpregnant and preterm guinea pigs demonstrate that both onapristone and RU 486 treatment substantially increased the ability of oxytocin to induce labor and delivery. Oxytocin receptors were also measured in these studies, the results of which will be considered below. A similar

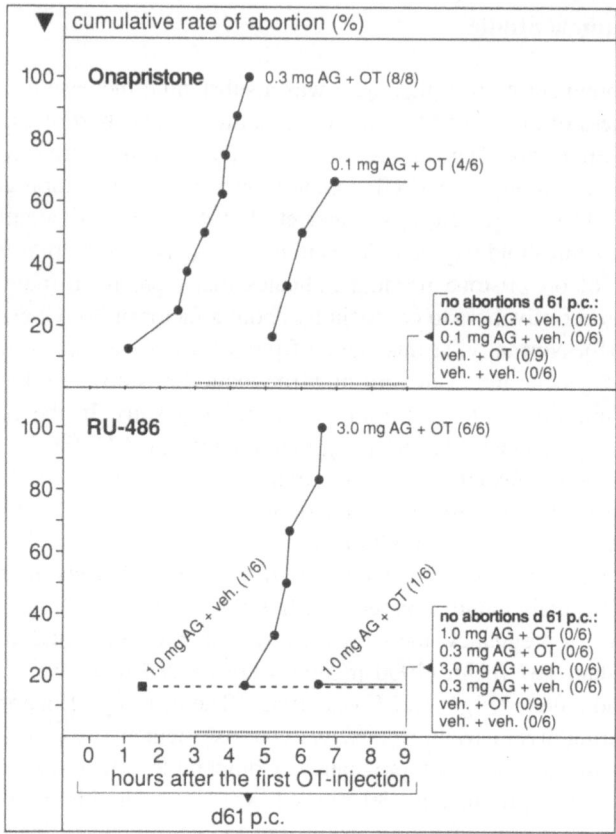

Fig. 15. Induction of preterm parturition with oxytocin *(OT)* after priming with the antiprogestins *(AG)* onapristone *(upper part)* and RU 486 *(lower part)*. Oxytocin was given on day 61 p.c. in serial injections of 25 mU at 1-h intervals (maximum six injections) 18 h after priming with onapristone, RU 486 or vehicle (controls), *n* = 6–9/group. Results are presented as cumulative rates of deliveries on d 61 p.c.

effect on uterine responsiveness to oxytocin has been shown in monkeys after RU 486 treatment (Wolf et al. 1989). Sequential treatment with RU 486 and oxytocin was much more effective in inducing preterm parturition in monkeys than treatment with oxytocin or the antiprogestin alone (see Sect. 6.5.3).

6.6.2 Human Studies

In humans, an increase in uterine sensitivity to sulprostone was demonstrated in 1985 by Bygdeman during the induction of abortion study with RU 486. When treatment with RU 486 was supplemented with intramuscular sulprostone, the incidence of complete abortion increased to 94% (Bygdeman and Swahn 1985). This observation was a significant breakthrough in the development of RU 486 for the induction of first- and second-trimester abortion. Meanwhile, a number of studies performed during the first, second, and third trimester of pregnancy have demonstrated an increase in myometrial responsiveness to both prostaglandins and oxytocin in women (see above). An increase in myometrial sensitivity to prostaglandins after RU 486 was observed in all stages of pregnancy. However, RU 486 did not elevate uterine responsiveness to oxytocin during early pregnancy in women (Swahn and Bygdeman 1988).

6.6.3 Mechanism of Action of Antiprogestins on the Myometrium

Effects of Antiprogestins on Gap Junctions. Gap junctions are intercellular channels connecting the interiors of two (myometrial) cells. These structures are sites of propagation and the basis for synchrony during labor (Garfield et al. 1977; Miller et al. 1989). The presence of gap junctions in the myometrium is essential for effective labor (see Chap. 1, this volume). The increase in gap junction could be the major mechanism of increased myometrial responsiveness to oxytocic agents following antiprogestin treatment and the most important mechanism of action of antiprogestins on the pregnant uterus. The observations that antiprogestins increase the myometrial responsiveness to oxytocin, prostaglandins, and even to mechanical stimuli (Elger et al. 1986,

1987) support this conclusion. An enhanced electrical coupling, due to the increase in the density of gap junctions in the myometrium following treatment with various antiprogestins, has been demonstrated in rats (Garfield et al. 1987; Garfield and Beier 1989) and guinea pigs (Chwalisz et al. 1991b; Sakai et al. 1992). In late pregnant guinea pigs, there was a substantial increase in electrical coupling (reflected in decreased input resistance) and gap junctions in the myometrium after onapristone treatment as evidenced by electrophysiological and immunocytochemical criteria (Chwalisz et al. 1991b). In this study a dramatic increase in connexin 43-binding has been found after onapristone treatment (Fig. 16). The results of intrauterine pressure measurements are also consistent with this view. Phasic and coordinated responses to oxytocin injections in onapristone-treated guinea pigs indicate that propagation of electrical events was improved by the treatment.

Effects on *Oxytocin Receptors.* Oxytocin is considered to play a major role in term labor. It is the most potent and specific natural uterotonic agent and uterine contractions induced with oxytocin are identical with those occurring during spontaneous labor, providing the uterus is in a reactive state (Csapo 1981). The myometrium of various species, including humans, is most reactive to oxytocin either near or at the time of parturition (Alexandrova and Soloff 1980; Fuchs et al. 1982, 1983). The enhanced myometrial responsiveness to oxytocin has been attributed to the increase in myometrial oxytocin receptor concentrations (Fuchs et al. 1982, 1983). Because in humans there is no increase in oxytocin concentrations in the peripheral blood before the onset of labor, it has been suggested that the increase in oxytocin receptor concentrations in the myometrium and decidua at term is one of the primary factors leading to the initiation of parturition (Fuchs et al. 1982).

Our studies performed in late-pregnant guinea pigs demonstrate that the onapristone-induced increase in the oxytocin response either during late pregnancy (day 42 p.c.) or at preterm (day 60 p.c.) was not due to the increase in the oxytocin receptor numbers in the myometrium (see Sect. 6.6.1). On the contrary, onapristone treatment led to a significant reduction in oxytocin receptor concentration on day 43 p.c. In this study, a continuous, approximately tenfold increase in oxytocin receptor concentrations from early to term pregnancy was found. The lack

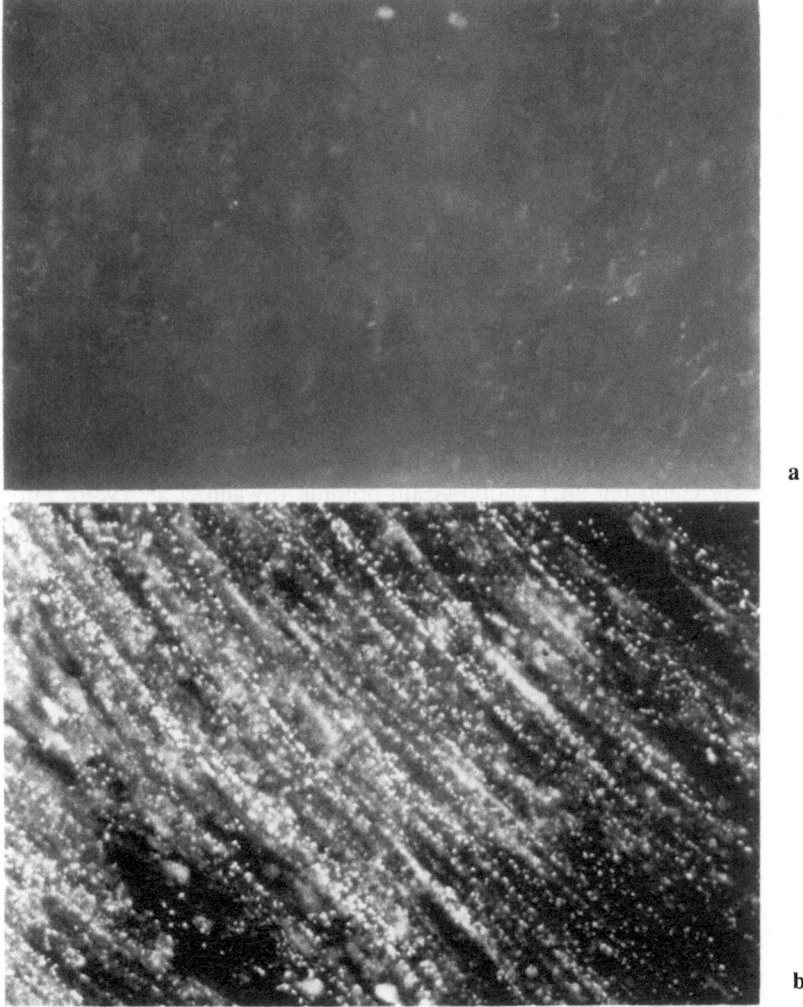

Fig. 16a,b. Effect of onapristone on gap junctions in late pregnant guinea pigs: Immunocytochemical staining with antibodies to connexin 43. Light micrograph (1000 ×) of myometrial tissue from a control animal on day 45 p.c. (**a**) and from an animal after 2 days of treatment with 10 mg onapristone s.c. on day 43-44 p.c.(**b**). **a** Note the lack of staining as compared to **b**. **b** Note the abundance of bright fluorescent spots which represent gap junctions

of an increase in oxytocin receptors after onapristone together with the observation that rising progesterone levels throughout pregnancy accompanied the increase in oxytocin receptors suggest that progesterone is not an inhibitor of myometrial oxytocin receptors in guinea pigs. Thus neither an acute increase in oxytocin receptors nor a change in oxytocin binding affinity is a prerequisite for the onset of labor in this species. This indicates that an increase in oxytocin sensitivity can occur at preterm in this species without an increase in oxytocin receptors. These data provide further evidence that a rise in oxytocin receptor concentrations alone may not be the only mechanism for oxytocin to be effective as suggested previously (Fuchs et al. 1982,1983).

These data are consistent with the results of studies with RU 486 performed in women during early pregnancy. An increase in uterine reactivity to sulprostone but not to oxytocin was found after RU 486 treatment in the study in which the effects of both uterotonic agents were compared after antiprogestin priming (Swahn et al. 1989). Furthermore, no increase in uterine responsiveness to oxytocin was found after treatment with the progesterone synthesis inhibitor epostane in early human pregnancy (Webster et al. 1985). However, the human uterus is not responsive to oxytocin during the first trimester of pregnancy, probably because oxytocin receptor concentrations are low at this time (Fuchs et al. 1982), whereas it is responsive to some degree to prostaglandins (Csapo 1981).

The data obtained in our guinea pig model conflict with the results of a study performed in pregnant rabbits which demonstrated an increase in myometrial oxytocin receptors after treatment with RU 486 (Jacobson et al. 1987). However, it is known that in this species, in contrast to guinea pigs, nonhuman primates and humans, progesterone withdrawal leads to the onset of labor. Therefore mechanisms regulating myometrial oxytocin receptors may be different in "progesterone-dependent" species such as rabbits.

6.7 Effects of Antiprogestins on the Uterine Cervix

The ripening of the uterine cervix occurring during late pregnancy seems to be the result of an active biochemical process within the cervix. This process represents a complex cascade with the involvement of degradative enzymes and changes in the synthesis of extracellular matrix proteins and glycoproteins (see Leppert 1992 for review). There is evidence that prostaglandins (Calder 1977; Hollingsworth et al. 1980) and relaxin (Downing and Sherwood 1985) are involved in spontaneous cervical ripening before birth. However, the exact mechanism which controls this process is still unknown. There is increasing evidence from animal and clinical studies with antiprogestins that progesterone is an important factor controlling both cervical ripening during pregnancy and even during the ovarian cycle.

6.7.1 Animal Studies

Studies in Pregnant Rats. In rats, a dose-dependent increase in extensibility and dilatation of the cervix were found 15 h after treatment with onapristone, lilopristone, and RU 486 in pregnant rats on day 14 p.c. These effects were seen long before the onset of labor, suggesting that cervical changes precede labor after antiprogestin treatment. Lilopristone was the most potent compound in this respect (Chwalisz et al. 1991c).

Studies in Guinea Pigs. The short duration of oxytocin-induced labor in antiprogestin-primed animals not only indicated increased uterine responsiveness but also suggested that the mechanical resistance of the genital tract was reduced. A dramatic softening and dilatation of the uterine cervix was found in onapristone-treated guinea pigs prior to the onset of labor (Chwalisz et al. 1987,1991c) (Fig. 17a). Electron microscopic (TEM) examinations revealed a dissolution, splitting up, and dissociation of collagen fibers and the expansion of the interfibrillar space due to edema after onapristone (Hegele-Hartung et al. 1989) (Fig. 17b). This was associated with the infiltration of the polymorphonuclear granulocytes, macrophages, and mast cells (Fig. 18). Furthermore, highly active fibroblasts with cytoplasmic components typical for secretory cells and polyploid nuclei were frequently found in onapristone-treated animals. Similar morphological changes were observed in control animals just before term. Biomechanical studies confirmed dose-dependent effects of both onapristone and RU 486 on the uterine cervix in guinea pigs (for review see Chwalisz et al. 1991c). We measured the mechanical properties of the cervix

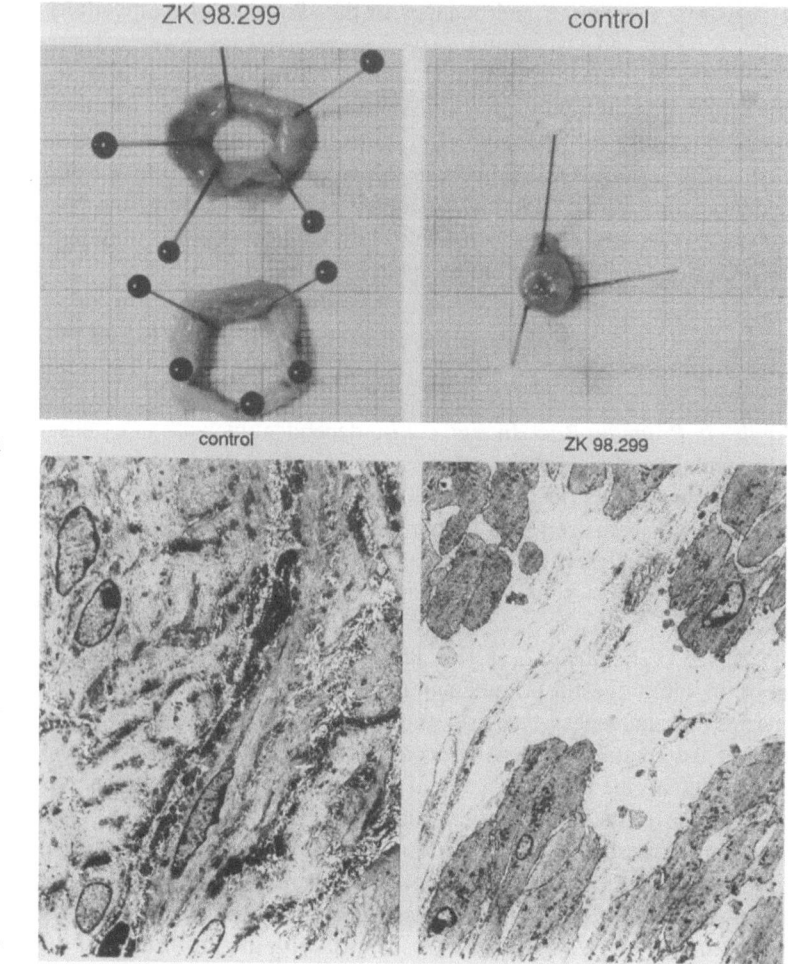

Fig. 17. a Effect of onapristone (3 mg s.c.) on the cervix in pregnant guinea pigs 15 h after treatment on day 59 p.c. **b** Transmission electron micrograph of the guinea pig cervix on day 60 p.c., 15 h after treatment with 1.0 mg onapristone s.c. *(right panel)* or vehicle *(left panel)*

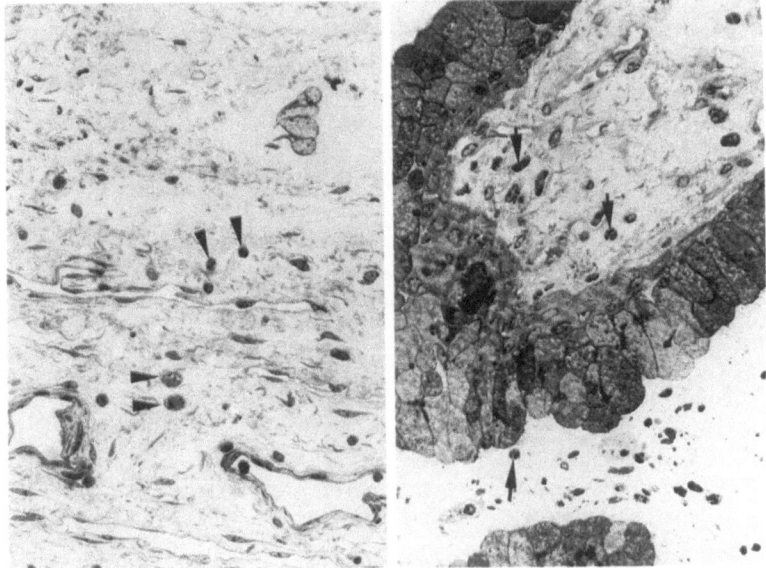

Fig. 18. Infiltration of polymorphonuclear leukocytes *(arrows)* into the guinea pig cervix on day 60 p.c., 15 h after onapristone treatment (1.0 mg s.c.)

by modifying the method described by Downing and Sherwood (1985) which allows for the quantification of cervical extensibility and dilatation of the uterine cervix under isometric conditions. The isolated cervix was incrementally stretched in steps of 0.2 mm at 2-min intervals and the resulting force (maximum values) digitally recorded. The slope of the regression line through the linear portion of the force curve was taken as a parameter of cervical extensibility. The reduction in slope in this relationship describes the increase in cervical extensibility (Chwalisz et al. 1991c). In nonpregnant guinea pigs, cyclical changes in the extensibility and dilatation of the cervix were found (Chwalisz et al. 1988). These changes were inversely related to the serum progesterone levels. During the latter half of pregnancy the cervix undergoes progressive softening and dilatation which starts around day 58 p.c. in guinea pigs. Maximum ripening was observed on day 65 p.c. (term day 67/68 + 3). There was no correlation between the mechanical properties of the cervix and ovarian steroid concentrations in the pe-

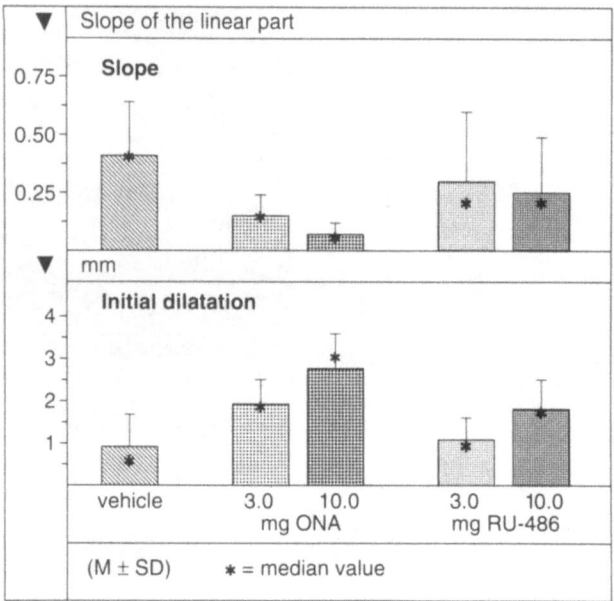

Fig. 19. Effect of onapristone *(ONA)* and RU 486 on cervical extensibility and dilatation in pregnant guinea pigs. The animals (n = 6/group) were treated sub-cutaneously on day 49 p.c. The measurements were performed 15 h after treatment. Reduction in the slope of the extensibility curve represents an increase in extensibility of the cervix. Note dose-dependent effects of onapristone ($p < 0.05$)

ripheral blood in pregnant animals (Chwalisz et al. 1991c). However, 15 h after treatment with onapristone (3 and 10 mg/animal s.c.) on day 49 p.c. there was a statistically significant dose-dependent increase in the extensibility and dilatation of the cervix. The effects of RU 486 administered at the same doses were less pronounced (Fig. 19). Time course studies revealed that the ripening effect of onapristone (3 mg/animal s.c., day 49 p.c.) was seen as early as 10 h after injection, but the maximum response came after 24 h (Chwalisz et al. 1987, 1991c). In late pregnancy (around day 60 p.c.), the softening effect of onapristone was much more pronounced and was evident after 30 times lower doses. Interestingly, both onapristone and RU 486 also induced cervical ripening in nonpregnant guinea pigs, indicating that conditions of pregnancy are not totally required for this response (Chwalisz et al. 1988).

The biomechanical studies performed in pregnant guinea pigs on day 50 of pregnancy shows that the "pure" progesterone agonist promegestone (R5020) fully reversed the onapristone effects on the cervix, which indicates that the observed effects were mediated via the progesterone receptor.

6.7.2 Human Studies

The clinical experience with RU 486 made during different stages of pregnancy and in nonpregnant women is consistent with the results of animal studies, demonstrating a ripening effect of antiprogestins on the uterine cervix. In all studies performed during the second and third trimester of pregnancy, RU 486 significantly reduced the induction–delivery interval and lowered the required prostaglandin (second trimester abortions) and oxytocin doses (term pregnancy). These results are an indirect indication that the cervix was prepared for delivery by RU 486 treatment. As presented in Table 1, the ripening effect of RU 486 was objectively assessed by biomechanical and morphological criteria in a number of studies performed during the first trimester and in one study performed during the second trimester of gestation (Frydman et al. 1988).

Moreover, a softening effect of RU 486 was demonstrated in nonpregnant women (Gupta and Johnson 1990), suggesting that the conditions of pregnancy (e.g., decidual or placental release of prostaglandins) are not required for this action. It is still unclear whether the ripening effect of RU 486 is dose related in women.

In the study performed during the second trimester, the cervical resistance was evaluated using Hegar dilators 48 h after pretreatment with 150 and 450 mg of RU 486, just prior to labor induction with a prostaglandin. A statistical improvement in cervical calibration was found after both doses, but the effect was independent of the dose (Frydmann et al. 1988). In another (WHO 1990) study performed during the first trimester of pregnancy comparing the cervical effects of different doses of RU 486 (25, 50, 100 mg, given twice at 24 and 12 h before vacuum aspiration), there was a significant although not dose-

142 Kristof Chwalisz

related increase in cervical dilatation. However, the ease of dilatation based on a subjective assessment by the operating surgeons was not only improved by RU 486, but was also dose related (WHO 1990). Further biomechanical studies employing improved assessment methods of cervical dilatation and extensibility are needed to clarify the question of dose dependency of the cervical effects of antiprogestins in humans.

6.7.3 Mechanism of Action of Antiprogestins on the Uterine Cervix

Biochemical studies revealed a decrease in the collagen and total glycosaminoglycan concentrations in the cervix after onapristone treatment in late pregnant guinea pigs, comparable to those seen in controls at term (Chwalisz et al. 1991c). These data suggest that an increased collagenolysis takes place within the cervix following onapristone treatment. Local edema seen in treated animals is probably the result of the qualitative changes in proteoglycans. Biochemical studies performed by other investigators are consistent with our findings, indicating that collagenase-mediated collagenolysis is the mechanism of cervical ripening in guinea pigs (Rajabi et al. 1991a,b,c). At spontaneous parturition (day 68+ p.c.), there was a marked increase in procollagenase, collagenase inhibitory activity, and net procollagenase activity (Rajabi et al. 1991a). Moreover, an increase in serum collagenase levels was found in normal pregnant women which correlated with the state of the cervix (Granström et al. 1992). In vitro studies employing monolayer cell cultures derived from cervices of day 50-prégnant guinea pigs demonstrated that at pharmacological concentrations progesterone ($10^{-6}M$) inhibited the procollagenase gene expression and blocked the estradiol-induced increase in collagenese production and secretion (Rajabi et al. 1991b). Recently, it was shown that the physiological concentrations of progesterone (10–200 nM) almost totally abolished the release of matrix metalloproteinases, collagenase, and related gelatinases in human endometrial explants, an effect which was antagonized by RU 486 (Marbaix et al. 1992). These studies suggest that progesterone may control directly or indirectly the collagenase activity in the uterine cervix.

It has been suggested that estrogens induce cervical ripening in humans (Gordon and Calder 1977), but this effect has never been convincingly demonstrated. We were also unable to demonstrate the ripening effects of estrogen in pregnant guinea pigs. In addition, estradiol did not influence the onapristone-induced ripening effect in this study (Chwalisz et al. 1991c). Since progesterone functionally opposes estrogen action in various tissues, an "unopposed estrogen action," cannot be excluded as a possible mechanism of antiprogestin-induced cervical ripening. Further studies are needed to answer this question.

Are Prostaglandins Involved in Antiprogestin-Induced Cervical Ripening? Since prostaglandins, in particular PGE$_2$, are widely believed to stimulate collagenase activity and eventually induce cervical ripening, we asked whether prostaglandins mediate antiprogestin-induced cervical ripening. In late-pregnant guinea pigs (day 42–43 p.c.) indomethacin had no effect on the onapristone-induced cervical ripening (Chwalisz et al. 1991c). However, the same dose of indomethacin fully inhibited cervical ripening induced with LPS in guinea pigs (Bukowski et al. 1993). These findings indicate that prostaglandins may not be mediators of antiprogestin action on the cervix in guinea pigs. Similar results were obtained with another cyclo-oxygenase inhibitor, diclofenac, in pregnant rats after treatment with RU 486 (Cabrol et al. 1991). Diclofenac inhibited RU 486-induced labor, but did not antagonize its ripening effect on the cervix (i.e., increased hydration, and hyaluronic acid concentration). Human studies with RU 486 are consistent with the animal data. It has been shown that pretreatment with RU 486 does not influence the capacity of human cervical tissue to produce prostaglandins from arachidonic acid (Rådestadt et al. 1990). In addition, treatment with the prostaglandin synthesis inhibitor naproxen did not inhibit the effect of RU 486 on the cervix in women (Rådestadt and Bygdeman 1992). In sum, all the data indicate that antiprogestins do not act on the cervix by means of stimulating the endogenous prostaglandin production.

The *"Inflammatory Reaction" as a Mechanism for Antiprogestin-Induced Cervical Ripening.* The theory of the role of inflammatory cells in the process of cervical ripening was first proposed by Liggins (1981). An infiltration of white blood cells into the cervix has been described in women at term (Junqueira et al. 1980). In this study there was a dissolution of the connective tissue matrix around polymorphonuclear leukocytes. Activated and degranulated eosinophils were reported in cervical biopsies in women at term (Knudsen et al. 1991). Neutrophilic granulocytes and macrophages are well-studied sources of collagenase and other enzymes that are capable of digesting the extracellular matrix proteins (Horwitz et al. 1987; Ito et al. 1987, 1988; Leppert 1992). It is likely that enzymes released from white blood cells play an important role during cervical ripening.

On the basis of the results of studies performed in pregnant guinea pigs which demonstrated an infiltration of polymorphonuclear granulocytes, macrophages, and mast cells (see above) and a lack of the effect of indomethacin, we proposed that chemotactic agents like cytokines could mediate onapristone effects in the cervix (Chwalisz et al. 1991c). There is increasing evidence that cytokines are involved in normal parturition during the ripening of the cervix. Uterine cervix explants from pregnant rabbits secrete IL-1-like factors in vitro (Ito et al. 1988). Recently, it was reported that an IL-8-like chemotactic factor is released from uterine cervical explants of rabbits at term pregnancy (Uchiyama et al. 1992). Furthermore, progesterone inhibits and RU 486 stimulates the IL-8 release from human choriodecidual cells in vitro (Kelly et al. 1992). It was reported that IL-8 concentrations increase during both preterm and term parturition (Romero et al. 1991). To evaluate the concept of the involvement of cytokines in cervical ripening we investigated the local effects of IL-1β, TNF-α, their combination and IL-8 on the extensibility and the morphology of the uterine cervix (Chwalisz et al. 1993b). Both biomechanical (Fig. 20) and morphological studies revealed a significant cervical softening and dilatation of the, cervix after local application of the cytokines in gel. Light and electron micrographs demonstrated a pronounced cervical ripening with the dissolution of collagen fibers, stromal edema, and infiltration by polymorphonuclear leukocytes in all groups treated with the cytokines. This study demonstrates that a local application in gel of the cytokines brought about cervical ripening without inducing labor in pregnant guinea pigs. The effects of IL-1β and IL-8 were similar to the physiological ripening at term and to the effects seen after onapristone.

The infiltration of inflammatory cells may be a fundamental part of the process of parturition. However, how the infiltration occurs during normal parturition is still not clearly defined (Leppert 1992). The experimental data (see above) suggest that the cytokines are involved in

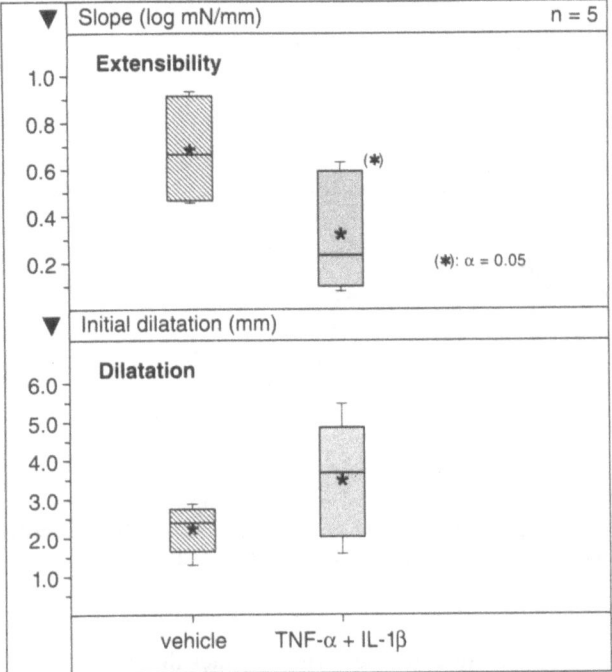

Fig. 20. Cervical ripening after local treatment with tumor necrosis factor-α *(TNF-α)* and interleukin-1β *(IL-1β)* combination in pregnant guinea pigs. Intracervically 2×10^5 U TNF-α + 2×10^6 U IL-1β (formulated in 3% hydroxycellulose gel in PBS buffer) were given on day 59 p.c. twice, at 9 a.m. and 5 p.m. The measurements were performed on day 60 p.c. at 9–10 a.m.

both normal and antiprogestin-induced cervical ripening. Morphologically and biochemically, antiprogestin-induced cervical ripening did not differ from the spontaneous process at term in guinea pigs (Hegele-Hartung et al. 1989; Chwalisz et al. 1991c). However, it is still not clear why spontaneous cervical ripening occur in this (and other) species in the presence of high serum progesterone levels.

Figure 21 presents the hypothetical mechanism of action of antiprogestins on the uterine cervix. Antiprogestins may act directly on the cervical fibroblast. The active fibroblasts may secrete collagenolytic enzymes which digest the extracellular matrix. Moreover, the active fi-

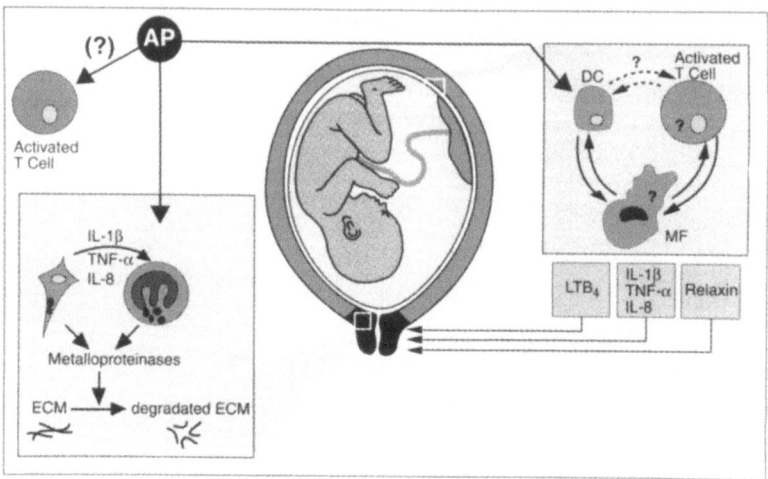

Fig. 21. Hypothetical mechanism of action of antiprogestins *(AP)* on the cervix. *IL*, interleukin; *TNF-α*, tumor necrosis factor-α; *ECM*, extracellular matrix; *DC*, decidual cells; *MF*, macrophages; *LTB₄*, leukotriene B₄

broblasts may secrete cytokines which attract the inflammatory cells after antiprogestin treatment. On the other hand, antiprogestins may act on the decidua, which contains a high amount of macrophages capable of releasing cytokines. It is also possible that antiprogestins act directly on the activated decidual or even peripheral T lymphocytes, which express progesterone receptors during pregnancy (Szekeres-Bartho et al. 1989). Alloactivated or pregnancy lymphocytes release at least two suppressor factors (PIBF and J6B7) which inhibit natural killer (NK) cells and TNF secretion by activated lymphocytes (see Chap. 9, this volume). The cytokines such as IL-1β, TNF-α, and IL-8, showing chemotactic activities, may be responsible for the infiltration of inflammatory cells, which in turn release metalloproteinases degradating the extracellular matrix. However, other factors such as leukotrienes and relaxin may also be involved in the antiprogestin action on the uterine cervix. .

6.8 General Discussion and Conclusions

The existing data collected with antiprogestins in various species, including nonhuman primates and humans, indicate that progesterone plays an essential role in the control of labor. Treatment with antiprogestins during advanced pregnancy induces myometrial and cervical effects in all species investigated to date. In species showing physiological progesterone withdrawal at term such as rats, rabbits, sheep, pigs, and cows, treatment with an antiprogestins during late pregnancy induces preterm parturition which is similar to normal parturition. In species in which there is no fall in progesterone concentrations prior to normal parturition such as guinea pigs, nonhuman primates, and humans, treatment with progesterone antagonists results in the increase in myometrial responsiveness and cervical ripening but not always in preterm birth. This observation suggests that both the myometrial responsiveness and the state of the cervix are controlled by progesterone during pregnancy in these species. In humans, the state of the cervix correlates very well with the myometrial sensitivity to oxytocin as assessed by the oxytocin infusion test. The manual evaluation of the cervix is a common method used to predict preterm birth in women.

The mechanism(s) that controls the release of uterine stimulants such as prostaglandins, oxytocin, endothelin (Suzuki 1990), etc. which eventually trigger the onset of labor might not be dependent on progesterone. It is possible that the release of "intrinsic uterine stimulants" is suppressed by factors of fetal and/or placental origin during pregnancy, similar to the way in which the antiluteolytic protein oTP-1 controls uterine prostaglandins in sheep. During early pregnancy, sheep conceptus secretes oTP-1, a protein from the interferon family, which suppresses the production of the luteolytic $PGF_{2\alpha}$ (see Bazer 1991 for review). The endogenous inhibition of uterine prostaglandins or other uterotonic factors may decrease during the progress of pregnancy and disappear at term. This concept of the release of endogenous inhibition occurring at term might explain why onapristone alone induces premature birth at preterm but not during mid-gestation in guinea pigs (see Sect. 6.5.2).

Overall, our studies performed in the guinea pig model indicate that the uterus (myometrium and cervix) goes through a conditioning (preparatory) step, which may be irreversible. This step can be induced with antiprogestins. After passing through this step the uterus is pre-

148   Kristof Chwalisz

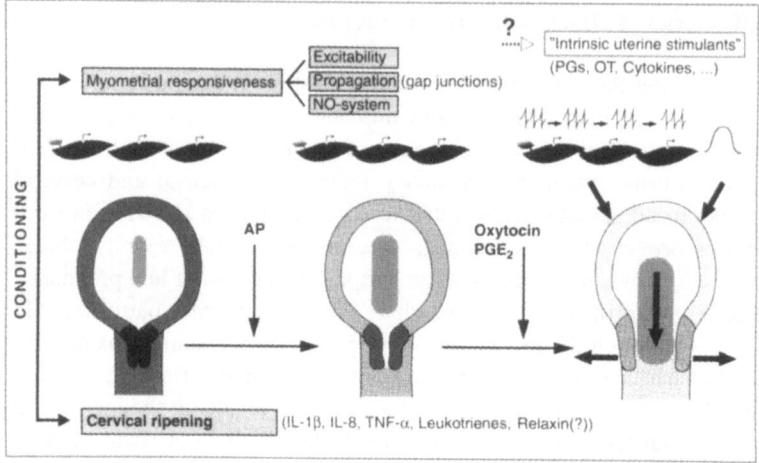

Fig. 22. Mechanisms of action of antiprogestins *(AP)* on the pregnant uterus during advanced pregnancy. *PG*, prostaglandin; *OT*, oxytocin; *IL*, interleukin; *TNF-α*, tumor necrosis factor-α

pared for labor and delivery and both exogenous or endogenous uterotonic agents may induce successful uterine contractions. The major mechanism of antiprogestin action on the late pregnant uterus may be the induction of gap junction protein synthesis which is essential for a successful labor (see Fig. 22). The substantial increase in electrical coupling following antiprogestin treatment results in the development of phasic and coordinated responses to injections of agonists (e.g., oxytocin or prostaglandins). However, antiprogestins may influence also other important factors responsible for the uterine quiescence during pregnancy. Antiprogestins may decrease the cGMP-effector system of nitric oxide, which is in part responsible for uterine relaxation during pregnancy (see Chap. 1, this volume). Furthermore, antiprogestins may result in the increase in ion channels and pumps (see Chap. 4, this volume), IP3-transduction mechanisms (see Chap. 2, this volume) etc. The second important component of the conditioning effect of antiprogestins is the cervical ripening. This action may be mediated by the cytokines (IL-1β, IL-8, TNF-α, etc). However, the involvement of leukotrienes and relaxin cannot be excluded.

Onapristone, which may be considered a "pure" progesterone antagonist shows high labor-inducing activity in both guinea pigs and *Tupaja belangeri* at preterm. On the contrary, RU 486 which is an antiprogestin with partial progesterone-agonistic activity, was not fully effective in inducing preterm parturition in these species. Therefore, the high labor-inducing activity of onapristone may be due to the absence of progesterone-agonistic activity in guinea pigs and *Tupaja belangeri*. In preterm guinea pigs, other 13α-configurated compounds are also very effective in inducing labor, which is consistent with this view. In guinea pigs, the efficacy of onapristone in inducing the uterine responsiveness to oxytocin and prostaglandins substantially increases during the course of pregnancy in spite of little changes in serum progesterone levels. At preterm (day 60 p.c.), 30 times lower doses of onapristone and a single treatment are required to achieve a similar response to both oxytocin and sulprostone compared to midpregnancy (see Sect. 6.5.2). This is remarkable in light of minor differences in serum progesterone concentrations at both stages of pregnancy (approximately 1500 nmol/l day 45 p.c. versus approximately 1000 nmol/l day 60 p.c.; Shi Shao Qing et al. 1988). Moreover, the spontaneous increase in both myometrial responsiveness and cervical ripening in guinea pigs (and humans) occurs in the presence of increasing progesterone concentrations in maternal blood. Why there is a decrease in progesterone action in the pregnant uterus at term without a progesterone decrease in peripheral blood remains a fundamental and still unanswered question. A hypothetical explanation of this phenomenon may be that the target organ sensitivity to progesterone (e.g., at receptor or postreceptor level) decreases in the course of pregnancy. Another possibility is that a "natural" antiprogestin is secreted in increasing amounts during late pregnancy. Recently, on the basis of in vitro experiments, transforming growth factor-β was proposed as a physiological antiprogestin (Casey and MacDonald 1993).

Antiprogestins may find a broad application in obstetrics as an adjunct to labor and delivery during second-trimester pregnancy termination, in intrauterine fetal death, and finally in the induction of labor at term. Termination of second-trimester pregnancy represents a serious clinical problem and this procedure is often associated with complications which are related to the gestational age and the method used. Although the availability of prostaglandins has greatly improved, procedures for second-trimester terminations and induction of abortion with intravenous, vaginal, extra- or intramniotic administrations of a prosta-glandin remains a long, distressing and painful experience for women because of prostaglandin-related side effects. The results of studies with a combination of RU 486 and prostaglandins indicate that this is an effective and safe method and a sequential antiprogestin–prosta-glandin treatment could eventually become a standard obstetric proce-dure during therapeutic second-trimester abortion and after intrauterine fetal death. The most exciting indication for antiprogestins seems to be induction of labor at term in combination with oxytocin. Our data dem-onstrate that the cervical ripening can be induced with antiprogestins without influencing uterine contractions. From an obstetrical point of view this is a favorable situation. If the uterus is prepared for delivery, subsequent treatment with oxytocin would permit better control of uterine contractility. Furthermore, the preparatory effect on the myometrium in terms of increased sensitivity to oxytocin is unique to antiprogestins.

Despite the fact that no acute or long-term unwanted effects of RU 486 have been reported in women or newborns to date and despite encouraging animal data, further studies are needed to establish the safety of RU 486 (and other antiprogestins) before these compounds can be routinely recommended as an adjunct to labor and delivery in pregnancies with living infants. To date there are no reports on the possible long-term effects of RU 486 in newborns.

Onapristone may be a preferable antiprogestin as an adjunct to labor and delivery at term. In the guinea pig model, onapristone is 10–30 times more effective than RU 486 in inducing both myometrial re-sponsiveness and cervical ripening (see Sect. 6.6.1). Moreover, onapris-tone alone induces preterm parturition in guinea pigs and *Tupaja belangeri*. The half-life of onapristone is about 2–4 h in women after oral administration (K. Zurth, unpublished data) (the half-life of RU 486 ranges between 24–48 h; Heikinheimo 1989). Furthermore, onapristone shows substantially lower antiglucocorticoid activity. However, further safety studies of labor induction with onapristone are needed in non human primates.

Acknowledgements. The initial phase of this research was conducted and supported by Dr. W. Elger. I would like to acknowledge this fruitful collaboration. Dis-cussions with Dr. R. Garfield, Dr. R. Romero, and Dr. R. Bukowski helped a lot in the development of ideas expressed here. I am grateful to Mrs. B. Kosub, Mrs. B. Bragulla, Mrs. H. Altmann, Mrs. G. Bauer, and Mr. S. Althof for their excellent technical assistance in performing experimental studies. I thank Mr. R. Kolberg, Mr. D. Niemeier, and Dr. Roth for the development of the equipment to assess both the cervical extensibility and intrauterine pressure.

References

Alexandrova M, Soloff MS (1980) Oxytocin receptors and parturition in the guinea pig. Biol Reprod 22:1106–11

Aubeny E, Baulieu EE (1991) Contragestive activity of RU 486 and oral active prostaglandin combinantion. Comptes Rendus de L'Academie des Science 312:539–545

Baulieu EE (1985) RU 486: an antiprogestin with contragestive activity in women. In: Baulieu EE, Segal SJ (eds) The antiprogesterone steroid RU 486 and human fertility control. Plenum Press, New York:1–25

Bazer FW (1991) Uterine-conceptus interactions during the peri-implantation period. In: Lavia LA (ed) Cellular signals controlling uterine function. Plenum Press, New York, London, pp 119–137

Beier HM, Mootz U, Hegele-Hartung C (1989) Studies on the establishment of mammalian pregnancy: synchronization of the maternal and the embryonic systems. In: Holstein AF, Voigt KD, Gräblin D (eds) Reproductive Biology and Medicine. Diesbach Verlag, Berlin, pp 210–223

Birgerson L, Ölund A, Odlind V, Somell C (1987) Termination of early human pregnancy with epostane. Contraception 35:111–20

Bosc MJ, Germain G, Nicolle A, Mouren M, Philibert D, Baulieu EE (1987) Control of birth in rats by RU 486, an antiprogesterone compound. J Reprod Fertil 79:1–8

Bukowski R (1993) Experimentelle Untersuchungen der Funktion von Zytokinen bei der Frühgeburt. Inaugural Dissertation, Free University of Berlin

Bukowski R, Scholz P, Hasan SH, Chwalisz K (1993) Induction of preterm parturition with the interleukin-1β, tumor necrosis factor-α (TNF-α) and with LPS in guinea pigs. Soc Gynecol Invest, Toronto, Ontario (abstract S26)

Bygdeman M, Swahn ML (1985) Progesterone receptor blockage. Effect on uterine contractility and early pregnancy. Contraception 32:45–51

Cabrol D, Bouvier D'Yvoire M, Mermet E, Cedard L, Sureau C, Baulieu EE (1985) Induction of labour with mifepristone after intrauterine fetal death. Lancet 1985 2:1019

Cabrol D, Dubois C, Cronje H, Gonnet JM, Guillot M, Maria B, Moodley J, Oury JF, Thoulon JM, Treisser A, Ulmann D, Correl S, Ulmann A (1990)

Induction of labor with mifepristone (RU 486) in intrauterine fetal death. Am J Obstet Gynecol 1990, 163:540–542

Cabrol D, Carbonne B, Bienkiewicz A, Dallot E, Alj AE, Cedard L (1991) Induction of labor and cervical maturation using mifepristone (RU 486) in the late pregnant rat. Influence of a cyclooxygenase inhibitor (diclofenac). Prostaglandins 42:71–79

Calder AA, Embrey MP Tait T (1977) Ripening of the cervix with extra-amniotic prostaglandin E_2 in viscous gel before induction of labour. Br J Obstet Gynecol 84:264

Casey ML, MacDonald PC (1993) Transforming growth factor-β (TGF-β) acts as a gene-specific antiprogestin. Society for Gynecologic Investigation. Scientific Program and Abstracts, 40th Annual Meeting, March 31-April 3, 1993 (Abstract S2)

Casey ML, Smith JW, Nagai K, Hersh LB, MacDonald PC (1992) Progesterone-regulated cyclic modulation of membrane metalloendopeptidase (enkephalinase) in human endometrium. J Biol Chem 34:23041–23047

Chwalisz K, Elger W (1986) Induction of labor with antigestagens in the rat and guinea pig. J Steroid Biochem [Suppl] 25:1315 (abstract)

Chwalisz K, Altmann H, Elger W (1986) Induction of premature parturition in the rat with antigestagens. Acta Endocrinol [Suppl] 111:12 (abstract)

Chwalisz K, Shi Shao Qing Neef G, Elger W (1987) The effect of the antigestagen ZK 89 299 on the uterine cervix. Acta Endocrinol [Suppl] 114:113–114 (abstract)

Chwalisz K, Shi Shao-Qing Esch A, Elger W (1988) Cervical softening in non-pregnant guinea pigs. Acta Endocrinol [Suppl] 117:194–195 (abstract)

Chwalisz K, Hegele-Hartung Ch, Fritzemeier KH, Beier HM, Elger W (1991a) Inhibition of the estradiol-mediated endometrial gland formation by the antigestagen onapristone in rabbits: relationship to uterine estrogen receptors. Endocrinology 129:312–322

Chwalisz K, Fahrenholz F, Hackenberg M, Garfield R, Elger W (1991b) The progesterone antagonist onapristone increases the effectiveness of oxytocin to produce delivery without changing the myometrial oxytocin receptor concentrations. Am J Obstet Gynecol 165:1760–1770

Chwalisz K, Hegele-Hartung, Schulz R, Shi Shao Qing, Louton PT, Elger W (1991c) Progesterone control of cervical ripening – experimental studies with the progesterone antagonists onapristone, lilopristone and mifepristone. In: Leppert P, Woessner F (eds) The extracellular matrix of the uterus, cervix and fetal membranes: synthesis, degradation and hormonal regulation, Perinatology Press, pp 119–131

Chwalisz K, Hsiu JG, Williams RF, Hodgen GD (1992) Evaluation of the antiproliferative actions of the progesterone antagonists mifepristone (RU 486) and Onapristone (ZK 98 299) on primate endometrium. Soc Gynecol Invest, Scientific, San Antonio, Texas, March 18–21, 1992 (abstract)

Chwalisz K, Stöckemann K, Fritzemeier KH, Gemperlein I, Fuhrmann U, Neef G (1993a) 13α-configurated progesterone antagonists: functional antiestrogenic action in the rat. Exp Clin Endocrinol 101; Suppl 1: 60 (abstract)

Chwalisz K, Scholz P, Hegele-Hartung Ch, Roth G, Bukowski R (1993b) Cervical ripening with interleukin 1β (IL-1β) and tumor necrosis factor-a (TNF-a) in pregnant guinea pigs. Soc Gynecol Invest, Toronto, Ontario (abstract S27)

Clark CL, Sutherland RL (1990) Progestin regulation of cellular proliferation. Endocrine Rev 11:266–310

Couzinet B, le Strat N, Silvestre L, Schaison G (1990) Late luteal administration of the antiprogesterone RU 486 in normal women: Effects on the menstrual cycle events and fertility control in a long-term study. Fertil Steril 54:1039–1044

Creange JE, Anzalone AJ, Potts GO, Schane HP (1981) WIN 32, 729 a new potent interceptive agent in rats and rhesus monkeys. Contraception 24:289–299

Creasy RK (1989) Preterm labor and delivery. In: Creasy RF, Resnik R (eds) Meternal-fetal medicine: Principles and practice. W.B. Saunders Company, Philadelphia, London, Toronto, Montreal, Sydney, Tokyo, pp 477–505

Croxatto HB, Salvatierra (1990) Cyclic use of antigestagens for fertility control. Third International Symposium on Contraception, Heidelberg, June 19–23, 1990

Csapo AI (1975) The "seesaw" theory of regulatory mechanisms of pregnancy. Am J Obstet Gynecol 121:578–581

Csapo AI (1981) Force of labor. In: Iffy L, Kaminetzky HA, (eds) Principles and practice of obstetrics and perinatology, Vol.2 John Wiley & Sons, New York, pp 761–799

Csapo AI, Wiest WG (1969) An examination of the quantitative relationship between progesterone and maintenance of pregnancy. Endocrinology 85:735–746

Downing SJ, Sherwood OD (1985) The physiological role of relaxin in the pregnant rat. II. The influence of relaxin on cervical extensibility. Endocrinology 116:1215–1220

Durlot F, Dubois C, Brunerie J, Frydman R (1988) Efficacy of progesterone antagonist RU 486 (mifepristone) for pre-operative cervical dilatation during first trimester abortion. Hum Reprod 3:583–584

Egarter CH, Husslein PW, Rayburn WF (1990) Uterine hyperstimulation after low-dose prostaglandin E_2 therapy: tocolytic treatment in 181 cases. Am J Obstet Gynecol 163:794–796

Elger W (1979) Pharmacology of parturition and abortion. Anim Reprod Sci 2: 133–148

Elger W, Hasan SG (1985) Studies on the mechanism of action of antifertile PG in animal models. Acta Physiol Hungarica 65:415–32

Elger W, Beier S, Chwalisz K, Fähnrich M, Hasan SH, Henderson D, Neef G, Rohde R (1986) Studies on the mechanism of action of progesterone antagonists. J Steroid Biochem 25:835–45

Elger W, Fähnrich M, Beier S, Shi Shao Qing, Chwalisz K (1987) Endometrial and myometrial effects of progesterone antagonists in pregnant guinea pigs. Am J Obstet Gynecol 157:1065–1074

Elger W, Neef G, Ottow E, Beier S, Chwalisz K, Fähnrich M, Kosub B, Hasan SH (1989) Studies on interactions of antigestagens with prostaglandins and sex hormone-related agents at the myometrial level in pregnant guinea pigs. In: Hormone antagonist for fertility regulation (Puri CP, Van Look PFA, eds.) Good Print, Bombay, pp 105–121

Elger W, Chwalisz K, Fähnrich M, Hasan SH, Laurent D, Beier S, Ottow E, Neef G, Garfield RE (1990) Studies on labor-conditioning and labor-inducing effects of antiprogesterones in animal models. In: Garfield RE (ed) Uterine contractility: mechanisms of control. Norwell, Ma.: Serono Symposia, USA, pp 153–157

Ellis ST, Heap B, Butchart AR, Rider V, Richardson NW, Wang MY, Taussig MJ (1988) Efficacy and specificity of monoclonal antibodies to progesterone in preventing the establishment of pregnancy in the mouse. J Endocrinol 118:69–80

Frydman R, Taylor S, Ulmann A (1985) Transplacental passage of mifepristone. Lancet ii:1252

Frydman R, Lelaidier C, Baton-Saint-Mleux C, Fernandez H, Vial M, Bourget P (1992) Labor induction in women at term with mifepristone (RU 486): a double-blind, randomized, placebo-controlled study. Obstet Gynecol 80:972–975

Frydman R, Fernandez H, Pons JC, Ulmann A (1988) Mifepristone (RU 486) and late pregnancy termination: a double-blind study of two different doses. Human Reprod 3:803–806

Frydman R, Baton C, Lelaidier C, Vial M, Bourget P, Fernandez H (1991) Mifepristone for induction of labour. Lancet 337:488–489

Fuchs AR, Mok E, Sundaram K (1974) Luteolytic effects of prostaglandins in rat pregnancy, and reversal by luteinizing hormone. Acta endocrinol (Kbh) 76:583–596

Fuchs AR, Fuchs F, Husslein P, Soloff MS, Fernstrom M (1982) Oxytocin receptors and human parturition: A dual role for oxytocin in the initiation of labor. Science 215:1396–1398

Fuchs AR, Periyasamy S, Alexandrova M, Soloff MS (1983) Correlation between oxytocin receptor concentration and responsiveness to oxytocin in pregnant rat myometrium. Effects of ovarian steroids. Endocrinology 113:742–749

Garfield RE, Beier S (1989) Increased myometrial responsiveness to oxytocin during term and preterm labor. Am J Obstet Gynecol 161:454–461

Garfield RE, Hertzberg EL (1990) Cell-to-cell coupling in the myometrium: Emil Bozler's prediction. In: Sperelakis N, Wood JD (eds) Frontiers in smooth muscle research. New York: Wiley-Liss, pp 673–681

Garfield RE, Sims S, Daniel EE (1977) Gap junctions: their presence and necessity in myometrium during gestation. Science 198:958–960

Garfield RE, Gasc JM, Baulieu EE (1987) Effect of the antiprogesterone RU 486 on preterm birth in the rat. Am J Obstet Gynecol 157:1281–1285

Gazol OS, Li Y, Schwabe C, Anderson LL (1993) Attenuation of antepartum relaxin surge and induction of parturition by anatiprogesterone RU 486 in sheep. J Reprod Fertil 97:233–240

Germain G, Philibert D, Pottier J, Mouren M, Baulieu EE, Sureau C (1985) Effect of the antiprogesterone agent RU 486 on the natural cycle and gestation in intact cynomolgus monkeys. In: Baulieu EE and Segal SJ (eds) The antiprogestin steroid RU 486 and human fertility control, Plenum Press, New York, pp 155–167

Glasier A, Thong KJ, Dewar M, Mackie M, Baird DT (1992) Mifepristone (RU 486) compared with high-dose estrogen and progestestogen for emergency postcoital contraception. N Engl J Med 327:1041–1044

Gordon AJ, Calder AA (1977) Oestradiol applied locally to ripen the unfavourable cervix. Lancet ii:1319–21

Granström LM, Gunvor EE, Malmström A, Ulmsten U, Woessner JF (1992) Serum collagenase levels in relation to the state of the human cervix during pregnancy and labor. Am J Obstet Gynecol 167:1284–1288

Gravanis A, Schaison G, George M, de Brux J, Satyaswaroop PG, Baulieu EE, Robel P (1985) Endometrial and pituitary responses to the steroidal antiprogestin RU 486 in postmenopausal women. J Clin Endocrinol Metab 60:156–63

Gottlieb C, Bygdeman M (1991) The use of antiprogestin (RU 486) for termination of second trimester pregnancy. Acta Obstet Gynecol Scand 70:199–203

Gronemeyer H, Benhamou B, Berry M, Bocquel MT, Gofflo D, Garcia T, Lerouge T, Metzger D, Meyer ME, Tora L, Vergezac A, Chambon P (1992) Mechanism of antihormone action. J Steroid Biochem Molec Biol 41:217–221

Gupta JK, Johnson N (1990) Effect of mifepristone on dilatation of the pregnant and non-pregnant cervix. Lancet 335:1238–40

Haluska GJ, Stanczyk FZ, Cook MJ, Novy MJ (1987) Temporal changes in uterine activity and prostaglandin response to RU 486 in rhesus macaques in late gestation. Am J Obstet Gynecol 157:1487–1495

Haluska GJ, West NB, Novy MJ, Brenner RM (1990) Uterine estrogen receptors are increased by RU 486 in late pregnant rhesus macaques but not after spontaneous labor. J Clin Endocrinol Metab 70:181–186

Haluska GJ, Cook MJ, Novy MJ (1991) Epostane but not RU 486 mimics spontaneous parturition in rhesus monkeys. Soc Gynecol Invest, San Antonio, TX (abstract 266)

Haluska, GJ, Cook MJ, Novy MJ (1993) Effect of onapristone (ZK 98 299) on uterine activity and cervical status in late gestation rhesus monkeys. Soc Gynecol Invest, Toronto, Ontario (abstract S24)

Healy DL (1990) Progesterone receptor antagonists and prostaglandins in human fertility regulation: a clinical review. Reprod Fertil Dev 2:477–490

Heap RB, Illingworth DV, Perry JS (1973) The secretory activity of the corpus luteum in the guinea pig and its role in the establishment and maintenence of pregnancy. In: Denamur, Netter (eds) Le Corps Jaune: 69–80

Hegele-Hartung C, Chwalisz K, Beier HM, Elger W (1989) Ripening of the uterine cervix of the guinea pig after treatment with the progesterone antagonist onapristone (ZK 98 299): an electron microscopic study. Hum Reprod 4:369–377

Heikinheimo O (1989) Pharmacokinetics of the antiprogesterone RU 486 in women during multiple dose administration. J Ster Biochem 32:21–25

Hill NCW, Selinger M, Ferguson J, López Bernal A, MacKenzie IZ (1990) The physiological and clinical effect of progesterone inhibition with mifepristone (RU 486) in the second trimester. Br J Obstet Gynaecol 97:487–492

Hill NCW, Selinger M, Ferguson J, MacKenzie IZ (1991) Transplacental passage of mifepristone and its influence on maternal and fetal steroid concentrations in the second trimester of pregnancy. Hum Reprod 6:458–462

Höfig A, Elger W, Chwalisz K, Hasan SH, Neef G, Ellendorff F (1988) Induction of parturition in the pig with the antigestagen ZK 112 993. Acta Endocrinol, Suppl 287, p 193

Hoff JD, Quigley ME, Yen SC (1983) Hormonal dynamics at midcycle: a reevaluation. J Clin Endocrinol Metab 57:792-796

Hollingsworth M, Gallimore S, Isherwood CNM (1980) Effects of prostaglandin $F2\alpha$ and E_2 on cervical extensibility in the late pregnant rat. J Reprod Fert 58:95–99

Horwitz AL, Hance AJ, Crystal RG (1987) Granulocyte collagenase: selective digestion of type I relative to type III collagen. Proc Natl Acad Sci USA 74:262–270

Horwitz KB (1992) The molecular biology of RU 486. Is there a role for anti-progestins in the treatment of breast cancer? Endocr Rev 13:146–163

Ito A, Hiro D, Sakyo K (1987) The role of leukocyte factors on uterine cervical ripening and dilatation. Biol Reprod 37:511–517

Ito A, Hiro D, Hioro D, Ojima Y, Mori Y (1988) Spontaneous production of interleukin1-like factors from pregnant rabbit uterine cervix. Am J Obstet Gynecol 159:261–265

Jacobson L, Riemer RK, Goldfien AC, Lykins D, Siiteri PK, Roberts JM (1987) Rabbit myometrial oxytocin and α_2-adrenergic receptors are increased by estrogen but are differentially regulated by progesterone. Endocrinology 120:1184–89

Johnson N, Bryce FC (1990) Could antiprogesterones be used as alternative cervical ripening agents? Am J Obstet Gynecol 162:688–90

Jungueira LCU, Zugaib M, Montes GS, Toledo OMS, Krisztan RM, Shigihara KM (1980) Morphologic and histochemical evidence for the occurence of collagenolysis and for the role of neutrophylic polymorphnuclear leukocytes during cervical dilatation. Am J Obstet Gynecol 138:273

Keirse MJN (1992) Therapeutic use of prostaglandins. In Elder MG (ed) Baillière's Clinical Obstetrics and Gynecology, Baillière Tindall, London, vol 6:787–809

Kelly RW, Leask R, Calder AA (1992) Choriodecidual production of interleukin-8 and the mechanism of parturition. Lancet 339:776–777

King JF, Grant A, Kierse MJN, Chalmers J (1988) β-Mimetics in preterm labour: an overview of the randomized controlled trials. Br J Obstet Gynecol 95:211–22

King RJB, Whithead MI (1986) Assessment of the potency of orally administered progestins in women. Fertil Steril 46:1062–1066

Klein-Hitpass L, Cato ACB, Henderson D, Ryffel GU (1991) Two types of antiprogestins identified by their differential action in transcriptionally active extracts from T47D cells. Nucleic Acid Res 19:1227–1234

Knudsen UB, Fredens K, Ulbjerg N (1991) Inflammatory cells in the cervix and their role during pregnancy. In: Leppert P, Woessner F (eds) The extracellular matrix of the uterus, cervix and fetal membranes: synthesis, degradation and hormonal regulation, Perinatology Press, Ithaca, New York, pp 141–145

Koering MJ, Healy DL, Hodgen GD (1986) Morphologic response of endometrium to a progesterone receptor antagonist, RU 486, in monkeys. Fertil Steril 45:280–287

Kozuka M, Ito T, Takahashi K, Hagiwara H (1989) Endothelin induces two types of contractions of rat uterus: phasic contractions by way of voltage-dependent calcium channels and developing contractions through a second type of calcium channels. Biochem Biochem Res Commun 159:317–323

158   Kristof Chwalisz

Lefebvre Y, Proulx L, Elie R, Poulin O, Lanza E (1990) The effect of
     RU 38486 on cervical ripening. Am J Obstet Gynecol 162:61–65
Lejeune B, Dehou MF, Leroy F (1986) Tentative extrapolation of animal data
     to human implantation. Ann NY Acad Sci 476:63–74
Leppert PC (1992) Cervical softening, effacement and dilatation: A complex
     biochemical cascade. J Mat Fet Med 1:213–223
Lessey BA, Damjanovich L, Coutifaris C, Castelbaum A, Albelda SM, Buck
     CA (1992) Integrin adhesion molecules in the human endometrium:
     correlation with the normal and abnormal menstrual cycle. J Clin Invest
     90:188–195
Li Y, Perezgrovas R, Gazal OS, Schwabe C, Anderson LL (1991) Antipro-
     gesterone, RU 486, fascilitates parturition in cattle. Endocrinology 129:
     765–770
Liggins GC (1981) Cervical ripening as an inflammatory reaction. In:Ellwood
     DA, Anderson ABM (eds) The cervix in pregnancy and labour, Edinburgh:
     Churchill Livingstone, pp 1–9
Lin TJ, Billiar RB, Little B (1972a) Metabolic clearence rate of progesterone
     in the menstrual cycle. J Clin Endocrinol Metab 35:879–892
Lin TJ, Lin SC, Erlenmayer F, Kline IT, Underwood R, Billiar RB, Little B
     (1972b) Progesterone production rates during the third trimester of preg-
     nancy in normal women, diabetic women, and women with abnormal glu-
     cose tolerance. J Clin Endocrinol 34:287–293
Luckett WP (1980) The suggested evolutionary relationships and classification
     of tree shrews. In: Luckett WP (ed) Comparative biology and evolutionary
     relationship of tree shrews. Plenum Press, New York, pp 3–29
MacDonald PC, Dombroski RA Casey L (1991) Recurrent secretion of pro-
     gesterone in large amounts: An endocrine/metabolic disorder unique to
     young women? Endocr Rev 12:372–401
Maggi M, Vannelli GB, Peri A, Brandi ML, Fantoni G, Giannini S, Torrisi C,
     Guardabasso V, Barni T, Toscano V, et al. (1991) Immunolocalization,
     binding, and biological activity of endothelin in rabbit uterus: Effect of
     ovarian steroids. Am J Physiol 260: E292-305
Marbaix E, Donnez J, Courrtoy PJ, Eeckhout Y (1992) Progesterone regulates
     the activity of collagenase and relateted gelatinases A and B in human en-
     dometrial explants. Proc Natl Acad Sci USA 89:11789–11793
Meyer ME, Pornon A, Jingwei J, Bocquel MT, Chambon P, Gronemeyer H
     (1990) Agonistic and antagonistic activities of RU 486 on the functions of
     the human progesterone receptor. EMBO J 12: 3923–3932
Miller SM, Garfield RE, Daniel EE (1989) Improved propagation in myome-
     trium associated with gap junctions during parturition. Am J Physiol 256
     (Cell Physiol 25):C130-C141

Moguilewski M, Philibert D (1985) Biochemical profile of RU 486. In: Baulieu EE, Segal SJ (eds) The antiprogestin steroid RU 486 and human fertility control. Plenum, New York, pp 87–99

Navot D, Scott RT, Droesch KD, Veeck LL, Hung-Ching Liu, Rosenvaks Z (1991) The window of embryo transfer and efficiency of human conception in vitro. Fertil Steril 55:114–118

Neef G, Beier S, Elger W, Henderson D, Wiechert R (1984) New steroids with antiprogestational and antiglucocorticoid activities. Steroids 44:349–372

Neulen J, Williams RF, Hodgen GD (1990) RU 486 (Mifepristone): induction of dose-dependent elevations of estradiol receptor in endometrium from ovariectomized monkeys. J Clin Endocrinol Metab 71:1074–1075

Padykula HA (1991) Regeneration in primate uterus. The role of stem cells. In Wynn, RM, Jollie WP (eds) Biology of the uterus. Plenum, New York, pp 279–288

Paislay LA, Mickelson WD, Anderson PD (1986) Mechanism and therapy for retained fetal membranes and uterine infections of the cow: a review. Theriogenology 25:353–381

Pasqualini JR, Kincl FA (eds) (1985) Hormones and the fetus, vol 1. Pergamon, Toronto, pp 194–268

Phillibert D, Moguilewsky M, Mary M, Lecaque D, Tournemine C, Secchi J, Deraedt R (1985) Pharmacological profile of RU 486 in animals. In: Baulieu EE, Segal SJ (eds) The antiprogesterone steroid RU 486 and human fertility control. Plenum, New York, pp 49–68

Porter DG (1970) The failure of progesterone to affect myometrial activity in the guinea pig. J Endocrinol 46:425–34

Psychoyos A (1986) Uterine receptivity for nidation. Ann NY Acad Sci 476:36–42

Puri CP, Van Look PFA (1991) Newly developed competitive progesterone antagonists for fertility control. In: Agarwal MK (ed) Antihormones in Health and Disease. Front Horm Res 19:127–167

Rodger MW, Baird D (1990) Pretreatment with mifepristone (RU 486) reduces interval between prostaglandin administration and expulsion in second trimester abortion. Br J Obstet Gynaecol 97:41–45

Spitz JM, Bardin CW (1993) Clinical application of the antiprogestin RU 486. Endocrinologist 3: 58–66

Qing SS, Fähnrich M, Hasan SH, Elger W (1988). PGFM and sex steroid concentration throughout the oestrus cycle and pregnancy in the guinea pig: effects of treatment with the progesterone antagonist ZK 98 299. In: Puri CP, Van Look PFA (eds) Hormone antagonists for fertility regulation. Good Print, Bombay, pp 87–99

Rajabi M, Solomon S, Poole R (1991a) Biochemical evidence of collagenase-mediated collagenolysis as a mechanism of cervical dilatation at parturition in the guinea pig. Biol Reprod 45:764–772

Rajabi M, Solomon S, Poole R (1991b) Hormonal regulation of interstitial collagenase in the uterine cervix of the pregnant guinea pig. Endocrinolgy 128:863–871

Rajabi M, Dodge G, Solomon S, Poole R (1991c) Immunochemical and immunohistochemical evidence of estrogen-mediated collagenolysis as a mechanism of cervical dilatation in the guinea pig at parturition. Endocrinology 128:371–378

Rådestadt A, Bygdeman M (1992) Cervical softening with mifepristone (RU 486) after pretreatment with naproxen. A double-blind randomized study. Contraception 45:221–227

Rådestadt A, Chistensen NJ, Strömberg L (1988) Induced cervical ripening with mifepristone in first trimester abortion: a double-blind randomized biomechanical study. Contraception 38:101–112

Rådestadt A, Bygdeman M, Green K (1990) Induced cervical ripening with mifepristone (RU 486) and bioconversion of arachidonic acid in human pregnant uterine cervix in the first trimester. Contraception 41:283–291

Roblero LS, Fernández O, Croxatto HB (1987) The effect of RU 486 on transport, development and implantation of mouse embryos. Contraception 36:549–555

Romero R, Ceska M, Avila CA, Mazor M, Behnke E, Lindley J (1991) Neutrophil attractant/activating peptide-1/interleukin-8 in term and preterm parturition. Am J Obstet Gynecol 165:813–20

Sakai N, Blennerhassett MG, Garfield RE (1992) Effects of antiprogesterones on myometrial cell-to-cell coupling in pregnant guinea pigs. Biol Reprod 46:385–365

Schweier A, Kuhn HJ, Hasan SH (1975) Serum progesterone levels during pregnancy in Tupaja belangeri correlated with ovarian and placental morphology. In: The laboratory animal in the study of reproduction, 6th ICLA Symposium, Thessaloniki, Gustav Fischer, Stuttgart

Shi Shao Qing, Fähnrich M, Chwalisz K, Hasan SH, Elger W (1988) PGFM and sex steroid concentrations throughout the oestrus cycle and pregnancy in the guinea pig: effects of treatment with the progesterone antagonist ZK 98 299. In: Puri P, Van Look PFA (ed) Hormone Antagonists for Fertility Regulation. Good Print, Bombay, pp 87–97

Slayden D, Hirst JJ, Brenner RM (1993) Estrogen action in the reproductive tract of rhesus monkeys during antiprogestin treatment. Endocrinology 132:1845–1856

Soloff MF (1989) Endocrine control of parturition. In: Wynn, RM, Jollie WP (eds), Biology of the uterus. Plenum, New York, pp 559–607

Swahn ML, Bygdeman M (1988) The effect of the antiprogestin RU 486 on uterine contractility and sensitivity to prostaglandin and oxytocin. Br J Obstet Gynaecol 95:126–134

Suzuki Y (1990) Properties of endothelin-induced contractions in the rabbit non-pregnant and pregnant myometria. Fukushima J Med Sci 36:29–40

Szekeres-Bartho J, Reznikoff EM, Varga P, Pichon MF, Varga Z, Chaouat G (1989) Lymphocytic progesterone receptors in human pregnancy. J Reprod Immunol 16: 239

Thong KJ, Baird DT (1992) A study of gemeprost alone, dilapan or mifepristone in combination with gemeprost for the termination of second trimester pregnancy. Contraception 46:11–17

Thorburn GD, Challis JRG (1979) Endocrine control of parturition. Physiol Rev 59:863–917

Uchiyama T, Ito A, Ikesue A, Nakagawa H, Mori Y (1992) Chemotactic factor in the pregnant rabbit uterine cervix. Am J Obstet Gynecol 167:1417–1422

Uldbjerg N, Ekman G, Malmström A, Sporrong B, Ulmsten U, Wingerup L (1981) Biochemical and morphological changes of human cervix after local application of prostaglandin E2 in pregnancy. Lancet:267–268

Urquhart DR, Templeton AA (1990a) Mifepristone (RU 486) for cervical priming prior to surgically induced abortion in the late first trimester. Contraception 42:191–199

Urquhart DR, Templeton AA (1990b) The use of mifepristone prior to prostaglandin-induced mid-trimester abortion. Hum Reprod 5:883–886

Van Look PFA, Bygdeman M (1989) Antiprogestational steroids: a new dimension in human fertility regulation. Oxf Rev Reprod Biol 11:1–60

van Uem JFHM, Hsiu JG, Chilik CF, Danforth DR, Ulmann A, Baulieu EE, Hodgen GD (1989) Contraceptive potential of RU 486 by ovulation inhibition: I Pituitary versus ovarian action with blockade of estrogen-induced endometrial proliferation. Contraception 40:171–184

Webb AMC, Russell J, Elstein M (1992) Comparison of Yuzpe regimen, danazol, and mifepristone (RU486) in oral postcoital contraception. BMJ 305:927–931

Webster MA, Phipps NS, Gillmer MDG (1985) Myometrial activity in first trimester human pregnancy after epostane therapy: Effect of intravenous oxytocin. Br J Obstet Gynaecol 92:957–962

Westphal U (1986) Progesterone binding protein (PBG). In: Westphal U (ed) Steroid-Protein Interactions II. Springer, Berlin Heidelberg New York, (Monographs on Endocrinology, vol 27), pp 138–194

Whitehead RJB, Townsend PT, Pryse-Davies, Ryder TA, King RJB (1981) Effects of estrogens and progestins on the biochemistry and morphology of the postmenopausal endometrium. N Engl J Med: 305:1599–605

Wolf JP, Chilik CF, Itskovitz J, Weyman D, Anderson TL, Ulmann A, Baulieu EE, Hodgen GD (1988) Transplacental passage of a progesterone antagonist in monkeys. Am J Obstet Gynecol 159:238–242

Wolf JP, Sinosich M, Anderson T, Ulmann A, Baulieu EE, Hodgen GD (1989) Progesterone antagonist (RU 486) for cervical dilatation, labor induction, and delivery in monkeys: Effectiveness in combination with oxytocin. Am J Obstet Gynecol 160:45–47

World Health Organization (1990) The use of mifepristone (RU 486) for cervical preparation in the first trimester pregnancy terminantion by vacuum aspiration. Br J Obstet Gynaecol 97:260–266

7 New Perspectives on Estrogen, Progesterone, and Oxytocin Action in Primate Parturition*

Miles J. Novy and George J. Haluska

7.1 Introduction

In studying the control of parturition in primates it is logical to concentrate on those endocrine and paracrine regulators which have proven to be important prerequisites for parturition in other species (i.e., estrogens, progesterone, prostaglandins, oxytocin, and cytokines). It has been shown in domestic ruminants that increased estrogen levels and falling progesterone levels at term (dependent upon the fetal adrenal

---

* This is publication no. 1890 from the Oregon Regional Primate Research Center.

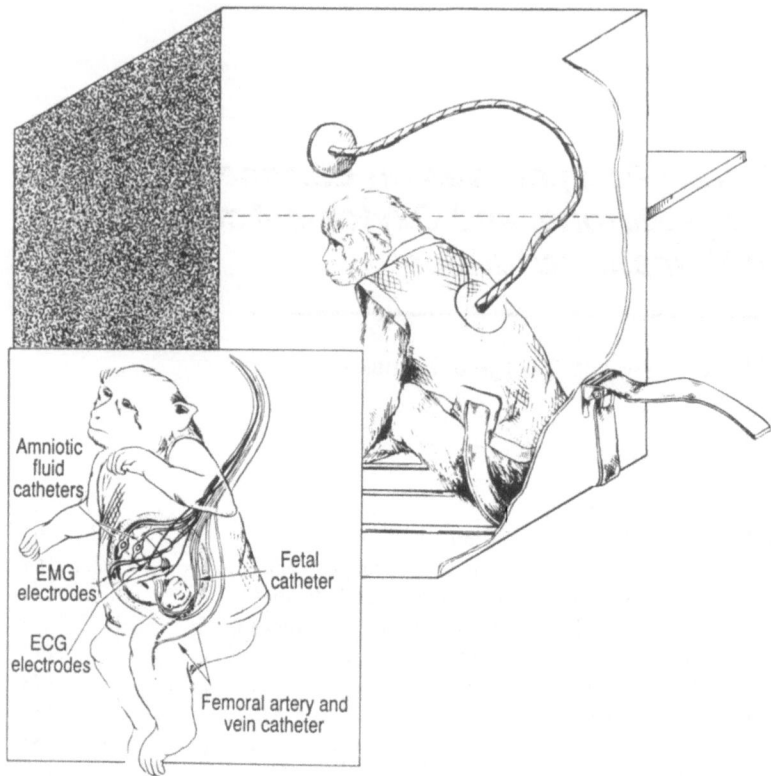

Fig. 1. The chronic jacket and tether system used to study physiological parameters in conscious, unrestrained rhesus macaques

secretion of cortisol) are pivotal for the initiation of parturition (Liggins et al. 1977; Challis and Olson 1988). Although the rising production of estrogens by the fetoplacental unit provides an attractive mechanism for the endocrine initiation of labor in primates, estrogens likely serve only a permissive function in parturition. There is no evidence for the lowering of progesterone concentrations in the maternal, fetal, or amniotic fluid compartments in higher primates prior to or during labor. However, a prepartum fall in progesterone is such a ubiquitous phenomenon across species that it is still tempting to search for evidence of a localized progesterone withdrawal mechanism in target tissues of primates. Indeed, the recent availability of effective antiprogestins has provided an efficient tool to facilitate such investigation.

Much of the recent research work in our laboratory at the Oregon Regional Primate Research Center has focused on the unique role of oxytocin in generating uterine biorhythms and on the importance of progesterone in maintaining uterine quiescence during pregnancy. The chronically catheterized fetal and maternal preparation in rhesus monkeys (*Macaca mulatta*) continues to provide a useful means to study in vivo endocrine and even paracrine interactions in the relatively inaccessible intrauterine space (Fig. 1). It has given us the opportunity to obtain measurements of maternal and fetal physiological variables and hormonal responses in conscious, unrestrained animals in a humane fashion.

7.2 The Role of Estrogens in Parturition

We have demonstrated in our laboratory that bound and "free" estrogens are increased in the maternal circulation of pregnant rhesus monkeys before spontaneous delivery (Stanczyk et al. 1986). Parturition is preceded by rising concentrations of dehydroepiandrosterone sulfate (DHEAS) in fetal blood, but not in maternal blood or amniotic fluid and by an increase in estrone in all three compartments (Siiteri and Serón-Ferré 1981; Walsh et al. 1984a). A rise in amniotic fluid estrone precedes or coincides with the rise in amniotic fluid prostaglandins, which begin to increase several days prior to the onset of labor near term (Novy and Haluska 1988). Concentrations of estrone and the $PGF_{2\alpha}$ metabolite in amniotic fluid were closely correlated after 134 days gestation ($r = 0.90$; $p < 0.01$) and both increased significantly before vaginal delivery (Walsh et al. 1984a).

The stimulus for the rise in fetal and maternal estrogens before parturition has not been unequivocally defined but it is likely to be fetal adrenocorticotropic hormone (ACTH) by analogy with sheep and because hypoxemic stress or infusions of ACTH to the fetus increase the production of estrogenic precursors such as DHEAS and androstenedione (Novy and Walsh 1981; Shepherd et al. 1992). A functioning fetal adrenal gland is clearly required because the prepartum estrogen trends are absent when the fetus is dead or anencephalic and parturition

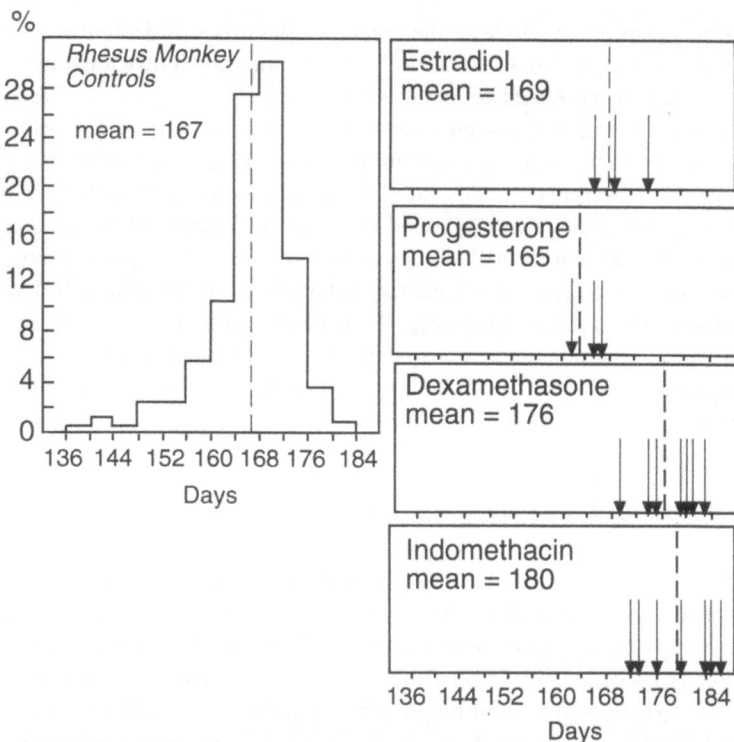

Fig. 2. Frequency distribution of normal live births in the colony of rhesus macaques at the Oregon Regional Primate Research Center *(top). Numbers* on the *abscissa* indicate gestational age in days (mean = 167 days; n = 317). The distribution of births after the maternal administration of estradiol benzoate (0.1–4.0 mg twice daily) or subcutaneous silastic implants containing progesterone were not significantly different. Maternal administration of dexamethasone or indomethacin significantly increased gestational length and shifted the distribution of births to the right (χ^2 52.6 and 60.5, respectively; $p < 0.001$). The *arrows* represent individual monkeys. Data from Novy and Walsh (1983)

is delayed in these situations (Walsh et al. 1979a). To test the effect of suppressing fetal and maternal adrenal secretion on parturition we treated rhesus monkeys with dexamethasone from day 130 of gestation to delivery (term = 167 days; Novy and Walsh 1983). Chronic treatment reduced estrogen biosynthesis during pregnancy, abolished the prepartum estrogen surge, and increased gestational length; i.e., 71% of the fetuses were born postmaturely (after day 175 of gestation, $p < 0.01$; Fig. 2). It is tempting to ascribe the dexamethasone effect on parturition to reduced estrogen biosynthesis by the fetoplacental unit but alternative mechanisms must also be considered. Contributory factors may include reduced prostaglandin production secondary to inhibition of phospholipase A_2 activity and/or prostaglandin synthetase (Mitchell et al. 1984). Recent data indicate that dexamethasone augments β-adrenergic receptor mediated inhibition of myometrial contractility (Kocan et al. 1993).

If estrogens facilitate parturition in primates then stimulation of adrenal estrogen precursors by ACTH should induce premature labor. However, long-term infusion of ACTH into the fetus either in a continuous or pulsatile fashion did not induce premature parturition in rhesus monkeys (Novy and Haluska 1988). Similarly, estradiol benzoate administered systemically to pregnant monkeys does not initiate labor before term (Fig. 2) unless the fetuses are dead (Novy and Walsh 1983). Taken together the evidence suggests that estrogens play only a supportive role in parturition and that stimulation of the rhesus fetal adrenals is not a sufficient stimulus to initiate labor either through cortisol or via androgenic precursors of estrogen. Nevertheless, there is a growing body of evidence which indicates that estrogens interact with the maternal circadian system to influence the timing of birth within the light/dark cycle.

7.3 Circadian Myometrial and Endocrine Rhythms

A 24-h biorhythm in uterine contractile activity during late gestation in rhesus monkeys was first described by Harbert (1977), Harbert and Spisso (1980), and later by Novy and coworkers (1980) more than a decade ago. Subsequent studies in our laboratory (Ducsay et al. 1983) and by others (Germain et al. 1982; Taylor et al. 1983; Honnebier et al.

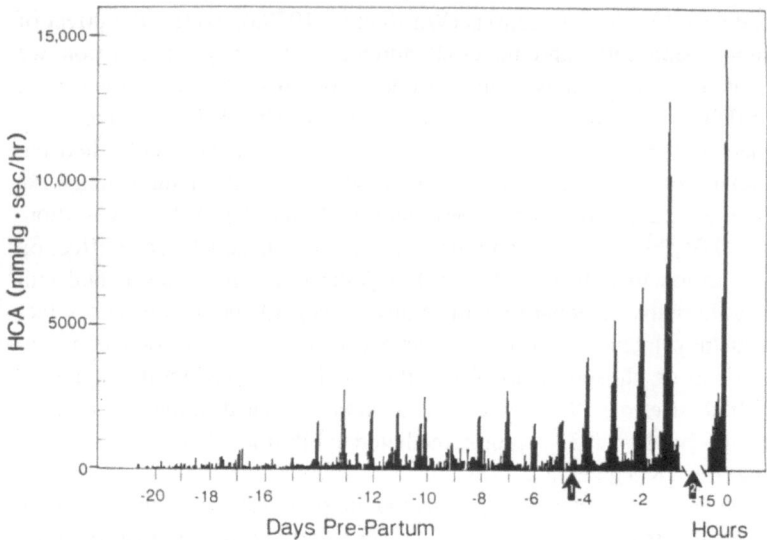

Fig. 3. Uterine activity expressed as hourly area under the contraction curve (*HCA*, mmHgs/h) for a group of rhesus monkeys (*n* = 5) delivering spontaneously at term. A progressive increase in nocturnal uterine activity is observed as term approaches. *Arrows* signify approximate times when accelerated changes occur in uterine activity and cervical ripening, respectively

1989b) have clearly established the nocturnal increase in uterine contractile activity which predominates in the last weeks of pregnancy and gradually culminates in labor and delivery (Fig. 3). Peak contractile activity occurs between 2100 hours and 0200 hours, given conventional lighting regimens of 12 h, 12 h or 16 h, 8 h light–dark photoperiods with lights on at 0600–0700 hours and light intensities of 350–500 lux. Evidence for a marked diurnal variation in uterine contractility which correlates with the sleep/wake cycle has also been reported during human pregnancy (Tambyraja and Hobel 1983; Germain et al. 1986, 1990; Main et al. 1991).

These biorhythms in uterine contractions help to explain why the onset of labor also shows circadian characteristics. Among day-active primates such as rhesus and squirrel monkeys, the majority of births occur during the late night or early morning hours (Jensen and Bobbitt

1967; Bowden et al. 1967; Brandt and Mitchell 1971). In nocturnal prosimians, the pattern is reversed (Doyle et al. 1967). In pregnant women, numerous studies have demonstrated that the peak hours of labor onset are between 2300 hours and 0400 hours (Longo and Yellon 1988). Additional evidence for photoperiod regulation of the timing of labor and delivery comes from the work of Jensen and Bobbitt (1967). By shifting the light–dark cycle 12 h they achieved a corresponding 12-h phase shift in the timing of normal births in pig-tailed macaques.

Similar findings have been reported by Figueroa and coworkers (1990) and by Ducsay and Yellon (1991) who subjected pregnant rhesus monkeys to photoperiod shifts of 6–11 h. These investigators noted that entrainment of uterine contractile activity (as judged by EMG episodes or intraamniotic pressure recordings) to the altered light/dark cycle occurred within 7 days. The mean time of peak contraction activity in pregnant monkeys always occurred 2–4 h after lights-off (Figueroa et al. 1990) and labor began in all animals during periods of darkness (Ducsay and Yellon 1991). These observations strongly suggest but do not prove a circadian rhythm. The term "circadian" (approximately 24 h) has a precise meaning in chronobiology but has often been used loosely to describe the contractile rhythms of the primate uterus. By definition, a biological rhythm is deemed circadian if in the absence of synchronizing cues (i.e., when exposed to constant environmental conditions) the endogenous free-running rhythm exhibits an approximate 24 h periodicity.

Recently, Honnebier and coworkers (1991) and Matsumoto et al. (1991) exposed pregnant rhesus monkeys to conditions of constant darkness or continuous light from midpregnancy until spontaneous delivery at or near term. In all animals a myometrial activity rhythm with a period of ca. 24 h was consistently present during the last week before delivery. The circadian phase of individual body temperature and uterine activity rhythms was idiosyncratic among these animals in the absence of a photoperiod and the temperature acrophase led the contraction acrophase by 6–7 h. Taken together these observations strongly support the hypothesis that the rhythm in preparturient myometrical activity depends upon the maternal circadian system.

Several groups of investigators have described 24-h variations in circulating maternal steroid hormone concentrations in pregnant non-human primates and pregnant women (Challis et al. 1983; Walsh et al.

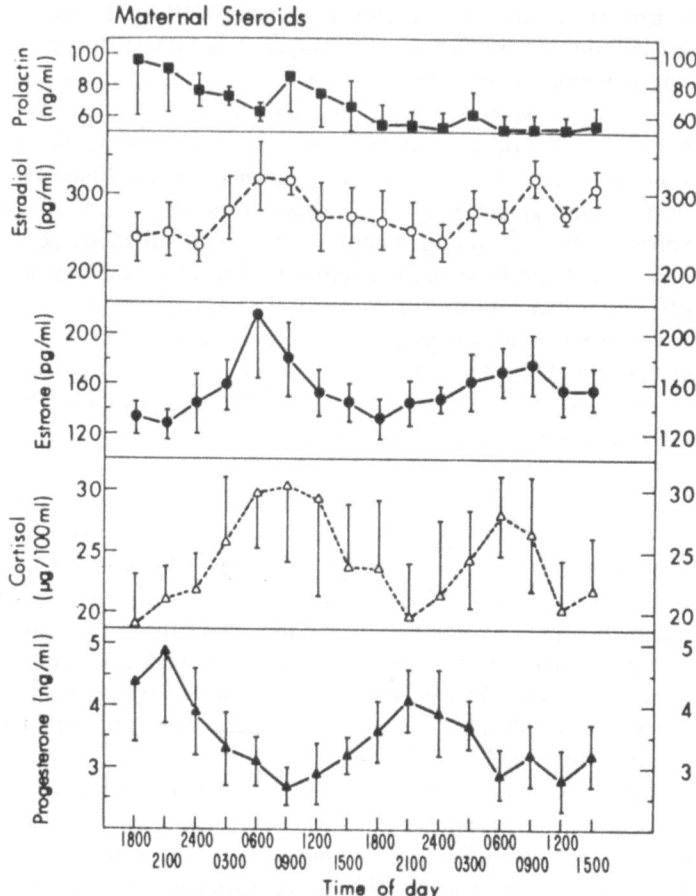

Fig. 4. Hormone fluctuations in the maternal peripheral plasma collected every 3 h during a 48-h period in long-term catheterized rhesus monkeys. Gestational age range = 127–138 days. Data represent means ± SE (*n* = 5). (From Walsh et al. 1984a)

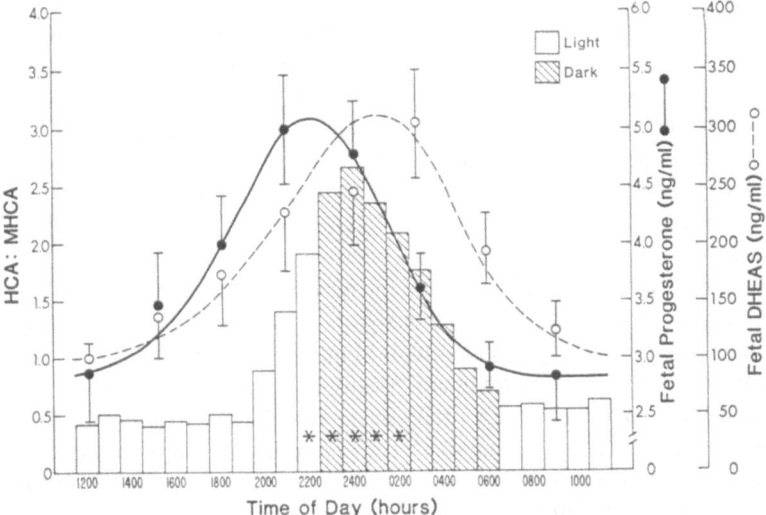

Fig. 5. Fetal plasma dehydroepiandrosterone sulfate *(DHEAS)* and progesterone in five long-term catheterized rhesus monkeys at 127–138 days gestation. Data collected over a 48-h sampling period were pooled and the mean ± SE hormone concentrations for each 3-h interval in a 24-h cycle were plotted and superimposed on the 24-h uterine activity pattern (hourly contraction area to mean hourly contraction area ratios, *HCA:MHCA*). Hours of elevated uterine activity are indicated by *asterisks* (*p* < 0.01). The circadian rhythms in fetal plasma DHEAS and progesterone closely parallel the biorhythm in uterine activity with peaks during periods of darkness between 2100 hours and 0300 hours (From Walsh et al. 1984a)

1984a,b). More recent studies in pregnant rhesus monkeys have established that the maternal rhythms in cortisol, estrogens, and progesterone depend upon the circadian oscillations of the maternal hypothalamic–pituitary–adrenal system (Honnebier et al. 1992). In the maternal circulation, cortisol, estrone, and estradiol were higher in the morning than in the evening (Walsh et al. 1984b; Matsumoto et al. 1991; Honnebier et al. 1992). The pattern observed for progesterone was reciprocal to that observed for cortisol (Fig. 4). Twenty-four hour oscillations in pregnancy-associated hormones are also demonstrable in the fetus but because glucocorticoids readily cross the primate placenta, the fetal pituitary is subject to feedback regulation by mater-

sult, there is a 180° phase difference between rhythms in fetal adrenal androgens and the maternal cortisol rhythm (Challis et al. 1983; Walsh et al. 1984b). Progesterone provides an interesting exception to the above since the nocturnal increase in maternal plasma progesterone is in phase with fetal progesterone – a consequence of the attenuated displacement of progesterone from corticosteroid binding globulin (CBG) by cortisol at night in the maternal circulation (Hess et al. 1981). The circadian rhythms in maternal progesterone and fetal DHEAS and progesterone (fetal adrenals have the biosynthetic capacity for pregnenolone and progesterone production) parallel the biorhythm in uterine contraction frequency and amplitude (Fig. 5). The seemingly paradoxical relationship of peak myometrial activity during a relatively low estradiol to progesterone (E/P) ratio in the maternal or fetal circulations has been observed by a number of different investigators in pregnant monkeys and women (Walsh et al. 1984b; Matsumoto et al. 1991; Honnebier et al. 1992; Fuchs et al. 1992).

Contrasting findings have been reported by Wilson and coworkers (1991) in the pregnant baboon near term. In this species the progesterone surge begins about 1900 hours and peaks at 0230 hours the following morning. The maternal plasma estradiol surge begins at 2200 hours and peaks at 0430 hours the next day. However, approximately 10 days before delivery the beginning of the estradiol surge shifts forward 6 h and precedes the progesterone surge by 2–3 h. These authors have proposed that the heightened E/P ratio increases myometrial oxytocin sensitivity and/or the production and release of hypothalamic-pituitary oxytocin which in turn coincides with nocturnal uterine activity.

According to this hypothesis, the forward shift in the estradiol surge prompts a rapid release of oxytocin presumably by nonreceptor-mediated mechanisms. Alternatively, if estrogens were acting through a classical genomic pathway to promote oxytocin-induced nocturnal uterine activity, a phase lag of 6–12 h would be appropriate and consistent with a morning (0800–0900 hours) peak in the maternal E/P ratio as has been observed in monkeys and women in late pregnancy (Walsh et al. 1984b; Matsumoto et al. 1991; Fuchs et al. 1992). In either case, it would not explain why the internal phase relationships between cortisol, estradiol, and progesterone rhythms in maternal plasma are quite different between baboons and monkeys and yet the nocturnal myometrial activity patterns have similar characteristics in the two species. A third possibility is that diurnal fluctuations in the E/P ratio are irrelevant to the development of nocturnal uterine activity patterns and that what is required is simply a threshold concentration of estrogen in the maternal circulation.

We favor the latter hypothesis, which is consistent with the view that estrogens serve a permissive function in generating nocturnal uterine activity (and in parturition as a whole) for the following reasons:

1. Stimulation of estrogen production by the fetoplacental unit (i.e., by intravenous Δ^4-androstenedione infusion) leads to the enhancement of nocturnal uterine activity (Figueroa et al. 1989), whereas the experimental suppression of estrogen production by dexamethasone attenuates or eliminates the 24-h rhythm (Ducsay et al. 1983).
2. Fetal death is associated with a loss of nocturnal uterine activity coincident with diminished maternal estrogen levels, but circadian activity is reestablished by maternal estradiol benzoate injections which produce sustained elevations in circulating estrogens (Novy and Haluska 1988).

Taken together, these observations demonstrate that rising estrogen concentrations during late pregnancy play a supportive role in establishing nocturnal uterine activity. Since estrogens are known to promote the development of myometrial oxytocin receptors, estrogens likely support nocturnal uterine activity by inducing a heightened myometrial sensitivity (Honnebier et al. 1989a). Estrogens also have a stimulatory effect at the level of the hypothalamus and pituitary, resulting in increased oxytocin synthesis and release into the circulation (Amico et al. 1981). Recent experiments demonstrate that oxytocin mRNA is synthesized by amnion, chorion, and decidua. Furthermore, oxytocin gene expression is increased in human intrauterine tissues around the time of labor onset and a similar increase can be achieved in vitro by estradiol (Chibbar et al. 1993). A potentiating effect of placental corticotrophin-releasing factor (CRF) and oxytocin on myometrial contractions has been demonstrated in human gestational myometrium in vitro (Quartero and Fry 1989). These findings support the notion of a paracrine system as well as an endocrine system involving oxytocin and sex steroids that may stimulate uterine contractions.

7.3.1 Importance of Maternal But Not Fetal Oxytocin

There is increasing evidence to indicate that oxytocin plays a key role in the generation of uterine activity rhythms. Honnebier et al. (1989b) and Hirst et al. (1991a) demonstrated that nocturnal uterine activity episodes are closely correlated with elevated concentrations of oxytocin, suggesting that these episodes are driven by oxytocin in the maternal plasma. The pivotal role of oxytocin in the generation of nocturnal uterine activity is supported by the observation that these episodes are abolished by oxytocin antagonist treatment in rhesus monkeys (Honnebier et al. 1989b; Hirst et al. 1991). We have reported that some late pregnant monkeys do not display a nocturnal increase in plasma oxytocin and these animals also characteristically lack a 24-h rhythm in uterine activity (Hirst et al. 1991). The proportion of animals that display a nocturnal peak in uterine activity increases steadily during late gestation particularly during the 10 days immediately preceding spontaneous labor.

Additional studies from our laboratory by Hirst et al. (1993) indicate a parallel and progressive rise in nocturnal uterine activity and plasma oxytocin concentrations with advancing gestation (Fig. 6). These observations suggest that the increasing levels of uterine activity result in part from rising nocturnal oxytocin levels in the maternal circulation as term approaches. Despite markedly higher levels of nocturnal uterine activity near parturition, there was no further increase in

Table 1. Comparison of maternal and fetal plasma oxytocin concentrations in rhesus monkeys at 0900 hours, 2100–0200 hours (during nocturnal uterine activity) and at labor and delivery (from Hirst et al. 1991,1993)

	Maternal oxytocin (pg/ml)	Fetal oxytocin (pg/ml)
Daytime	2.9 ± 0.3	3.1 ± 0.2
Nighttime	20.0 ± 3.4	2.9 ± 0.7
Early labor	18.2 ± 5.0	3.0
Advanced labor	30.0 ± 7.4	n.d.
Delivery	62.5 ± 5.5	

Values represent mean $\pm$ SE
n.d., not detectable.

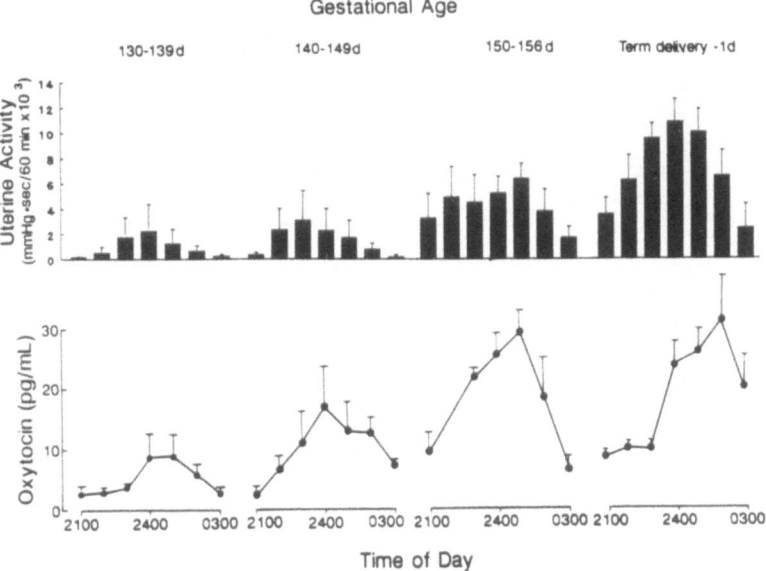

Fig. 6. Maternal plasma oxytocin (OT) concentrations and uterine activity during late gestation in four rhesus monkeys. For all animals, OT concentrations and uterine activity were measured at hourly intervals between 2100 hours and 0300 hours on one night in each of three gestational age ranges (130–139 days, 140–149 days, and 150–156 days) as well as on the night before term delivery (term delivery, 1 day). Uterine activity and OT concentrations increased during the hours of darkness and were maximal around 2400 hours. The magnitude of the nocturnal rise in OT concentrations ($p < 0.005$) and uterine activity ($p < 0.001$) increased with advancing gestational age. *Vertical bars* indicate SE. (From Hirst et al. 1993)

oxytocin concentrations on the night preceding vaginal delivery or in the 1- to 3-h period before delivery compared to the levels attained between 150–156 days of gestation (about 1 week before delivery). These observations suggest that the overall myometrial sensitivity to oxytocin also increases substantially during the days just around delivery and are in agreement with the previously reported rises in the number of myometrial oxytocin receptors near term in women and in rhesus monkeys (Fuchs 1986; Hirst et al. 1991). Uterine responsiveness to oxytocin also exhibits a 24-h rhythm which coincides with the

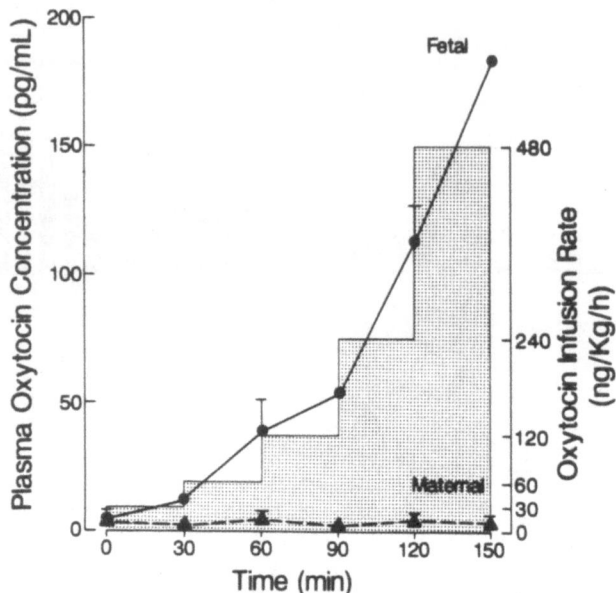

Fig. 7. Effect of infusion on OT into the fetus on maternal (σ) and fetal (λ) plasma OT concentrations. Infusions were via catheters chronically implanted into the fetal jugular vein and three infusions were conducted in different animals at 134, 146, and 151 days of gestational age. The rate of infusion was increased in steps at 30-min intervals (indicated by *shading*) and the total length of the infusions was 150 min. Maternal and fetal arterial blood samples were obtained prior to the commencement of the infusions and at the end of each 30-min period immediately before increasing to the next infusion rate. Means of three OT determinations are shown with *vertical bars* indicating SE; otherwise the means are of two OT determinations. Fetal plasma OT concentrations increased significantly during the infusions (60 ng/kg per hour, $p < 0.05$; 240 ng/kg per hour, $p < 0.0001$), whereas there was no change in maternal OT concentrations. There was also no increase in uterine activity which remained at basal levels (150–300 mmHgsec/h). (From Hirst et al. 1993)

nocturnal rhythm in uterine activity (Honnebier et al. 1989b). Fuchs and coworkers (1992) have recently demonstrated that plasma oxytocin concentrations exhibit a nocturnal peak and a morning nadir in women during late pregnancy. Therefore, the possibility that increased nocturnal oxytocin secretion contributes to the timing of spontaneous labor must also be considered in women.

We found no change in fetal plasma oxytocin concentrations during the hours of darkness, indicating that unlike the mother, the fetus does not display cyclic changes in oxytocin levels (Hirst et al. 1993). The absence of a rise in fetal oxytocin levels during episodes of nocturnal uterine activity or during labor also indicates that oxytocin secretion by the fetus is unlikely to contribute to the generation of circadian uterine activity or to the initiation or maintenance of labor (Table 1). In our studies, graded infusions of oxytocin into the fetal circulation raised fetal plasma levels more than 50-fold but had no effect on maternal oxytocin concentrations or on uterine activity (Fig. 7), suggesting that fetal oxytocin does not cross the primate placenta during late gestation.

The renewed interest in oxytocin should not detract from the importance of prostaglandins as necessary components in the process of parturition. Husslein et al. (1981) have shown that oxytocin stimulates prostaglandin $F_{2\alpha}$ ($PGF_{2\alpha}$) production by the decidua and induction of labor with oxytocin depends in part on an adequate prostaglandin response (Fuchs et al. 1983). PGE_2 is also an important component during parturition since it promotes cervical ripening. We have previously demonstrated that amniotic fluid prostaglandins begin to increase several days prior to the onset of labor (Walsh et al. 1984b; Haluska et al. 1987; Novy and Haluska 1988). Suppression of prostaglandin biosynthesis by long-term administration of indomethacin during rhesus pregnancy (Novy and Walsh 1983) prevents the timely onset of parturition and dramatically prolongs gestation (Fig. 2).

7.4 Contrasting Effects of Progesterone Antagonism and Inhibition of Biosynthesis

The ability of progesterone to maintain uterine quiescence during pregnancy is clearly acknowledged in those species in which maternal peripheral concentrations of progesterone decrease before labor and de-

livery. However, the role of progesterone in late pregnancy maintenance in primates has been questioned since maternal, fetal, and amniotic fluid levels of progesterone are sustained or increase just before
parturition (Novy and Liggins 1980; Walsh et al. 1984b). Recent
studies in nonhuman primates indicate there is also no change in the
concentration or distribution of myometrial or decidual progesterone
receptors with the onset of spontaneous parturition (Hirst et al. 1990).
This does not preclude the possibility that parturition is initiated by an
interference with progesterone-receptor binding to DNA or by some
other block in the transcriptional apparatus. It is conceivable that labor
is triggered by a natural "antiprogestin" which appears at term. This
substance could be a steroid metabolite which competes with progesterone for its receptor or, more likely, a protein which decreases the
number of accessible nuclear binding sites (Novy and Haluska 1988).
Precedent for an inhibitory substance which may interfere with progesterone–receptor–acceptor binding is "processin", a protein of 56 kDa
molecular mass found in rat placenta (Ogle 1986).

Recent work has focused on mechanisms which may effect a local
tissue withdrawal of progesterone particularly at the level of the fetal
membranes and decidua while peripheral progesterone levels remain
elevated (Novy and Haluska 1988). A number of studies demonstrate
that the fetal membranes and decidua are fully capable of metabolizing
estrogens and progesterone and that the formation of these steroids
may be regulated locally (Gibb 1977; Mitchell et al. 1982). However,
whether any of these factors are related to in vivo prostaglandin production by intrauterine tissues or to the physiologic changes associated
with parturition remains to be conclusively demonstrated. The available synthetic progesterone receptor antagonists and inhibitors of progesterone synthesis provide useful tools to investigate the endocrine
and paracrine effects of functional progesterone withdrawal in the control of primate parturition.

7.4.1 Myometrial Contractility

RU486 is a potent synthetic 19-norsteroid which interacts with the progesterone receptor and antagonizes progesterone action in nearly all
species studied so far. We studied the effect of RU486 on uterine ac-

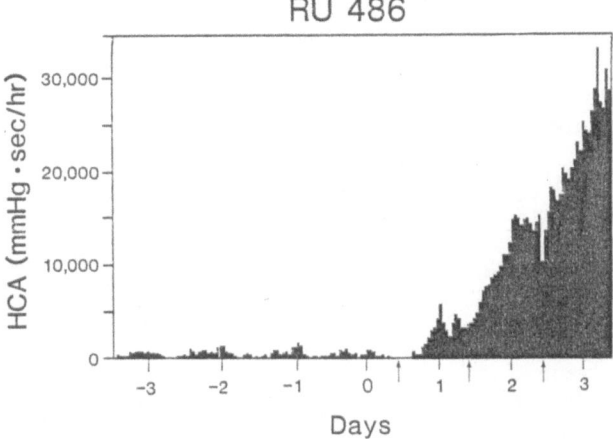

Fig. 8. Uterine activity expressed as hourly contraction area *(HCA)* in mmHgsec/h in rhesus monkeys ($n = 7$) at 130–135 days gestation treated with RU486 in three consecutive doses of 20 mg/kg per day. Cumulative uterine activity equals and eventually exceeds the HCA observed during spontaneous parturition

tivity and amniotic fluid prostaglandins in rhesus monkeys during the mid-third trimester of pregnancy. RU486 was administered orally (20 mg/kg per day) for 3 days. Uterine activity increased from basal conditions within 8 h after the first dose of RU486 reaching levels of 12 000 mmHgsec/h. A sustained increase in uterine activity (13 000–30 000 mmHgsec/h) was observed for 48–72 h prior to cesarean section with little of no cervical effacement or dilatation (Fig. 8). Increases in amniotic fluid $PGF_{2\alpha}$, PGF metabolite, and the bi-cyclo derivative of PGE metabolite occurred 40 h after the onset of increased uterine activity, In contrast, amniotic fluid prostaglandins in control animals which deliver at term increase 24–48 h before prelabor increases in uterine activity (Haluska et al. 1987).

The individual contractions which resulted from RU486 treatment were qualitatively different from the contractions observed in control animals either at night or during spontaneous labor although the force of the contractile activity (area under the contraction curve) was equal to or greater than observed in normal labor. Figure 9 demonstrates

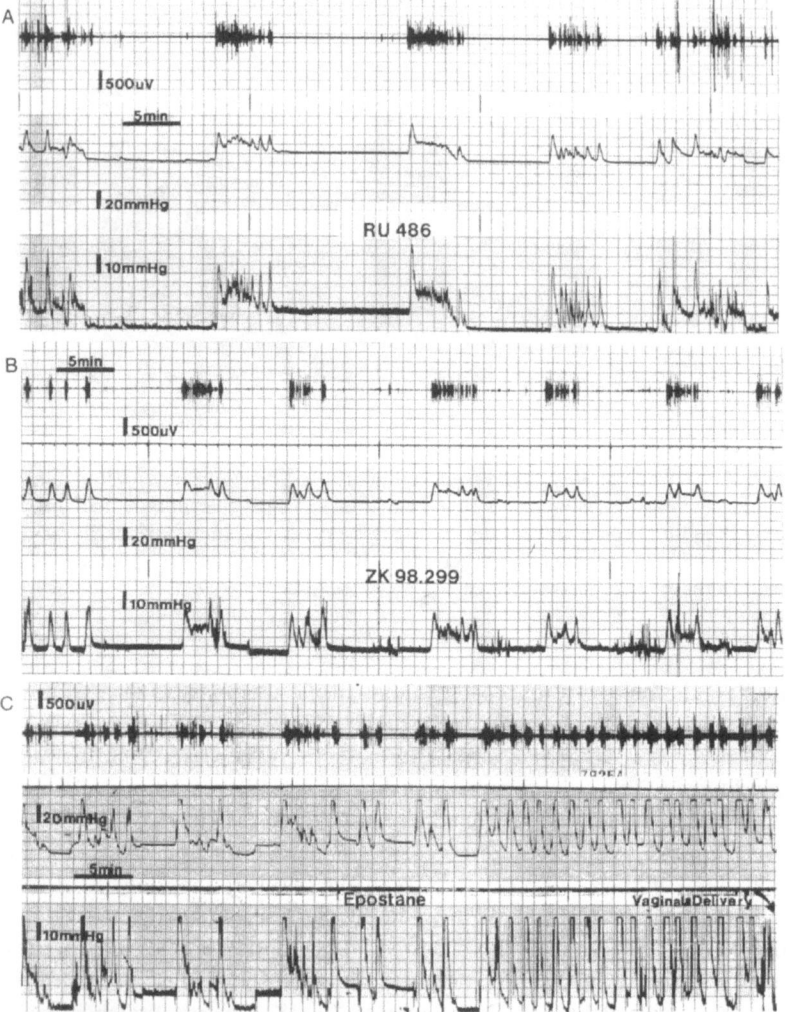

Fig. 9A–C. Uterine activity recordings for a period of 1 h from individual rhesus monkeys. In each *panel*, the *top recording* is electromyographic activity (EMG), the *middle section* is filtered intra-amniotic pressure (IAP), and the *lower portion* is unfiltered IAP. **A** Section of a recording demonstrating contractility at 60+ h after RU486 administration. **B** Demonstrates the uterine contractility after 60+ h of ZK98299 administration. **C** Uterine contractility after 40+ h of epostane administration and just before vaginal delivery

GJ/1000µm

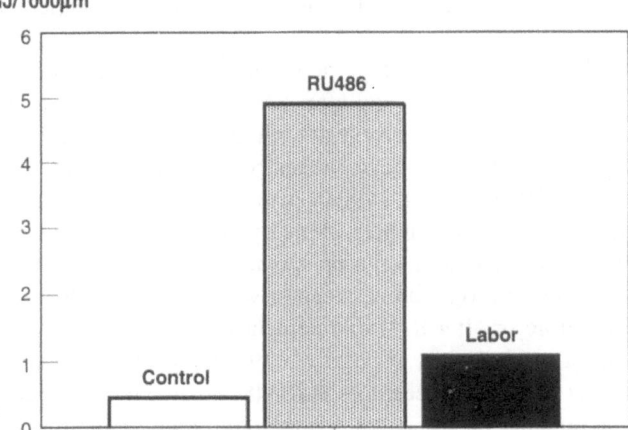

Fig. 10. The number of gap junctions/1000 µm of myometrial smooth muscle cell membrane. Tissues were collected from pregnant control rhesus monkeys who received no treatment and were not in labor, animals treated with RU486 at 20 mg/kg per day for 3 days, and tissues collected at night from animals who were in the final stages of spontaneous labor

EMG and intra-amniotic pressure (IAP) recordings from a representative monkey treated with RU486. Coupled to a prolonged electrical event is a protracted increase in intraamnionic pressure. Therefore, as a result of RU486 administration, myometrial excitability, propagation of action potentials, and excitation–contraction coupling all appear to be increased but the reason for the hypertonus is not clear.

Evidence gathered primarily in rodents indicates that progesterone promotes myometrial relaxation by increasing the membrane potential of the resting smooth muscle cell and by inhibiting the formation of gap junctions between myometrial cells (Garfield et al. 1990). Our findings in rhesus monkeys treated with RU486 are consistent with the earlier findings of Garfield et al. (1987) in rats. We found that RU486 treatment resulted in a dramatic increase in gap junctions in the rhesus myometrium (Haluska et al. 1989) which was even greater than that observed after spontaneous labor and parturition (Fig. 10). This result should not be surprising if one considers that progesterone inhibits and estrogens promote increases in both gap junctions and nuclear estrogen receptors in the myometrium (Garfield et al. 1980; Brenner et al. 1990).

We obtained samples of myometrium, decidua, and fetal membranes from monkeys before and after RU486 treatment in late gestation and after spontaneous delivery at term. Nuclear estrogen receptors were localized using immunocytochemistry and quantified by sucrose gradient assay (Haluska et al. 1990a). A marked and dramatic increase in estrogen receptor was noted in myometrium and decidua after RU486 administration but not during spontaneous term labor (Fig. 11). Neither estrogen or progesterone receptor were detected in the amnion or chorion before or after RU486 treatment or at spontaneous parturition (Haluska et al. 1990a; Hirst et al. 1990). Despite the unusually large increase in estrogen receptors in RU486-treated myometrium and decidua, it remains to be shown whether this high level of receptor contributes to the biological effects. However, the high level of uterine contractility and the atypical contraction pattern generated by RU486 may be related to the unmasking of large quantities of estrogen receptor and to the formation of unusually high numbers of gap junctions. Since there was no change in the concentration of estrogen or progesterone receptor between pregnant and parturient uterine tissues, we have concluded that an increase in myometrial or decidual estrogen receptors or a decrease in myometrial or decidual progesterone receptors is not part of normal parturition in primates (Haluska et al. 1990a; Hirst et al. 1990).

Steroid hormones, particularly estrogens, progesterone, and glucocorticoids, exert important effects on the myometrial response to β-agonists (Roberts et al. 1989). Recent reports suggest that progesterone and glucocorticoids promote myometrial relaxation also by increasing the number of functional β-adrenergic receptors and by protecting against agonist-induced desensitization (Kocan et al. 1993). Theoretically, therefore, the antiglucocorticoid effect of RU486 could synergize with the antiprogesterone effect in stimulating myometrial contractions.

Although RU486 administration to rhesus monkeys during late pregnancy stimulated intense uterine activity, it did not produce the orderly sequence of changes observed during normal parturition (i.e., sinusoidal contraction patterns, diurnal rhythms in uterine activity or cervical changes) (Haluska et al.1987). We elected to perform a ce-

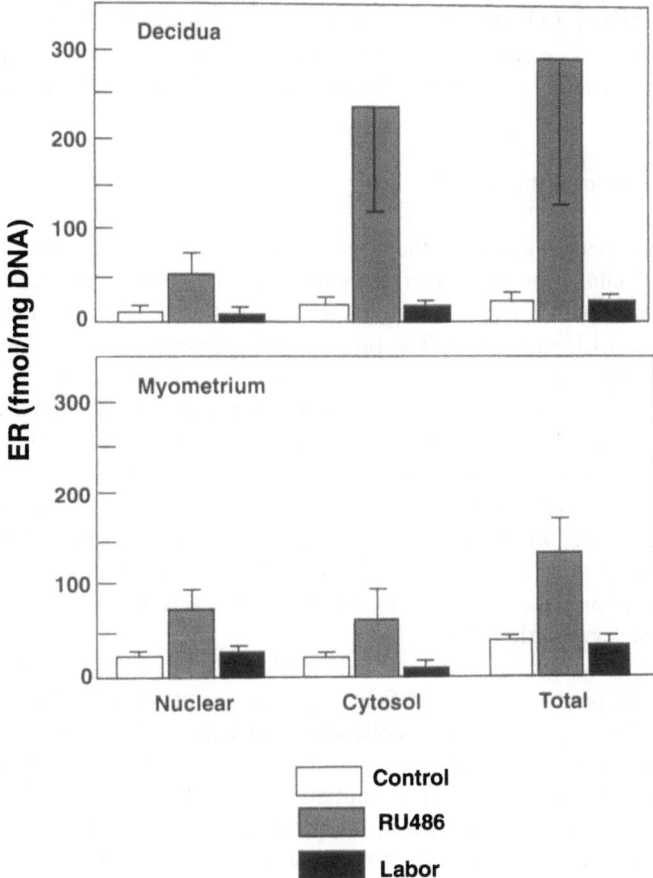

Fig. 11. Nuclear, cytosolic and total estrogen receptors in decidua and myometrium measured by sucrose gradient assay. Tissues were collected from pregnant control animals *(open bars)* who received no treatment, animals treated with RU486 at 20 mg/kg per day for 3 days, and tissues collected at night from animals who were in the final stages of normal labor *(black bars)*. The increase in nuclear estrogen receptors from animals treated with RU486 *(stippled bars)* was significantly greater than both pregnant control and in labor animals at $p < 0.05$. From Haluska et al. (1990a)

sarean section 72 h after the start of RU486 treatment because the cervix did not dilate and there was evidence of progressive fetal asphyxia (inferred by fetal heart rate decelerations and prolonged bradycardia).

7.4.2 Cervical Ripening

Cervical ripening at term is thought to be regulated by a complex interaction of endocrine and paracrine factors including estrogens and prostaglandins, with PGE_2 as a major contributor to cervical softening (Novy and Liggins 1980). The presence of inflammatory cells during cervical ripening has also been noted. More recent studies (Ito et al. 1988) have focused on various cytokines (IL-1 and IL-8) which are derived from the maternal inflammatory cells and which function as stimulators of collagenase and initiators of the ripening process (A.A. Calder, D. Lincoln, personal communication). The fact that amniotic fluid prostaglandins and lipoxygenase products (Haluska et al. 1990b) rose slowly after RU486 administration suggested to us that retarded eicosanoid production by fetal membranes could account in part for the deficient cervical ripening.

Therefore, in related experiments (Haluska et al. 1988), we determined the production rate of PGE_2 by fetal membranes in vitro in a superfusion system. Amnion collected from untreated animals during spontaneous term labor showed a 20-fold increase in PGE_2 production compared to untreated control animals not in labor (Fig. 12). There was no significant difference in the production rate of amnion PGE_2 between controls and RU486 treated animals. This lack of increase in amnion PGE_2 production with RU486 may partially explain the lower amniotic fluid PGE_2 concentrations and the minimal cervical change after antiprogestin treatment. Calder (1990) has speculated that the region of the internal cervical os and adjacent fetal membranes may represent a critical area for the process of cervical ripening because the fetal membranes could transmit PGE_2 to the adjacent cervical tissue.

There is in vitro evidence that dexamethasone increases PGE_2 output from human amnion cells in culture and that RU486 (acting as a glucocorticoid receptor antagonist) attenuates dexamethasone-enhanced PGE_2 output (Potestio et al. 1988). Since glucocorticoid binding sites (unlike estrogen or progesterone receptors) are present in am-

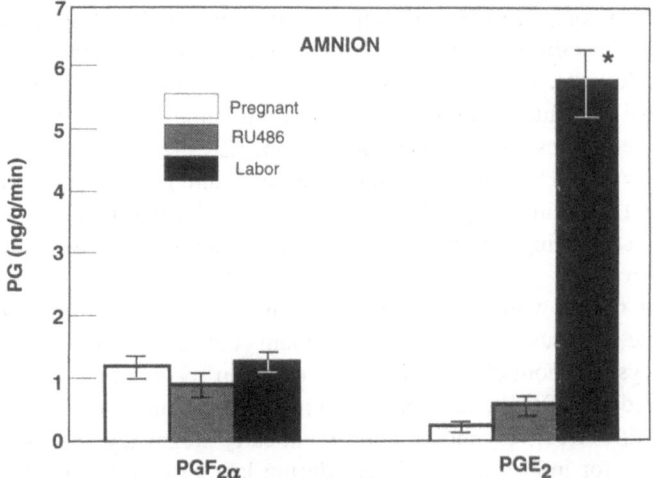

Fig. 12. $PGF_{2\alpha}$ and PGE_2 production from amnion in an in vitro superfusion system. Tissues were collected from pregnant control animals *(open bars)* who received no treatment, and were not in labor, from animals treated with RU486 *(stippled bars)* at 20 mg/kg per day for 3 days, and from animals who were in the final stages of spontaneous labor *(black bars)*. The increase in production rate of amnionic PGE_2 is significant at $p < 0.01$

nion and chorion (Giannopoulos et al. 1983). RU486 may inhibit prostaglandin production by fetal membranes even as it increases myometrial contractility.

RU486 is effective when used to abort first trimester pregnancies in humans (Couzinet et al. 1986); however, the success rate of abortion is increased when RU486 is used in conjunction with synthetic PGE_2. Treatment with RU486 and PGE_2 has also been effective in terminating human pregnancy in the second trimester (Rodger and Baird 1990). The additive effect of PGE_2 with RU486 suggests that RU486, by itself, may not adequately increase endogenous PGE_2 concentrations. Hill et al. (1990) indicate that RU486 increases uterine contractility and sensitivity to prostaglandins but does not increase endogenous PGE_2 production.

We, therefore, studied and compared the effects of a novel type of antiprogestin (ZK98299, type I; Schering AG, Berlin, Germany) which, in contrast to type II antiprogestins such as RU486, interferes with the specific binding of progesterone receptor to the DNA progesterone response elements and exhibits no progestomimetic effects and minimal antiglucocorticoid activity (Klein-Hitpass et al. 1991). At equivalent doses, the effects of onapristone (ZK98299) on uterine contractility (Fig. 9), amniotic fluid prostaglandin levels, and cervical ripening were indistinguishable from those of RU486 (Haluska et al. 1993), suggesting that antiglucocorticoid properties play a negligible role here.

Our results with type I and type II antiprogestins confirm and extend the previous observations of Germain et al. (1985) in cynomolgus monkeys in whom obstructed labors, cesarean section, and fetal death occurred after RU486 administration in late pregnancy. Although antiprogestins effectively induce abortion in early pregnancy, they are less effective for induction of delivery during late gestation in nonhuman primates. The single exception appears to be the study by Wolf et al. (1989) but in this study effective delivery was achieved only in combination with oxytocin. In addition to its uterotonic effects, oxytocin is known to stimulate decidual PG release particularly at term (Fuchs 1986; Husslein et al. 1981). In contrast to our observations in animals with unripe cervices in the mid-third trimester of pregnancy, Wolf and coworkers (1989) conducted their experiments in monkeys at term when the endogenous cervical ripening process may already have been underway. Similar considerations apply to the clinical study of RU486 for labor induction in women at term conducted by Frydman et al. (1992).

In contrast to RU486 or ZK98299, epostane, a potent 3β-hydroxy-steroid–dehydrogenase (3β-HSD) blocker, produced a very different result in monkeys at the same gestational age (Haluska et al. 1991). In all animals tested, epostane caused an 80–90% decrease in maternal, fetal, and amniotic fluid progesterone concentrations and by 24 h a dramatic increase in cervical ripening and delivery occurring within 48 h. The pattern of uterine contractility showed a diurnal rhythmicity and there was a transition to the short duration, high amplitude type contractions seen in spontaneous term labor. Effective cervical ripening and the transition to "labor-like" contractions did not occur in our experiments with either RU486 or ZK98299. The results of our work with antiprogestins on cervical status (using a modified Bishop's score) are summarized in Table 2.

Table 2. Cervical scores based on a modified Bishops Score for animals
treated with RU486, ZK98299 and Epostane

	0	24	48	72
RU486 (20 mg/kg) ($n = 7$)	2.3 ± 0.4	2.6 ± 0.2	3.3 ± 0.5	4.0 ± 0.3
RU486 (1 mg/kg) ($n = 3$)	2.0 ± 0.4	2.3 ± 0.5	3.3 ± 1.1	3.5 ± 1.3
ZK98299 (10 mg/kg) ($n = 5$)	2.4 ± 0.4	2.4 ± 0.4	2.8 ± 0.6	3.4 ± 0.9
Epostane (10 mg/kg) ($n = 3$)	1.6 ± 1.2	8.4 ± 0.3	(Delivery 36–48 h)	

A score of 10 represents a completely dilated and effaced cervix.

Since circulating estrogens and cortisol decrease along with progesterone after epostane administration, we performed additional experiments with epostane and progesterone implants in order to demonstrate
that progesterone withdrawal was specifically responsible for inducing
parturition (Haluska et al. 1992). While epostane was administered for
3 days, circulating progesterone levels were maintained at higher than
normal levels by the silastic implants even though estrogens and cortisol tended to decrease slightly. When the progesterone implants were
removed, delivery occurred within 48 h in a fashion similar to our
original experiments with epostane alone. Clearly, exogenous substitution of progesterone was able to prevent delivery in the face of epostane administration. In contrast to RU486 and ZK98299, 3β-HSD inhibition mimics spontaneous labor and results in vaginal delivery
during late pregnancy. Taken together, our results could mean that
nonreceptor-mediated progesterone withdrawal activates chorioamnion
and/or bone marrow-derived cells in decidua to release cytokines and
stimulatory eicosanoids. However, we are left with a conundrum because pharmacologic progesterone withdrawal induces preterm labor
and delivery (which can be blocked by progesterone substitution) but
exogenous progesterone, even in substantial quantities (Fig. 13) does
not prevent parturition at term.

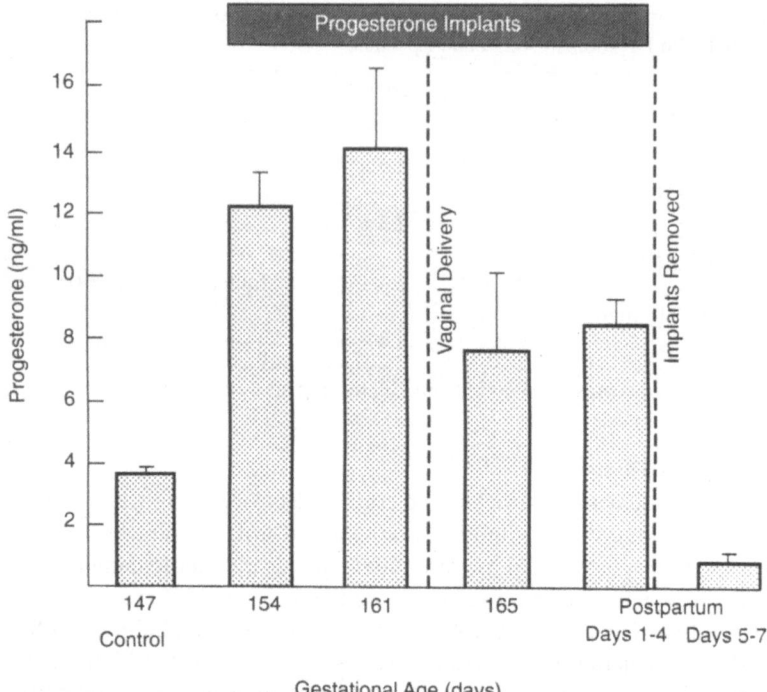

Fig. 13. Maternal plasma progesterone concentrations in pregnant rhesus monkeys ($n = 3$) who received subcutaneous implants of silastic capsules containing progesterone beginning at 148 days gestation and continuing postpartum after spontaneous vaginal delivery. Despite elevated plasma concentrations of progesterone which greatly exceeded normal values at term, animals delivered at the expected gestational age (mean = 165 days)

7.5 Concluding Remarks

We have reviewed the evidence that estrogens play an important supportive role in regulating the prepartum trends in nocturnal uterine activity which are present in nonhuman primates and women. There is persuasive evidence that this phenomenon is mediated by a heightened myometrial sensitivity to oxytocin and is driven by the maternal circadian system via the pituitary release of oxytocin. Other findings support the hypothesis that oxytocin synthesized in decidua may act in a paracrine fashion to stimulate intrauterine prostaglandin synthesis.

The contrasting effects of antiprogestins and inhibitors of progesterone biosynthesis on rhesus parturition suggest that if progesterone withdrawal is a prerequisite for normal parturition, it is not expressed by an obvious receptor-mediated mechanism. However, we remain cautious in extrapolating our results to signify that spontaneous labor and delivery at term is necessarily the result of a decrease in progesterone production by intrauterine tissues. We have been unable to block normal parturition by the exogenous administration of progesterone. It is possible that labor is triggered by a natural "antiprogestin" which appears at term. Clearly, we know more about the biochemistry and electrophysiology of myometrial contraction than about the endocrine and paracrine signals which initiate uterine activity and how they are coordinated with cervical ripening.

Acknowledgements. These investigations were supported in part by grants from the national Institutes of health (RR-00163) and from the National Institute of Child Health and Human Development (HD06159 and HD 18185). We wish to acknowledge the generosity of Dr. K. Chwaliscz and Schering AG in providing samples of onapristone (ZK98299).

References

Amico JA, Seif SM, Robinson AG (1981) Oxytocin in human plasma: correlation with neurophysin and stimulation with estrogen. J Clin Endocrinol Metab 52:988–993

Bowden D, Winter P, Ploog G (1967) Pregnancy and delivery behavior in the squirrel monkey (*Saimri sciureus*) and other primates. Folia Primatol (Basel) 5:1–42

Brandt EM, Mitchell G (1971) Parturition in primates: behavior related to birth. In: Rosenblum LA (ed) Primate Behavior Development in Field and Laboratory Research. Academic Press, New York, pp 177–183

Brenner RM, West NB, McClellan MC (1990) Estrogen and progestin receptors in the reproductive tract of male and female primates. Biol Reprod 42:11–19

Calder AA (1990) Control of Parturition: Scientific and Clinical Aspects. In: Garfield RE (ed) Uterine Contractility, Serono Symposia, USA, Massachusetts, pp 283–293

Challis JRG, Olson DM (1988) Parturition. In: Knobil E, Neill J (eds) The Physiology of Reproduction, Raven Press, New York, pp 2177–2216

Challis JRG, Sprague C, Patrick JE (1983) Relation between diurnal changes in peripheral plasma progesterone, cortisol, and estriol in normal women at 30–31, 34–35, and 28–29 weeks of gestation. Gynecol Obstet Invest 16:33–44

Chibbar R, Miller FD, Mitchell BF (1993) Synthesis of oxytocin in amnion, chorion, and decidua may influence the timing of human parturition. J Clin Invest 91:185–192

Couzinet B, Le Strat N, Ulmann A, Baulieu EE, Schaison G (1986) Termination of early pregnancy by the progesterone antagonist RU486 (mifepristone). N Engl J Med 315:1656–1670

Doyle GA, Pelletied A, Bekkeer T (1967) Courtship, mating and parturition in the lesser bushbaby (*Balago senegalensis moboli*) under semi-natural conditions. Folia Primatol (Basel) 7:169–197

Ducsay CA, Cook MJ, Walsh SW, Novy MJ (1983) Circadian patterns and dexamethasone-induced changes in uterine activity in pregnant rhesus monkeys. Am J Obstet Gynecol 145:389–396

Ducsay CA, Yellon SM (1991) Photoperiod regulation of uterine activity and melatonin rhythms in the pregnant rhesus macaque. Biol Reprod 44:967–974

Figueroa JP, Honnebier MBOM, Binienda Z, Wimsatt J, Nathanielsz PW (1989) Effect of 48-hour intravenous D4-androstenedione infusion on the pregnant rhesus monkey in the last third of gestation: Changes in maternal plasma estradiol concentrations and myometrial activity. Am J Obstet Gynecol 161:481–486

Figueroa JP, Honnebier MBOM, Jenkins S, Nathanielsz PW (1990) Alteration of 24 h rhythms in myometrial activity in the chronically catheterized pregnant rhesus monkey following a 6 h shift in the light-dark cycle. Am J Obstet Gynecol 163:648

Frydman R, Lelaidier C, Baton-Saint-Mleux C, Fernandez H, Vial M, Bourget P (1992) Labor induction in women at term with mifepristone (RU486): A double-blind, randomized, placebo-controlled study. Obstet Gynecol 80:972–975

Fuchs AR (1986) The role of oxytocin in parturition. In: Huszar G (ed) The Physiology and Biochemistry of the Uterus in Pregnancy and Labor. CRC Press, Boca Raton, FL pp 163–183

Fuchs AR, Behrens O, Liu H-C (1992) Correlation of nocturnal increase in plasma oxytocin with a decrease in plasma estradiol/progesterone ratio in late pregnancy. Am J Obstet Gynecol 167:1559–1563

Fuchs AR, Goeschen K, Husslein P, Rasmussen AB, Fuchs F (1983) Oxytocin and the initiation of human parturition. III. Plasma concentrations of oxytocin an d13,14-dihydro-15-keto-prostaglandin F2a in spontaneous and oxytocin-induced labor at term. Am J Obstet Gynecol 147:497–502

Garfield RE, Gasc JM, Baulieu MD (1987) Effects of the antiprogesterone RU486 on preterm birth in the rat. Am J Obstet Gynecol 157:1281–1285

Garfield RE, Kannan MS, Daniel EE (1980) Gap junction formation in myometrium: control by estrogens, progesterone, and prostaglandins. Am J Physiol 238:C81–C89

Garfield RE, Tabb T, Thilander G (1990) Intercellular coupling and modulation of uterine contractility. In: Garfield RE (ed) Uterine Contractility, Serono Symposia, USA

Germain AM, Valenzuela GJ, Ivanhovic M, Ducsay CA, Gabella C, Seron-Ferre M (1993) Relationship of circadian rhythms of uterine activity with term and preterm delivery. Am J Obstet Gynecol 168:1271–1277

Germain G, Cabrol D, Visser A, Sureau C (1982) Electrical activity of the pregnant uterus in the cynomolgus monkey. Am J Obstet Gynecol 142:513–518

Germain G, Lopes P, Cabrol D, Barbe MP, Huneau D, le Houezec R, Sureau C (1986) A comparison of uterine motility in the pregnant and nonpregnant cynomolgus monkesy (*Macaca fascicularis*) and pregnant women: A manometric and electromyographic study. Acta Physiol Hung 67:95–115

Germain G, Philibert D, Pottier J, Mouren M, Baulieu EE, Sureau C (1985) Effects of the antiprogesterone agent RU486 on the natural cycle and gestation in the intact cynomolgus monkeys. In: Baulieu EE, Segal SJ (eds) The Antiprogesterone Steroid RU486 and Human Fertility Control. Plenum Press, New York, pp 155–168

Giannopoulos G, Jackson K, Tulchinsky D (1983) Specific glucocorticoid binding in human uterine tissues, placenta and fetal membranes. J Steroid Biochem 19:1375

Gibb W (1977) Steroid modulation by the human fetal membranes. In: Second Symposia publications, The Endocrine Physiology of Pregnancy and the Peripartal Period, Raven Press, New York, p 135

Haluska GJ, Cook MJ, Novy MJ (1988) Prostaglandin production by intrauterine tissues of the rhesus monkey after RU486 treatment. Biol Reprod 38 [suppl 1]: abstract 266

Haluska GJ, Cook MJ, Novy MJ (1991) Epostane but not RU486 mimics spontaneous parturition in rhesus monkeys. Soc Gynecol Invest, San Antonio TX, abstract 11

Haluska GJ, Cook MJ, Novy MJ (1993) Effect of onapristone (ZK98299) on uterine activity an cervical status in late gestation rhesus monkeys. Soc Gynecol Invest, Toronto, Ontario, abstract S24

Haluska GJ, Cook MJ, Witkin SS, Novy MJ (1992) Epostane induces parturition in rhesus monkeys by lowering endogenous progesterone concentrations. In: Soc Gynecol Invest, San Antonio TX, abstract 244

Haluska GJ, Jay IC, Novy MJ (1989) Gap junctions in the primate myometrium: effect of RU486. Biol Reprod 40 [Suppl 1]:abstract 250

Haluska GJ, Mitchell MD, Novy MJ (1990b) Amniotic fluid lipoxygenase metabolites during spontaneous labor and after RU486 treatment during late pregnancy in rhesus macaques. Prostaglandins 40:99–105

Haluska GJ, Stanczyk FZ, Cook MJ, Novy MJ (1987) Temporal changes in uterine activity and progesterone response to RU486 in rhesus macaques in late gestation. Am J Obstet Gynecol 157:1487–1495

Haluska GJ, West NB, Novy MJ, Brenner RM (1990a) Uterine estrogen receptors are increased by RU486 in late pregnant rhesus macaques but not after spontaneous labor. J Clin Endocrinol Metab 70:181–186

Harbert GM (1977) Biorhythms of the pregnant utereus (Macaca mulatta). Am J Obstet Gynecol 129:401–408

Harbert GM, Spisso KR (1980) Biorhythms of the primate uterus (Macaca mulatta) during labor and delivery. Am J Obstet Gynecol 138:686–692

Hess DL, Spies HG, Hendrickx AG (1981) Diurnal steroid patterns during gestation in the rhesus macaque: Onset, daily variation and the effects of dexamethasone treatment. Biol Reprod 24:609–616

Hill NCW, Selinger M, Ferguson J, Lopez-Bernal A, MacKenzie IZ (1990) The physiological and clinical effects of progesterone inhibition with mifepristone (RU486) in the second trimester. Br J Obstet Gynecol 97:487–492

Hirst JJ, West NB, Brenner RM, Novy MJ (1990) Myometrial and decidual progesterone receptors during gestation, in spontaneous labor and after RU486 treatment in rhesus monkeys. In: Proceedings of the Serono Symposium on Uterine Contractility: Mechanisms of Control, St. Louis, Missouri, p 41, abstract 17

Hirst JJ, Haluska GJ, Cook MJ, Hess DL, Novy MJ (1991a) Comparison of plasma oxytocin and catecholamine concentrations with uterine activity in pregnant rhesus monkeys. J Clin Endocrinol Metab 73:804–810

Hirst JJ, Haluska GJ, Cook MJ, Novy MJ (1991b) Maternal plasma oxytocin (OT) and oxytocin receptor concentrations during late pregnancy in rhesus monkeys. In: Proceedings of the 38th annual meeting of the Society for Gynecologic Investigation, San Antonio, TX, p 146

Hirst JJ, Haluska GJ, Cook MJ, Novy MJ (1993) Plasma oxytocin and nocturnal uterine activity: Maternal but not fetal concentrations increase progressively during late pregnancy and delivery in rhesus monkeys. Am J Obstet Gynecol 169:414–422

Honnebier MBOM, Figueroa JP, Nathanielsz PW (1989a) Variation in myometrial response to pulsatile intravenous oxytocin administration – a pulsatile oxytocin challenge test at different times of the day in the pregnant rhesus monkey at 121 to 138 days gestational age. Endocrinology 125:1498–1503

Honnebier MBOM, Figueroa JP, Rivier J, Vale W, Nathanielsz PW (1989b) Studies of the role of oxytocin in late pregnancy in the pregnant rhesus monkey: plasma concentrations of oxytocin in the maternal circulation throughout the 24-h day and the effect of the synthetic oxytocin antagonist [1-B-Mpa (B-(CH2)5)1,(Me(Tyr2,Orn8] oxytocin. J Dev Physiol 12:225–232

Honnebier MBOM, Jenkins SL, Wentworth RA, Figueroa JP, Nathanielsz PW (1991) Temporal structural of delivery in the absence of a photoperiod: Preparturient myometrial activity of the rhesus monkey is related to maternal body temperature and depends on the maternal circadian system. Biol Reprod 45:617–625

Honnebier MBOM, Jenkins SL, Nathanielsz PW (1992) Circadian timekeeping during pregnancy: endogenous phase relationships between maternal plasma hormones and the maternal body temperature rhythm in pregnant rhesus monkeys. Endocrinology 131:2051–2058

Husslein P, Fuchs AR, Fuchs F (1981) Oxytocin and the initiation of human parturition. I. Prostaglandin release during induction of labor by oxytocin. Am J Obstet Gynecol 141:688–393

Ito A, Hiro D, Ojima Y, Mori Y (1988) Spontaneous production of interleukin-1-like factors from pregnant rabbit uterine cervix. Am J Obstet Gynecol 159:261–265

Jensen GD, Bobbitt RA (1967) Changing parturition time in monkeys (*Macaca nemistrina*) from night to day. Lab Anim Care 17:379–381

Klein-Hitpass L, Cato ACB, Henderson D, Ryffel GU (1991) Two types of antiprogestins identified by their differential action in transcriptionally active extracts from T47D cells. Nucleic Acid Res 19:1227–1234

Kocan LH, MacLusky NJ, Lye SJ (1993) Dexamethasone reverses the labor-associated myometrial desensitization to β-adrenergic agonists in the rat. Am J Obstet Gynecol 168:961–968

Liggins GC, Fairclough RJ, Grieves SA, Forster CS, Knox BS (1977) Parturition in the sheep. In: Knight J, O'Connor M (eds) The fetus and birth. Ciba Found Symp 47:5

Longo LD, Yellon SM (1988) Biological timekeeping during pregnancy and the role of circadian rhythms in parturition. In: Kunzel W, Jensen A (eds) Endocrine Control of the Fetus. Springer, Berlin Heidelberg New York, pp 173–192

Main DM, Grisso JA, Wold T, Snyder ES, Holmes J, Chiu G (1991) Extended longitudinal study of uterine activity among low-risk women. Am J Obstet Gynecol 165:1317–1322

Matsumoto T, Hess DL, Kaushal KM, Valenzuela GY, Yellon SM, Ducsay CA (1991) Circadian myometrial and endocrine rhythms in the pregnant rhesus macaque: Effects of constant light and timed melatonin infusion. Am J Obstet Gynecol 165:1777–1784

Mitchell B, Cruickshank B, McLean D, Challis JRG (1982) Local modulation of progesterone production in human fetal membranes. J Clin Endocrinol Metab 55:1237–1239

Mitchell MD, Brennecke SP, Novy MJ (1984) Prostaglandin synthease inhibitor activity in the plasma of rhesus monkeys during late pregnancy: Effect of dexamethasone. Prostaglandins Leukotrienes Med 14:199–204

Novy MJ, Haluska GJ (1988) Endocrine and paracrine control of parturition in rhesus monkeys. In: McNellis D, Challis JRG, MacDonald PC, Nathanielsz PW, Roberts JM (eds) The Onset of Labor: Cellular and Integrative Mechanisms, Perinatology Press, Ithaca, NY pp 321–337

Novy MJ, Liggins GC (1980) Role of prostaglandins, prostacyclin, and thromboxanes in the physiologic control of the uterus and in parturition. Sem Perinatol 4:45–66

Novy MJ, Walsh SW (1981) Regulation of fetoplacental steriodogenesis in rhesus macaques. In: Novy MJ, Resko JA (eds) Fetal endocrinology. Academic, New York, pp 65–94

Novy MJ, Walsh SW (1983) Dexamethasone and estradiol treatment in pregnant rhesus macaques: Effects on gestational length, maternal plasma hormones, and fetal growth. Am J Obstet Gynecol 145:920–931

Novy MJ, Walsh SW, Cook MJ (1980) Chronic implantation of catheters and electrodes in pregnant nonhuman primates. In: Nathanielsz PW (ed) Animal Models in Fetal Medicine, Elsevier/North-Holland Biomedical Press, Amsterdam, p 133

Novy MJ, Haluska GJ, Cook MJ (1989) Circadian uterine activity is abolished by fetal death but restored by maternal estradiol (E2) administration. In: Soc Gynecol Invest, San Diego CA, abstract 342

Ogle TF (1986) Evidence for nuclear processing of progesterone receptors in rat placenta. J Steroid Biochem 25:183

Potestio FA, Zakar T, Olson DM (1988) Glucocorticoids stimulate prostaglandin synthesis in human amnion cells by a receptor-mediated mechanism. J Clin Endocrinol Metab 67:1205–1210

Quartero HWP, Fry CH (1989) Placental corticotrophin-releasing factor may modulate human parturition. Placenta 10:439–443

Rodger MW, Baird DT (1990) Pretreatment with mifepristone (RU486) reduces interval between prostaglandin administration and expulsion in second trimester abortion. Br J Obstet Gynecol 97:41–45

Shepherd RW, Stanczyk FZ, Bethea CL, Novy MJ (1992) Fetal and maternal endocrine responses to reduced uteroplacental blood flow. J Clin Endocrinol Metab 75:301–307

Siiteri PK, Serón-Ferré (1981) Some new thoughts on the fetoplacental unit and parturition in primates. In: Novy MJ, Resko JA (eds) Fetal Endocrinology, Academic Press, New York, pp 1–34

Stanczyk FZ, Hess DL, Namkung PC, Senner JW, Petra PH, Novy MJ (1986) Alterations in sex steroid-binding protein (SBP), corticosteroid-binding globulin (CBG), and steroid hormone concentrations during pregnancy in rhesus macaques. Biol Reprod 35:126–132

Tambyraja RL, Hobel CJ (1983) Characterization of 24 hr uterine activity (UA) in the second half of the human pregnancy. Proceedings of the 30th annual meeting of the Society for Gynecologic Investigation, Washington DC, abstract 516

Taylor NF, Martin MC, Nathanielsz PW, Serón-Ferré M (1983) The fetus determines circadian oscillation of myometrial electromyographic activity in the pregnant rhesus monkey. Am J Obstet Gynecol 146:557–567

Walsh SW, Kittinger GW, Novy MJ (1979) Maternal peripheral concentrations of estradiol, estrone, cortisol, and progesterone during late pregnancy in rhesus monkeys (Macaca mulatta) and after experimental fetal anencephaly and fetal death. Am J Obstet Gynecol 135:37–42

Walsh SW, Ducsay CA, Novy MJ (1984a) Circadian hormonal interactions among the mother, fetus, and amniotic fluid. Am J Obstet Gynecol 150:745–753

Walsh SW, Norman RL, Novy MJ (1979) In utero regulation of rhesus monkey fetal adrenals: Effects of dexamethasone, adrenocorticotropin, thyrotropin-releasing hormone, prolactin, human chorionic gonadotropin, and α-melanocyte-stimulating hormone on fetal and maternal plasma steroids. Endocrinology 104:1805–1813

Walsh SW, Stanczyk FZ, Novy MJ (1984b) Daily hormonal changes in the maternal, fetal and amniotic fluid compartments before parturition in a primate species. J Clin Endocrinol Metab 58:629–639

Wilson L, Parsons MT, Flouret G (1991) Forward shift in the initiation of the nocturnal estradiol surge in the pregnant baboon: Is this the genesis of labor. Am J Obstet Gynecol 165:1487–1498

Wolf JP, Sinosich M, Anderson TL, Ulmann A, Baulieu EE, Hodgen GD (1989) Progesterone antagonist (RU486) for cervical dilation, labor induction, and delivery in monkeys: Effectiveness in combination with oxytocin. Am J Obstet Gynecol 160:45–47

8 The Role of the Infection and Cytokines in Preterm Parturition

Roberto Romero, Ricardo Gomez, Peter Baumann,
Moshe Mazor, and David Cotton

8.1 Introduction

A growing body of evidence suggests that infection plays a key role in
the pathogenesis of preterm labor and delivery. This chapter reviews
this evidence and also the proposed mechanisms by which infection
leads to preterm parturition.

Three lines of evidence support a role for infection in the onset of
preterm labor: (1) the administration of bacteria or bacterial products
to animals results in either abortion or labor [1–8]; (2) systemic maternal infections such as pyelonephritis, pneumonia, malaria, and typhoid
fever are associated with the onset of labor [9–21]; and (3) localized
intrauterine infection is associated with preterm labor and delivery.

8.2 Evidence Derived from Animal Experimentation

In 1943, Zahal and Bjerknes[2] demonstrated that the injection of *Shi-
gella* and *Salmonella* endotoxin into mice and rabbits is capable of inducing abortion. Takeda and Tsuchiya [3,4] confirmed this observation
using *Escherichia coli* endotoxin in pregnant mice and rabbits. Subsequently, several investigators have replicated these findings using
different animal species [5–7]. Furthermore, immunization of animals
with an antiendotoxin antibody ameliorates this biological effect [22].
The mechanisms of endotoxin-induced abortion appear to be mediated
by prostaglandins (PGs) as the concentration of prostaglandin F (PGF)
increased in serum, endometrium, and urine after the administration of
10 μg of *Salmonella* endotoxin to pregnant mice on day 16 [8]. Fur-

thermore, pretreatment of the animals with indomethacin reduced the endotoxin-induced abortion rate [8]. Recently, an animal model of ascending infection has been developed by placing bacteria through a hysteroscope into the uterine cavity in rabbits [23].

8.3 Association Between Systemic Maternal Infection and Preterm Labor and Delivery

Systemic maternal febrile infections such as pneumonia, pyelonephritis, malaria, and typhoid fever have been associated with preterm labor and delivery [9–21]. The rate of preterm delivery associated with maternal pneumonia ranges from 15% to 48% [9–12]. Although the advent of antibiotic treatment has dramatically reduced maternal mortality from this condition, it has not affected the rate of preterm delivery [11,12]. In contrast, pyelonephritis was associated with preterm delivery in the preantibiotic era but not after the introduction of antibiotics in clinical medicine. Currently, pyelonephritis is associated with preterm labor but not preterm delivery [13–16]. Similarly, typhoid fever in the preantibiotic era carried a 60–80% risk of abortion and preterm labor, but this risk decreased after the introduction of antibiotic therapy [17–19]. Malaria has also been associated with a 50% rate of preterm delivery [20]. However, chemoprophylaxis seems to protect patients from preterm delivery [21]. Collectively, these data support the concept that severe untreated systemic maternal infection is associated with preterm labor and delivery and that treatment may decrease the rate of preterm delivery in some cases (e.g., pyelonephritis, typhoid fever) but not in others (e.g., pneumonia). The mechanisms involved in the initiation of labor in the setting of systemic infections have not been studied in humans. However, wide clinical experience indicates that maternal fever is associated with increased uterine activity. This effect has been demonstrated with parenteral administration of endotoxin to women at term. A two- to threefold increase in uterine activity was noted during the chill period (15–60 min), and uterine activity gradually diminished [24]. Since parenteral administration of endotoxin to animals and humans results in the production and release to cytokines and this, in turn, can stimulate prostaglandin production, we have proposed that these products mediate the increase in uterine activity in the setting of febrile maternal infection [25].

8.4 Association Between Intrauterine Infection and Preterm Delivery

Intrauterine infections can be classified, according to the location of the microorganisms, into two broad categories: intraamniotic and extra-amniotic infections. The gold standard for the diagnosis of an intrauterine infection is a microbiologic culture. The technical difficulties in obtaining a culture from the extra-amniotic compartment make this type of infection difficult to study. Thus, most literature refers to intraamniotic infection.

The amniotic cavity is normally sterile and therefore the isolation of any microorganism from the amniotic fluid constitutes strong evidence of microbial invasion. This condition can exist even in the absence of clinical signs and symptoms of infection [26]. The method of amniotic fluid collection for microbiologic studies is critical. The two techniques that have been used for microbiological studies are transabdominal amniocentesis and transcervical retrieval either by needle puncturing of the membranes or by aspiration through an intrauterine catheter. Transcervical amniotic fluid collection is associated with an unacceptable risk of contamination with vaginal flora. Therefore, when analyzing the prevalence of microbial invasion of the amniotic cavity in preterm labor, we will consider only studies in which amniotic fluid was obtained by transabdominal amniocentesis.

The term "clinical chorioamnionitis" refers to the clinical syndrome associated with microbial invasion of the amniotic cavity. Manifestations include maternal fever, uterine tenderness, foul-smelling vaginal discharge, fetal tachycardia, and maternal leukocytosis [27]. This clinical syndrome appears only in a small fraction of women with microbiologically proven intraamniotic infections. In a recent study, we found that only 12.5% of women with preterm labor and intact membranes with a positive amniotic fluid culture had clinical chorioamnionitis [28]. The presence and severity of clinical chorioamnionitis are probably related to both microbial and host factors. Microbial factors include the type and virulence of the microorganism, inoculum size, and pathway of infection (hematogenous or ascending infection). Host factors include the magnitude of the local inflammatory response and the local and systemic production of cytokines.

8.4.1 Pathways of Intraamniotic Infection

Microorganisms may gain access to the amniotic cavity and fetus using any of the following pathways: (1) ascending from the vagina and the cervix; (2) by hematogenous dissemination through the placenta (transplacental infection); (3) retrograde seeding from the peritoneal cavity through the fallopian tubes; and (4) accidental introduction at the time of invasive procedures such as amniocentesis, percutaneous fetal blood sampling, chorionic villous sampling, or shunting [28–33].

Indirect evidence indicates that the most common pathway of intrauterine infection is the ascending route. This evidence includes: (1) histologic chorioamnionitis is more common and severe at the site of membrane rupture than in other locations, such as the placental chorionic plate or umbilical cord; (2) in virtually all cases of congenital pneumonia (stillbirths or neonatal), inflammation of the chorioamniotic membranes is present; (3) the bacteria identified in cases of congenital infections are similar to those found in the genital tract; and (4) in twin gestation, histologic chorioamnionitis is more common in the firstborn twin and has not been demonstrated in only the second twin. As the membranes of the first twin are generally apposed to the cervix, this is taken as evidence in favor of an ascending infection [28–33]. This is consistent with our observations in the microbiology of amniotic fluid in twin gestations. Indeed, in all our cases of microbial invasion of the amniotic cavity diagnosed by transabdominal amniocentesis in twin gestation, the presenting sac was involved. When both amniotic cavities were infected, the inoculum size was larger in the presenting sac [34].

We have proposed a four-stage process leading to intrauterine infection (Fig. 1) [26]. The first stage consists of an overgrowth of facultative organisms or the presence of pathologic organisms (i.e., *Neisseria gonorrhoeae*) in the vagina and/or the cervix. Bacterial vaginosis may be one of the manifestations of stage I. Once microorganisms gain

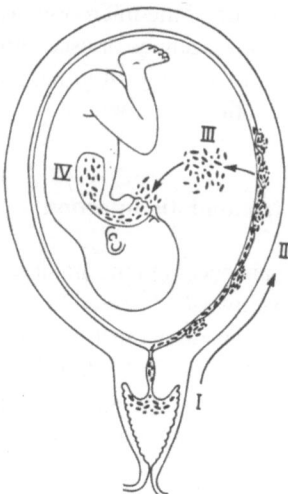

Fig. 1. The stages of ascending infection

access to the intrauterine cavity, they reside in the decidual (stage II). A localized inflammatory reaction leads to deciduitis and further extension to chorionitis. The infection may invade the fetal vessels (choriovasculitis) or proceed through the amnion (amnionitis) into the amniotic cavity, leading to an intraamniotic infection (stage III). Rupture of the membranes is not a prerequisite for intraamniotic infection, as microorganisms are capable of crossing intact membranes [35]. Once in the amniotic cavity, the bacteria may gain access to the fetus by different ports of entry (stage IV). Aspiration of the infected fluid by the fetus may lead to congenital pneumonia. Otitis, conjunctivitis, and omphalitis are localized infections that occur by direct spreading of microorganisms from infected amniotic fluid. Seeding from any of these sites to the fetal circulation leads to bacteremia and sepsis. Another possible pathway for fetal sepsis is the spread of an infection located in the decidua parietalis to the decidua basalis, and from there directly to the fetal villous circulation.

The mechanisms responsible for preterm premature rupture of the membranes (PROM) may also be associated with ascending infection. A localized infection in the choriodecidual junction can lead to rupture of the membranes. Microbial invasion of the amniotic cavity may result from the spread of microorganisms from the localized choriodecidual nidus or by direct spread from the vagina through the site of rupture. Rupture of membranes can also result when ascending infection, as described in the previous paragraph, reaches stage III. The effect of bacterial protease and host products secreted in response to bacterial infection from both sides of the membranes may lead to weakening of the membranes [36]. This model explains why women with PROM can have either a positive or a negative amniotic fluid culture at the time of presentation. Over time, a progressive increase in the incidence of positive amniotic fluid cultures is expected.

8.4.2 Microbiology of Microbial Invasion of the Amniotic Cavity

The most common microbial isolates from the amniotic cavity from women with preterm labor and intact membranes are *Ureaplasma urealyticum*, *Fusobacterium* species, and *Mycoplasma hominis* [26]. Fifty percent of patients with microbial invasion have more than one microorganism isolated from the cavity. The inoculum size varies considerably, and in 71% of the cases more than 10^5 colony forming units per milliliter (cfu/mL) are found [28]. Our observations are consistent with those of other studies supporting a role for *Fusobacterium* [37,38] and mycoplasma species [39] in preterm labor. It is noteworthy that the most common microorganisms responsible for neonatal sepsis are not frequently isolated from amniotic fluid.

The role of *Chlamydia trachomatis* as an intrauterine pathogen has not been elucidated. This microorganism is an important cause of cervicitis and has been recently isolated from amniotic fluid [40,41]. A case of congenital pneumonia caused by *C. trachomatis* suggests that this microorganism may be capable of causing ascending intraamniotic infection [40]. The uncertainty about the role of *C. trachomatis* in the etiology of microbial invasion and intrauterine infection may be related to difficulties in isolating the microorganisms from amniotic fluid with standard culture techniques. The use of polymerase chain reaction to detect specific sequences for this microorganism should help resolve this question [42].

8.4.3 Microbial Invasion of Amniotic Cavity in Preterm Delivery

Studies examining the clinical circumstances surrounding preterm delivery indicate that a third of all patients present with preterm labor and intact membranes, a second third are associated with preterm PROM, and the remaining third results from delivery because of maternal or fetal indications (i.e., preeclampsia, growth retardation etc.) [43].

To examine the role of microbial invasion of the amniotic cavity in the etiology of preterm delivery, we will review the evidence supporting an association between intrauterine infection and spontaneous preterm labor (with or without intact membranes).

8.4.4 Microbial Invasion of the Amniotic Cavity in Patients with Preterm Labor and Intact Membranes

Table 1 displays the results of 13 studies in which amniocentesis was performed in women with preterm labor and intact membranes [28,44–54]. The mean rate of positive amniotic fluid cultures was 11.9% (90/758). Women with positive amniotic fluid cultures generally did not have clinical evidence of infection at presentation, but they were more likely to develop chorioamnionitis subsequently [42.2% (38/90) vs. 3.9% (13/328)], to be refractory to tocolysis [62.5% (35/56) vs. 13% (36/276)], and to rupture their membranes spontaneously [19.6% (9/46) vs. 5.1% (15/292)] than were women with negative amniotic fluid cultures.

8.4.5 Microbial Invasion of the Amniotic Cavity in Patients with Preterm PROM

Table 2 displays the results of amniotic fluid cultures in women with preterm PROM in seven published studies [53,55–60]. Positive amniotic fluid cultures occurred in 27.9% (113/404). This figure probably underestimates the true prevalence of intraamniotic infection. Recent evidence gathered with ultrasound indicates that women with PROM and severely reduced amniotic fluid volume have a higher incidence of intraamniotic infection [59,61]. Since these women are less likely to

Table 1. Intraamniotic infection in women with preterm labor and intact membranes as determined by amniotic fluid studies obtained by transabdominal amniocentesis (from [26])

Author	Year	No. of Patients	Positive cultures (n). (%)	Clinical chorioamnionitis (n) (%)	PROM	Refractory to tocolysis (n) (%)
Miller et al. [47]	1980	23	11 (47.8)	8 (72.7)	2/7 (28.5)	
Bobbitt et al. [48]	1981	31	8 (25.8)	6 (75.0)		7/8 (87.5)
Wallace and Herrich [49]	1981	25	3 (12.0)	1 (33.3)		
Hameed et al. [50]	1984	37	4 (10.8)	3 (75.0)		3/4 (75.0)
Wahbeh et al. [51]	1984	33	7 (21.2)	2 (28.5)		4/7 (57.1)
Wieble and Randall [52]	1985	35	1 (2.9)	1 (100.0)		
Leigh and Garite [41]	1986	59	7 (11.8)	4 (57.1)	4/7 (57.1)	7/7 (100.0)
Gravett et al. [53]	1986	54	13 (24.0)	5 (38.5)		5/13 (38.5)
Iams et al. [54]	1987	5	0 (0.0)			
Duff and Kopelman [55]	1987	24	1 (4.2)	0	0/1	0/1
Romero et al. [56]	1988	41	4 (9.8)			
Skoll et al. [57]	1989	127	7 (5.5)	1 (14.3)	1/7	9/16 (56.2)
Romero et al. [32]	1989	264	24 (9.1)	3 (12.5)	2/24	35/56 (62.5)
Totals		758	90 (11.9)	38 (42.2)	9/46 (19.6)	35/56 (62.5)

Table 2. Intraamniotic infection in women with preterm PROM as determined by amniotic fluid studies obtained by transabdominal amniocentesis (From [26])

Author	Year	No. of Patients	Positive culture (n) (%)	Success rate (%)	Clinical chorioamnionitis infection (n) (%)	Neonatal infection (n) (%)
Garite et al. [58]	1979	59	9/30 (30.0)	51	6/9 (66.6)	2/9 (22.2)
Garite et al. [59]	1982	207	20/86 (23.2)	49	11/20 (55.0)	5/20 (25.0)
Cotton et al. [60]	1984	61	6/41 (14.6)	69	6/6 (100.0)	1/6 (16.6)
Broekhuizen et al. [61]	1985	79	15/53 (28.3)	66	3/15 (20.0)	8/15 (53.3)
Vintzileos et al. [62]	1985	54	12/54 (22.2)	–	2/12 (16.6)	4/12 (33.3)
Feinstein et al. [63]	1986	73	12/50 (20.0)	68	6/12 (50.0)	5/12 (41.6)
Romero et al. [64]	1988	90	39/90 (43.3)	95	– –	
Totals		623	113/404 (27.9)	59	34/74 (45.9)	25/74 (33.7)

have an amniocentesis, the bias in these studies is to underestimate the prevalence of infection. Another bias in these studies is that women with preterm PROM admitted in labor did not undergo amniocentesis. Therefore, such studies provide information about the prevalence of intraamniotic infection in women with preterm PROM who were not in labor. We have documented that patients who were in preterm labor on admission had a tendency to have higher incidence of positive amniotic fluid cultures in comparison to women admitted with PROM who were not labor [39% (24/61) vs. 25% (41/160), $p < 0.05$]. Furthermore, of patients who were not in labor on admission, 60% had a positive amniotic fluid culture when entering in active labor [62].

8.4.6 Microbial Invasion of the Amniotic Cavity in Patients Presenting with Acute Cervical Incompetence

We have recently determined that 51.5% (17/33) of patients presenting with cervical dilatation of 2 cm or more an intact membranes between 14 and 24 weeks have a positive amniotic fluid culture for microorganisms [63]. The outcome of patients with microbial invasion was uniformly poor as they developed complications (rupture of membranes, clinical chorioamnionitis, or pregnancy loss). Microbial invasion of the amniotic cavity in this setting may be primary or secondary. In instances of primary cervical disease (i.e., cervical hypoplasia), dilatation and effacement of the cervix result in exposure of the chorioamniotic membranes to the lower genital tract flora. Microorganisms may cross intact membranes [35]; therefore they can gain access to the amniotic cavity. Under these circumstances infection would be a secondary phenomenon. It is unknown why microbial invasion develops in some patients and not in others. Factors that may play an important role include magnitude of membrane surface exposed to the vaginal flora, duration of the exposure, microbiologic characteristics of the genital flora, and host response to microbial invasion. An alternative possibility is that an ascending infection in the midtrimester of pregnancy produces myometrial activation and subsequently cervical dilatation. Because uterine contractions are usually painless and are often undetected by patients in this stage of pregnancy, the clinical picture of an infection-induced spontaneous abortion may be indistinguishable from that of an incompetent cervix.

8.4.7 Microbial Invasion of the Amniotic Cavity in Patients with Spontaneous Labor and Intact Membranes at Term

The high frequency of microbial invasion of the amniotic cavity in patients with preterm labor and intact membranes and in patients' preterm PROM data (Tables 1,2) has been often invoked to support a causal relationship between infection and preterm labor. However, until recently the prevalence of microbial invasion in normal spontaneous labor at term and in patients with term PROM – the appropriate controls groups – was unknown. We have recently completed such studies. These results are important to understand the biology and significance of microbial invasion in both term and preterm parturition.

To establish the prevalence of microbial invasion of the amniotic cavity in women in spontaneous labor at term with intact membranes, amniotic fluid was retrieved transabdominally in a group of women undergoing primary or repeat cesarean section in active labor or who were suspected to have preterm labor but subsequently delivered a term infant by weight and pediatric examination. We found that 18.8% (17/90) of these patients had a positive amniotic fluid culture. The frequency of positive amniotic fluid cultures in patients with term PROM was 34% (11/32) [64,65]. These data indicate that the prevalence of microbial invasion of the amniotic cavity is similar in women with term and preterm labor and leading to preterm delivery (both with intact membranes) and also in patients with term and preterm PROM. Consequently, the argument could be made that microbial invasion is a phenomenon associated with labor or rupture of membranes per se rather than preterm labor and preterm PROM. Microorganisms gain access to the sterile amniotic cavity when cervical dilatation exposes intact membranes to the normal vaginal flora or immediately after membrane rupture. Therefore, microbial invasion may be the consequence of labor or rupture of the membranes rather than the cause of preterm labor and preterm PROM.

There are, however, striking differences between the findings at term and preterm gestation. First, the inoculum size in term labor is much smaller than in preterm labor [64]. Second, the concentrations of several macrophage derived cytokines interleukin-1 (IL-1), tumor necrosis factor (TNF), IL-6 and IL-8 are severalfold higher in samples obtained during preterm labor with infection than during term labor [66]. Third, the incidence and severity of histopathologic chorioamnionitis is much higher in patients with preterm labor and delivery than in patients with term gestation (30% vs. 10%), as well as the incidence of clinical chorioamnionitis [67]. Collectively, these data indicate that there are fundamental differences between microbial invasion of the amniotic cavity in the context of term and preterm gestation. Moreover, recently we have determined that among patients with preterm PROM who are not in labor on admission, those with microbial invasion of the amniotic cavity have a shorter amniocentesis to delivery interval than patients with sterile amniotic fluid [68].

We believe that microbial invasion of the amniotic cavity can probably be both a cause and a consequence of labor. Ascending microbial invasion may lead to macrophage activation and initiation of parturition when present for an extended period of time; this is the likely sequence of events in preterm labor associated with microbial invasion. When labor has begun, secondary microbial invasion may also occur. We believe that is the most likely sequence of events associated with microbial invasion during spontaneous term labor.

8.4.8 Histologic Chorioamnionitis in Preterm Delivery

Further evidence that intrauterine infection is associated with preterm delivery is derived from histopathologic studies of the placenta. Inflammation of the placenta is a host–response mechanism to a variety of stimuli such as infection and immune injury. Traditionally, acute inflammation of the chorioamniotic membranes has been considered an indicator of amniotic fluid infection [27,29–32,69–72]. This view has been based upon indirect evidence. Previous studies have demonstrated an association between acute inflammatory lesions of the placenta and the recovery of microorganisms from the subchorionic plate [73,74] and from the chorioamniotic space [75]. Bacteria have been recovered from the subchorionic plate from 72% of placentas with histologic chorioamnionitis [74–76]. Furthermore, we have recently found that there is an excellent correlation between positive amniotic fluid cultures for microorganism and histologic chorioamnionitis [77].

Several studies have examined the prevalence of inflammation in placentas from women delivering preterm infants. We have critically reviewed these studies elsewhere [63]. Collectively the evidence indicates that there is association between preterm birth and the occurrence of acute chorioamnionitis.

8.4.9 Clinical Evidence of Infection (Chorioamnionitis, Endometritis, and Neonatal Sepsis) in Preterm Delivery

The prevalence of endometritis is higher in women delivering preterm than in women delivering at term [preterm PROM: 18.7% (38/203) vs. term PROM: 8.4% (38/454), $p < 0.001$; preterm intact membranes: 13.1% (36/274) vs. term intact membranes: 6.4% (120/1881) $p < 0.001$]. Furthermore, the prevalence of endometritis is the same after preterm delivery with intact membranes as after delivery with preterm PROM. These data suggest that there is an association between postpartum infection and preterm delivery [78]. The prevalence of neonatal sepsis is 4.3 per 1000 live births in premature infants, in contrast to 0.8 per 1000 live births for term infants [79]. Furthermore, the lower the birth weight, the higher the prevalence of sepsis (164/1000 for 1001–1500 g; 91/1000 for 1501–2000 g; and 23/1000 for 2001–2500 g [80]. The conventional interpretation of these data is that premature newborns are more susceptible to infection. The observation that at least half of the cases of sepsis are diagnosed within 48 h after delivery, together with the high incidence of microbial invasion of the amniotic cavity in women with preterm labor and PROM, calls for a reappraisal of this traditional view. We would suggest that the higher incidence of sepsis in the preterm newborn is partially attributable to higher incidence of intrauterine infection in women with preterm labor. Furthermore, we propose that the onset of preterm labor in this subpopulation may be part of the repertoire of host defense against infection.

8.5 Cellular and Biochemical Mechanisms Proposed to Mediate Parturition in the Setting of Infection – The Role of Arachidonic Acid Metabolites

8.5.1 Prostaglandins

Until recently, prostaglandins have been considered the universal mediators of parturition in mammalian species [81–85]. Traditional evidence that supports the participation of prostaglandins in the mechanism of labor in human includes the following: (1) Administration of prostaglandin results in abortion or labor; (2) treatment with prostaglandins inhibitors delays the process of midtrimester abortion and the onset of labor and can arrest preterm labor; (3) parturition at term is associated with elevated amniotic fluid and maternal plasma concentrations of prostaglandins; (4) arachidonic acid (prostaglandin precursor) concentrations in the amniotic fluid increase during labor; (5) intraamniotic administration of arachidonic acid results in labor.

The evidence supporting a role for prostaglandins in the mechanisms responsible for preterm labor is less firm than for term labor [85–94]. In both plasma and amniotic fluid, levels of prostaglandins have been reported to be either normal [84–86] or increased [87–88] in women in preterm labor. The discrepancies between these studies may be attributed to the heterogenous nature of disease causing preterm labor. Patients with preterm labor and microbial invasion of the amniotic cavity have significantly higher amniotic fluid concentrations of prostaglandin E_2 and $F_{2\alpha}$ and their stable metabolites (bicycloprostaglandin E_2 and 13,14-dihydro-15-keto-prostaglandin $F_{2\alpha}$) than women in preterm labor with negative amniotic fluid cultures [89–94]. In contrast, amniotic fluid concentrations of prostaglandins E_2 and $F_{2\alpha}$ are not different between women in preterm labor with negative amniotic fluid cultures and women without labor with negative cultures at similar gestational ages [91]. These findings are consistent with the observation that the production of prostaglandin by amnion and chorion decidua is higher in patients with preterm labor and histologic chorioamnionitis than in patients with no evidence of placental inflammation [94]. These data suggest an increased bioavailability of prostaglandins in preterm parturition associated with infection. It remains to be proved whether prostaglandins play a role in preterm labor not associated with infection.

8.5.2 Arachidonic Lipoxygenase Metabolites

Metabolites of arachidonic acid derived through the lipoxygenase pathway involving leukotrienes and hydroxyeicosatraeonic acids have also been implicated in the mechanisms of spontaneous parturition at term [95–100]. Arachidonate lipoxygenase products are inflammatory mediators, so it is possible that they also participate in the mechanisms of preterm labor associated with infection. Concentrations of 5-hydroxyeicosatetraenoic acid, leukotriene B_4 and 15-hydroxyeicosatetraenoic acid are increased in the amniotic fluid of women with preterm labor and microbial invasion of the amniotic cavity [101–103]. Similarly, amnion from patients with histologic chorioamnionitis releases more leukotriene B_4 in vitro than amnion from women delivered preterm without inflammation [104].

The precise role of arachidonate lipoxygenase products in parturition in association with infection remain to be determined. 5-Hydroxyeicosatetraenoic acid and leukotriene C_4 may stimulate uterine contractility, and leukotriene B_4 may recruit neutrophils to the site of infection and participate in the regulation of the cyclooxygenase pathway [105–106]. Leukotriene B_4 has been shown to act as a calcium ionophore [107] and thus may increase phospholipase activity and enhance the rate of prostaglandin synthesis by intrauterine tissues.

8.6 Mechanisms Stimulating the Bioavailability of Arachidonic Acid Metabolites in Preterm Labor Associated with Infection: The Role of Microbial Products and of Cytokines

8.6.1 Bacterial Products

Prostaglandin biosynthesis may be stimulated by either bacterial or host factors secreted in response to microbial presence. The traditional explanation for the onset of labor in the presence of infection has been that bacterial products directly stimulate prostaglandin biosynthesis.

Fig. 2A–C. The effects of bacterial endotoxin on the production of PGE_2 by human amnion cells. **A** 026 E coli, **B** 055 E coli, and **C** *S. typhosa*, respectively. *White bars*,control; *striped bars*,lipopolysaccharide (LPS)

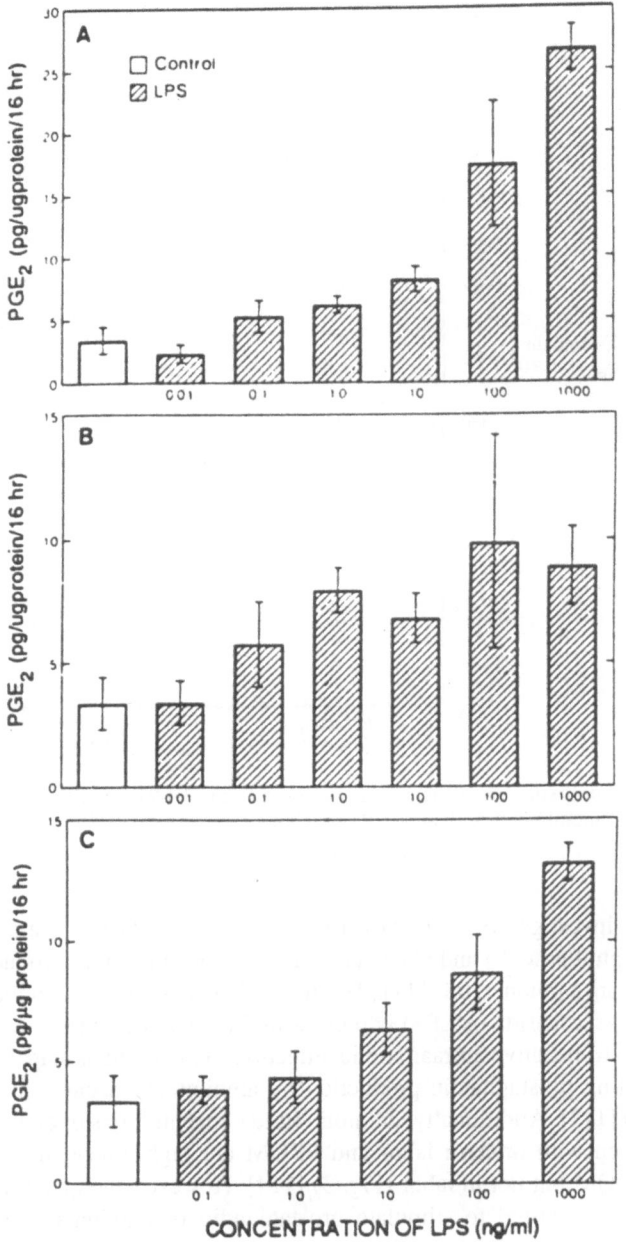

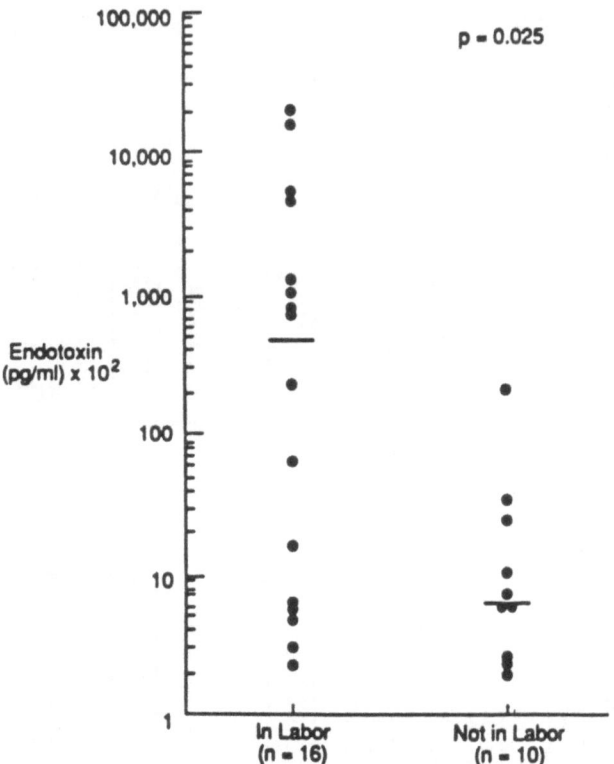

Fig. 3. Endotoxin concentrations in amniotic fluid of women with premature rupture of membranes with and without labor

Several investigators have shown that bacterial products are a source of phospholipase A_2 and C and can stimulate prostaglandin production by human amnion [108–111]. We have also reported that endotoxin (lipopolysaccharide or LPS) is present in the amniotic fluid of women with gram-negative intraamniotic infections [112] and is capable of stimulating prostaglandin production by amnion and decidua in vitro (Fig. 2) [113]. Additionally, amniotic fluid concentrations of endotoxin in women with preterm labor and PROM are higher than in women with PROM but not in labor (Fig. 3) [114]. However, the quantities of endotoxin required to stimulate prostaglandin production by human amnion are not generally found in the amniotic fluid of women with intraamniotic infection and preterm labor. Moreover, the overlap in endotoxin concentrations between the two groups (with and without labor) suggests that factors other than endotoxin are involved in triggering the onset of premature labor in the presence of intraamniotic infection. Therefore, it is possible that other bacterial products may be responsible for the stimulation of arachidonic acid metabolites by intrauterine tissues or, alternatively, that host–defense mechanisms are operative.

The observation that 28–33% of women with preterm PROM have an intraamniotic infection without labor suggests that the mere presence of microorganisms in the amniotic cavity is not sufficient to lead to the onset of labor. Furthermore, there is now evidence that the effects of microbial products on prostaglandin production by intrauterine tissues are more variable than previously thought [115–117].

8.6.2 The Role of Cytokines

Until recently, it was widely accepted that microorganisms alone were responsible for the ill effects and metabolic derangements associated with infection. It has now been established, however, that many of these ill effects are mediated by endogenous host products. A typical example of this is the pathophysiology of endotoxic shock. Bacterial endotoxin exerts its deleterious effects through the release of endogenous mediators such as tumor necrosis factor.

The onset of labor in the presence of infection may be considered the pathophysiologic counterpart of endotoxin shock and thus a host-mediated response. In view of the pivotal role of the macrophage–monocyte system in the host response against infection and tissue injury, we propose that secretory products of macrophage activation play a key role in the mechanisms of parturition in the presence of infection.

Macrophages are ubiquitous cells present in the maternal (decidua), fetal, and placental compartments. These cells are activated by microbial products to secrete a wide variety of mediators including IL-1 and TNF, IL-6, IL-8, and IL-10.

8.6.2.1 Interleukin-1

IL-1 is produced by activated monocyte/macrophage cells in response to bacterial products such as endotoxin [118]. IL-1 is pleotropic cytokine which along with TNF and IL-6 has been shown to mediate host responses to infection and injury. The biologic properties of IL-1 include the mediation of fever, activation of T and B lymphocytes, induction of collagenase activity, and prostaglandin biosynthesis.

Two biochemically related but distinct forms of IL-1 have been isolated: IL-1α and IL-1β. These two cytokines are the product of two different genes. They have the same molecular weight but a different isoelectric point (pI for IL-1α = 5; pI for IL-1β = 7). Despite sharing only a 25% aminoacid sequence homology, IL-1α and IL-1β bind to the same receptor and have the same spectrum of biologic activities.

In 1985 we postulated that IL-1 produced by the host (fetus or mother) could serve as a signal for initiation of labor. The evidence to support this view includes: (1) IL-1 stimulates prostaglandin production by amnion, decidua, and myometrium [119–120]; (2) human decidua can produce IL-1 in response to bacterial products [121]; (3) amniotic fluid IL-1 bioactivity and concentrations are elevated in women with preterm labor and intraamniotic infection. In contrast, amniotic fluid from patients with preterm labor but without intraamniotic infection does not contain IL-1 (Fig. 4) [122–123]; (4) in women with preterm PROM and intraamniotic infection, IL-1 bioactivity is higher in the presence of labor (Fig. 5) [122]; (5) IL-1 can induce preterm parturition when administered systematically to pregnant mice [124]; and (6) intrauterine administration of IL-1 (and also TNF) to pregnant guinea pigs can reliably induce preterm labor and delivery[125]. These data suggest that the host response to microbial invasion of the uterine cavity plays an important role in the onset of labor in cases of intrauterine infection.

The intracellular cellular sources of IL-1 have been studied with the use of immunohistochemistry and in situ hybridization. Increased IL-1b gene expression has been demonstrated in fetal membranes and decidua from patients with clinical and histologic chorioamnionitis using northern blot analysis. IL-1β mRNA expression has been preferentially localized within macrophages present in the chorion and adherent decidua, and to a lesser extent to neutrophils and occasional decidual cells.

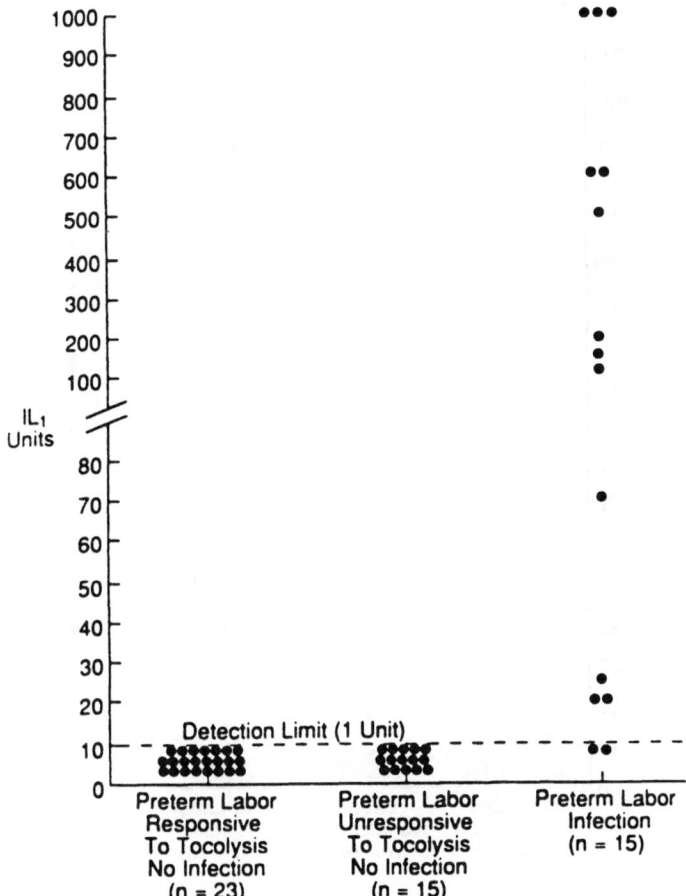

Fig. 4. IL-like bioactivity in the amniotic fluid of women with preterm labor. IL-1-like bioactivity is expressed in units because samples have been run in different assays. One unit of IL-1 is defined as that amount of IL-1 required to double the proliferative response of thymocytes stimulated with concanavalin A

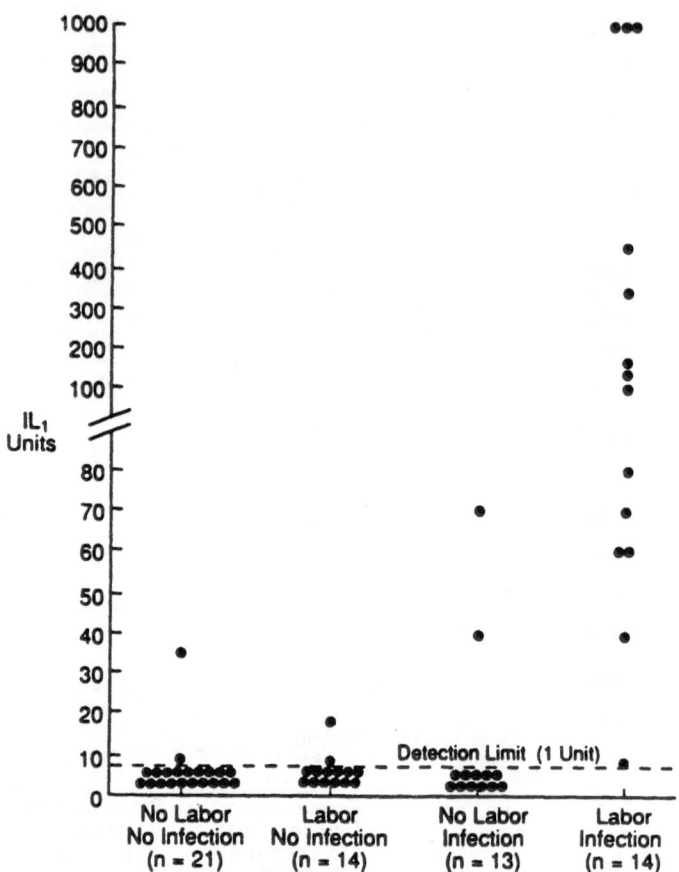

Fig. 5. IL-1-like bioactivity in the amniotic fluid of women with preterm premature rupture of membranes. A significantly greater proportion of women with microbial invasion of the amniotic cavity than those without infection had detectable amniotic fluid IL-1-like activity ($p < 0.001$)

These findings suggest that most of IL-1β production in chorioamnionitis is from infiltrating inflammatory cells and not from the local decidual cells [126]. Indeed, we have found that decidual cells (endometrial stromal cells) are capable of producing IL-1β but that they do not export this cytokine (R. Romero et al., Unpublished observations).

Paulesu et al., using immunohistochemical techniques, found that fetal trophoblast cells in direct contact with maternal blood or tissue (villous syncytiotrophoblast and extravillous trophoblast, respectively) but not villous cytotrophoblast and cytotrophoblast column stained positive for IL-1α and IL-1β. The patterns of distribution for IL-1α and IL-1β were similar. However, IL-1β was more intense than IL-1α [127]. Increased staining for IL-1β is observed in the fetal membranes of patients with documented microbial invasion of the amniotic cavity (R. Romero, unpublished observations). Taniguchi et al. 1991 have recently reported that placental tissue obtained from patients with labor had produced larger amounts of IL-1 than those obtained from women not in labor, with a predominant production of IL-1β [128]. Moreover, placental tissue obtained from women with labor and chorioamnionitis produced 17-fold higher concentrations of both IL-1α and IL-1β than those obtained from women with labor but without evidence of chorioamnionitis [128].

Specific bindings sites for IL-1 have been demonstrated in human myometrial cells Hertelandy et al. [129] and amnion cells in culture [130].

8.6.2.2 Tumor Necrosis Factor

TNF is secreted by activated macrophages and has properties similar to IL-1 (Table 3) [131]. Evidence suggesting a role for TNF in the onset of labor associated with infection includes: (1) TNF stimulates prostaglandin production by human amnion, decidua, and chorion [132]; (2) TNF is produced by human decidua in response to bacterial products [133–134]; and (3) TNF is absent from normal amniotic fluid but present in the amniotic fluid of women who have intraamniotic infection and preterm labor (Fig. 6) [135–136]; (4) Systemic and intrauterine administration of TNF to pregnant animals can induce preterm labor and delivery [137]. Moreover, Bry has recently demonstrated a

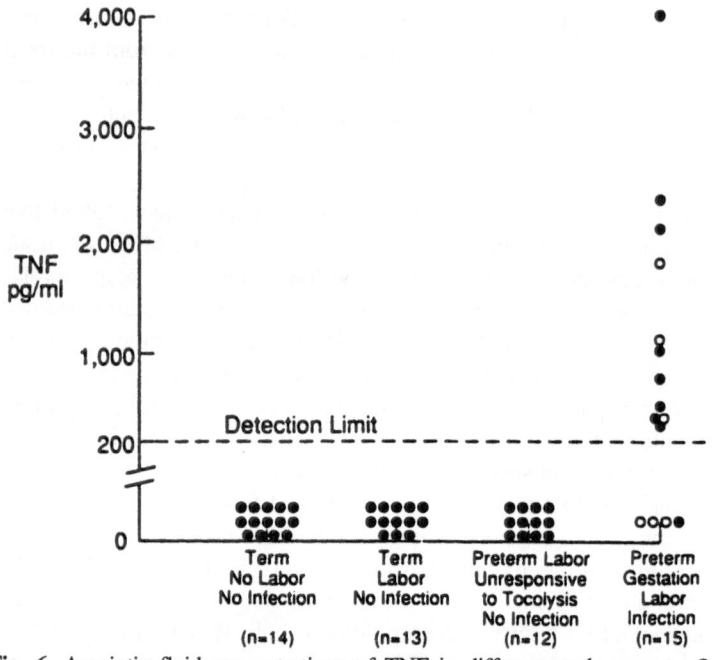

Fig. 6. Amniotic fluid concentrations of TNF in different study groups. *Open circles* represent women with premature rupture of membranes. *Solid circles* represent women with preterm labor and intact membranes

Table 3. Biologic properties of IL-1 and TNF/cachectin (from [173])

Biologic property	IL-1	TNF/cachectin
Endogenous pyrogen, causes fever	+	+
Induces slow wave sleep	+	+
Elicits hemodynamic shock	+	+
Increases hepatic acute phase protein synthesis	+	+
Decreases albumin synthesis	+	+
Activates endothelium	+	+
Decreases lipoprotein lipase	+	+
Decreases cytochrome p450	+	+
Decreases plasma Fe/Zn	+	+
Increases fibroblast proliferation	+	+
Increases synovial cell collagenase and PGE_2	+	+
Induces formation of IL-1	+	+
Elicits T/B cell activation	+	−

synergistic effect of TNF and IL-1 on prostaglandin production by in-trauterine tissues [138].

8.6.2.3 Interleukin-6

IL-6 has been implicated as a major mediator of the host response to infection and tissue damage. This cytokine is produced by a wide variety of cells such as fibroblasts, monocyte/macrophages, endothelial cells, keratinocytes, and endometrial stromal cells. IL-6 consists of a group of differentially modified phosphoglycoproteins ranging in size from 23 to 30 kDa and additional complexes of 43–45 kDa derived from a single gene located at 7p21 in the human genome. IL-6 expression is induced by several inflammation-associated cytokines including IL-1, TNF, and interferons, bacterial products, RNA- and DNA-containing viruses, and second messenger agonists (diacyglycerol, cAMP, and Ca^{2+}) that activate any of the three major signal transduction pathways. IL-6 elicits major changes in the biochemical, physiologic, and immunologic status of the host, including the acute phase plasma protein response, activation of T and natural killer cells, and stimulation and proliferation of immunoglobulin production by B cells. IL-6 induces the production of corticotrophin-releasing factor (CRP) by liver cells. This may be important in the context of intraamniotic infection, as clinical studies have indicated that an increase in maternal serum CRP often precedes the development of clinical chorioamnionitis and the onset of preterm labor in women with preterm PROM [139–144]. In addition, it has been demonstrated that IL-6 stimulates prostaglandin production by human amnion and decidual cells in primary cultures [145]. We have measured amniotic fluid IL-6 bioactivity in women in preterm labor with and without intact membranes using the hepatocyte stimulating factor assay (in Hep3B2) and sodium dodecyl sulfate-polyacrylamide gel electrophoresis SDS-PAGE/immunoblot assay. Low levels of IL-6 were detected in the amniotic fluid of normal women in the second and third trimester of pregnancy. However, women with preterm labor with intraamniotic infection had higher amniotic fluid levels of IL-6 than women in preterm labor without intraamniotic infection (Figs. 7,8) [146].

These findings have been confirmed by measuring amniotic fluid IL-6 concentration using a two-site immunoassay in 146 consecutive patients admitted with the diagnosis of preterm labor [147].

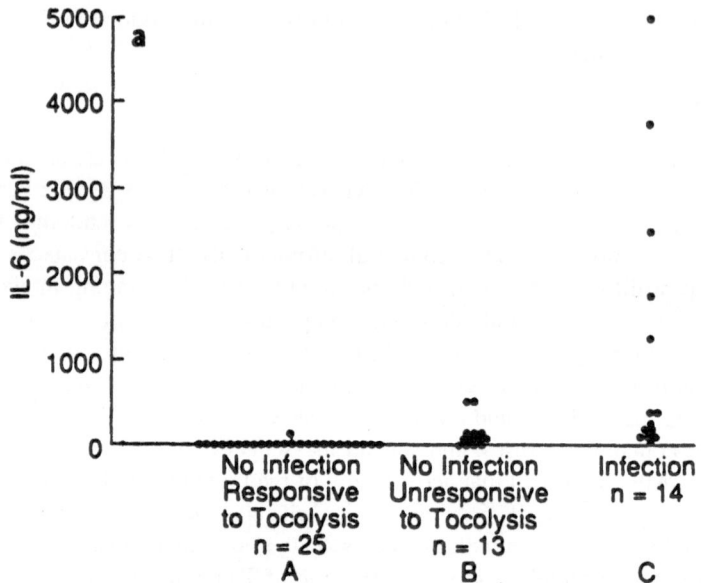

Fig. 7. Amniotic fluid concentrations of IL-6 in women with preterm labor. Amniotic fluid culture was negative, positive response to tocolysis (**A**). Amniotic fluid culture was negative, no response to tocolysis (**B**). Amniotic fluid culture was positive, no response to tocolysis (**C**). Differences between groups were significant (Kruskal Wallis, $p < 0.0001$)

Recently, Matsuzaki and coworkers examined the fetal production of IL-6 both in vivo and in vitro, measuring IL-6 concentrations in cord sera from normal infants and from infants with intraamniotic infections with and without antibiotic treatment and production of IL-6 by placental explants. IL-6 activity was only detected at delivery in those infants with intraamniotic infection who did not receive antibiotic treatment. The major cell population producing IL-6 was the macrophages, as compared with T cells and non-T-cell fractions from mononuclear cells [148]. On the other hand, Liechty et al. studied IL-6 production by fetal and maternal cells before and after stimulation with IL-1. Very low concentrations were found in nonstimulated cells, but higher concentrations were found after stimulation. Interestingly, IL-1 stimulated in vitro production of IL-6 in adult and term neonatal cells

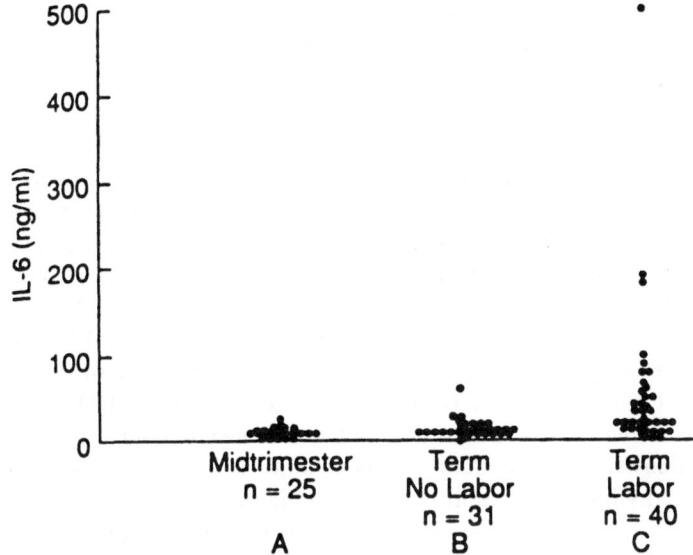

Fig. 8. Amniotic fluid concentrations of IL-6 in pregnant women not in labor and in spontaneous labor at term. Midtrimester (**A**), at term but not in labor (**B**), in spontaneous labor at term (**C**). Differences between groups were significant (Kruskal Wallis, $p < 0.0001$)

but not in those from preterm neonates. After 18–24 h, IL-6 concentrations in supernatants were similar to those from term neonates but were lower in those from preterm cells [149].

The determination of amniotic fluid IL-6 concentrations has been found to be of diagnostic and prognostic value in women with preterm labor and intact membranes and in patients with preterm PROM. An elevated amniotic fluid IL-6 concentrations is more sensitive than other rapid test (including the Gram stain examination of amniotic fluid, amniotic fluid white blood cell count, and glucose determinations) in the detection of microbial invasion of the amniotic cavity. Moreover, patients with an elevated amniotic fluid IL-6 concentration are at risk for impending preterm delivery, despite tocolysis, and have a higher rate of neonatal morbidity and mortality. In women with preterm PROM, those with an elevated amniotic fluid IL-6 determination have a shorter amniocentesis to delivery interval than patients with low amniotic fluid IL-6 concentrations [147,150–151].

8.6.2.4 Interleukin-8

IL-8, or neutrophil attractant/activating peptide-1, is a cytokine that is capable of inducing selective neutrophil chemotaxis and activation. This cytokine has been purified from conditioned media obtained from bacterial endotoxin (LPS)-stimulated human mononuclear cells, LPS-stimulated lung macrophages, mitogen-stimulated lymphocytes, and virus-infected fibroblasts. The IL-8 gene is placed at position 4q12-q21 on the human genome in a gene cluster with other members of the platelet factor-4 superfamily. We have studied IL-8 in the context of preterm labor and term labor and found that elevated amniotic fluid concentrations of IL-8 are found in both term and preterm parturition (with and without infection) [152]. Microbial invasion was associated with elevated concentrations of this cytokine in amniotic fluid.

Recent evidence indicates that the sex steroid hormones can regulate the production of IL-8 by intrauterine tissues. Specifically, progesterone inhibits and antiprogestins (mifepristone) stimulate the production of this cytokine by choriodecidual explants [153].

The precise role of IL-8 in human parturition remains to be established. Since cervical ripening is associated with an influx of inflammatory cells into the cervix, it is possible that cytokines with chemotactic properties play a physiologic role in this process. Recently, Chwalitz et al. have demonstrated that the local administration of IL-8 in the cervix can induce cervical ripening in the guinea pig [154]. We have also been able to induce similar changes in the rabbit by injecting this cytokine into the cervix (R. Romero et al., unpublished observations). Although IL-8 does not stimulate prostaglandin production by intrauterine tissues, it potentiates the stimulatory effect of other cytokines such as IL-1 in prostaglandin production (M.D. Mitchell, personal communication).

8.6.2.5 Colony Stimulating Factors

Another group of cytokines which may be important in parturition are the colony stimulating factors (CSFs). These peptides are a family of glycoprotein growth factors which regulate the survival, proliferation, and differentiation of hematopoietic stem cells as well as the functional activities of mature cells. According to the cell type that they stimulate, CSFs are classified into four different groups: (1) macrophage CSFs (M-CSF or CSF-1) stimulate the differentiation of stem cells along the monocyte/macrophage lineage; (2) granulocyte–macrophage CSF (GM-CSF) stimulates granulocyte and monocyte colony formation; (3) granulocyte CSF (G-CSF) stimulates only the formation of granulocytic colonies, and (4) multi-CSF (or IL-3) stimulates pluripotent stem cells leading to a mixture of cell types. We have demonstrated that human decidual explants are capable of producing CSF-1 and GM-CSF. CSF-1 is normally present in amniotic fluid and its concentration increases with gestational age. Spontaneous parturition at term is associated with an increased CSF-1 concentration in amniotic fluid and its concentration is also increased during preterm labor with intraamniotic infection [155]. We have proposed that one of the roles of CSF-1 in pregnancy is to regulate the proliferation and functional state of decidual macrophages.

8.6.2.6 Platelet Activating Factor

Other noncytokine bioactive agents secreted during the inflammatory process may also participate in parturition associated with infection. Platelet activating factor (PAF) is produced by the fetus and is present in the amniotic fluid from women with preterm labor; this lipid is capable of stimulating prostaglandin E_2 production by amnion and stimulating myometrial contractions directly [156]. The secretion of PAF-acetylhydrolase (the enzyme responsible PAF inactivation) by decidual macrophages is regulated by bacterial endotoxin and cytokines. Specifically, bacterial endotoxin, IL-1α, IL-1β and TNF inhibit PAF–acethylhydrolase secretion by decidual macrophages. The effect of endotoxin could be partially block by the natural IL-1 receptor antagonist and this cytokine could abolish the effect of IL-1α and IL-1β on PAF–acethylhydrolase secretion by decidual cells (H. Narahara and J.M. Johnston, personal communication).

8.6.2.7 Summary of the Role of Cytokines in Parturition

We have proposed a model in which the initiation of human parturition in the presence of infection is controlled by the host. Systemic maternal infections, such as pyelonephritis, or localized infections, such as deciduitis, could trigger parturition via the monocyte/macrophage system in peripheral blood and human decidua. Preterm labor could, according to this, be considered an event occurring when the intrauterine or maternal environment is hostile to the well-being of the fetus. From this point of view, the initiation of preterm labor may have survival value.

Although, this chapter has focused on parturition associated with infection, cytokines may also play a role in physiologic parturition. For example, we and others have found IL-6 gene expression is enhanced in fetal membranes and myometrium obtained from women in active labor at term without demonstrable infection in the amniotic cavity (R. Romero, P. Shegal, unpublished observations, and reference [157]). Moreover, cervical ripening which is a requirement for both term and preterm parturition has been likened to an inflammatory reaction and thus can be regulated by cytokines. Changes in the biophysical properties of the cervix are associated with modifications in collagen and glycosaminoglycans. IL-1 have been shown to stimulate collagenase activity and also has effects on the metabolism of glycosaminoglycans. In vitro experiments have shown that rabbit uterine cervix from term pregnancy produces more IL-1-like activity than the uterine cervix from nonpregnant animals [158]. IL-8 and other chemotactic cytokines may play a central role in the regulation of the influx of inflammatory cells into the cervical stroma at the time of parturition/cervical ripening.

Cytokines may also participate in physiologic and pathologic rupture of membranes. The pathophysiology of PROM may be similar to that of preterm labor. Indeed, the membranes consist mainly of connective tissue. Bacterial infection, directly or indirectly (via IL-1 and TNF), may induce the release of proteases (collagenase, elastases, etc.) from macrophages or other cell types which degrade the fetal membranes and lead to rupture. The reason why some intraamniotic infections result in preterm labor and others in PROM remains to be determined. We consider them as two different expressions of the same basic phenomenon; activation of the host–defense macrophage system.

Infection seems an important, but not the only, etiology for this chain of events.

8.7 Potential Role of Cytokines in the Inhibition of Human Parturition

Although much of the interest in cytokines has focused on the potential role of molecules on prostaglandin production, myometrial contractility, cervical ripening and membrane rupture, it is likely that cytokines which down regulate the inflammatory process (i.e., IL-1 receptor antagonist, transforming growth factor β, TGF-β) may also play an important role in the control of parturition. Moreover, they may represent a novel approach to the control of preterm parturition. This section will review the available data on the natural IL-1 receptor antagonist and transforming growth factor β.

8.7.1 The IL-1 Receptor Antagonist

This cytokine is a member of the IL-1 gene family that binds to the IL-1 receptors but has no agonist IL-1 activity. Exogenous administration of the IL-1 receptor agonist (IL-1ra) blocks virtually all IL-1 properties both in vitro and in vivo, including IL-1-induced prostaglandin production by several cell types [159–164]. Since this property may have potential therapeutic implications for blocking IL-1-induced prostaglandin production by intrauterine tissues associated with intrauterine infection, we have recently studied the concentrations of the IL-1ra in human amniotic fluid [165]. We found that IL-1ra is a physiologic constituent of human amniotic fluid, an observation that is in keeping with our previous work which demonstrated that human amniotic fluid contained an inhibitor(s) to IL-1 bioactivity. Moreover, the concentrations of IL-1ra detected in normal amniotic fluid are among the highest that have been observed in any biologic fluid examined thus far (even in pathologic states), including human plasma concentrations after experimental endotoxemia and in critically ill patients [166,167]. Interestingly, no changes in amniotic fluid concentrations of IL-1ra was found in those patients with intraamniotic infections, des-

228 Roberto Romero et al.

pite a dramatic increase in IL-1α and IL-1β. The IL-1ra to IL-1 ratio in women with preterm labor and intraamniotic infection was 2.7:1, a ratio that clearly is insufficient to prevent IL-1-induced prostaglandin production by intrauterine tissues in vitro. Similar observation have been recently reported in monkeys with experimental intraamniotic infection [168].

We also studied the effect of IL-1ra on IL-1-induced prostaglandin E$_2$ production by human amnion and chorion primary cultures and found that the IL-1ra reduced IL-1-prostaglandin production in a dose-dependent manner and a complete blocking effect for amnion cells was achieved at a ratio of 1000:1. These results support a role for the IL-1ra in term and preterm human parturition, and that adjunctive administration of anticytokine agents may be of value in the treatment of preterm labor [165]. Recently, we have reported that the IL-1 ra can block IL-1 induced preterm parturition in pregnant mice [169].

8.7.2 Transforming Growth Factor-β

TGF-β is a multifunctional peptide with potent immunosuppressive activities, including inhibition of immunostimulatory properties of IL-1β. TGF-β and TGF-β mRNA have been isolated from term human placenta and fetal membranes [170]. Recently, Bry and Hallman have demonstrated that TGF-β is able to reduce IL-1 and TNF prostaglandin production by human amnion [171]. Moreover, TGF-β was able to reduce also the IL-1 plus TNF synergistic effect in amnion cells by 80%. Moreover, treatment with TGF-β can block IL-1-induced preterm parturition in rabbits [172].

More studies are required to determine if naturally occurring cytokines (i.e., IL-1ra and TGF-β) or anticytokine agents can play a role in the treatment of preterm parturition.

References

1. Bang B. J Comp Path and Therap 1897; 10:125, as quoted in Thiersch JB (1962) Effect of lipopolysaccharides of gram negative bacilli in the rat litter in utero. Proc Soc Exper Biol Med 109:429–437
2. Zahl PA, Bjerknes C (1943) Induction of decidua-placental hemorrhage in mice by the endotoxins of certain gram-negative bacteria. Proc Soc Exper Biol Med 54:329–332
3. Takeda Y, Tsuchiya I (1953) Studies on the pathological changes caused by the injection of the Shwartzman filtrate and the endotoxin into pregnant rabbits. Jap J Exper Med 21:9–16
4. Takeda Y, Tsuchiya I (1953) Studies on the pathological changes caused by the injection of the Shwartzman filtrate and the endotoxin into pregnant animals. II. On the relationship of the constituents of the endotoxin and the abortion-producing factor. Jap J Exper Med 23:105–110
5. Rieder RF, Thomas L (1960) Studies on the mechanisms involved in the production of abortion by endotoxin. J Immunol 84:189–193
6. McKay DG, Wong T-C (1963) The effect of bacterial endotoxin on the placenta of the rat. Am J Pathol 42:357–377
7. Kullander S (1977) Fever and parturition: An experimental study in rabbits. Acta Obstet Gynecol Scand Suppl 66:77–85
8. Skarnes RC, Harper MJK (1972) Relationship between endotoxin-induced abortion and the synthesis of prostaglandin F. Prostaglandins 1:191–201
9. Finland M, Dublin TD (1939) Pneumococcic pneumonias complicating pregnancy and the puerperium. JAMA 112:1027–1032
10. Oxorn H (1955) The changing aspects of pneumonia complicating pregnancy. Am J Obstet Gynecol 70:1057–1063
11. Benedetti TJ, Valle R, Ledger WJ (1976) Antepartum pneumonia in pregnancy. Am J Obstet Gynecol 144:413–507
12. Madinger NE, Greenspoon JS, Ellrodt AG (1989) Pneumonia during pregnancy: Has modern technology improved maternal and fetal outcome. Am J Obstet Gynecol 161:657–662
13. McLane CM (1939) Pyelitis of pregnancy: A five-year study. Am J Obstet Gynecol 38:117–123
14. Kass E (1962) Maternal urinary tract infection. New York State J Med 1:2822–2826
15. Cunningham FG, Morris GB, Mikal A (1973) Acute pyelonephritis of pregnancy: A clinical review. Obstet Gynecol 42:112–117
16. Fan YD, Pastorek JG, Miller JM, Mulvey J (1987) Acute pyelonephritis in pregnancy. Am J Perinatol 4:324–326
17. Wing ES, Troppoli DV (1930) The intrauterine transmission of typhoid. JAMA 95:405

18. Diddle AW, Stephens RL (1938) Typhoid fever in pregnancy: probable intrauterine transmission of the disease. Am J Obstet Gynecol 38:300
19. Stevenson CS, Glasko AJ, Gillespie EC (1951) Treatment of typhoid in pregnancy with chloramphenicol (chloromycetin). JAMA 146:1190
20. Herd N, Jordan T (1981) An investigation of malaria during pregnancy in Zimbabwe. C Afr J Med 27:62–68
21. Gilles HM, Lawson JB, Sibelas M, Voller A, Allan N (1969) Malaria, anaemia and pregnancy. Ann Trop Med Pharmacol 63:245
22. Rioux-Darrieulat F, Parant M, Chedid L (1978) Prevention of Endotoxin-Induced Abortion by Treatment of Mice with Antisera. J Infect Dis 137:7
23. Dombrowski RA, Woodward DS, Harper JK et al (1990) A rabbit model for bacterial-induced preterm pregnancy loss. Am J Obstet Gynecol 163:1938–1943
24. Sereno JA, Poseiro JJ, Sica-Blanco Y, Pose SV (1985) Stimulatory effects of chills on the contractility of the pregnant human uterus. Int Congr Phys Sci Buenos Aires, Argentina, August 9–5, 1959
25. Romero R, Mitchell MD, Duff GW, et al. A possible mechanism for premature labor in gram-negative maternal infection: A monocyte product stimulates prostaglandin release by the amnion. Presented at the 32nd Meeting of the Society for Gynecologic Investigation, Phoenix, AZ, March 20–23, 1985
26. Romero R, Mazor M. Infection and preterm labor. Clin Obstet Gynecol 1988;31:553–584
27. Gibbs RS, Blanco JD, St Clair PJ, Castaneda YS (1982) Quantitative bacteriology of amniotic fluid from women with clinical intraamniotic infection at term. J Infect Dis;145:1–8
28. Romero R, Sirtori M, Oyarzun E, et al. (1989) Infection and labor. V. Prevalence, microbiology, and clinical significance of intraamniotic infection in women with preterm labor and intact membranes. Am J Obstet Gynecol;817–824
29. Blanc WA (1953) Infection amniotique et neonatal. Gynaecologia;136:101–104
30. Blanc WA (1964) Pathways of fetal and early neonatal infection: Viral placentitis, bacterial and fungal chorioamnionitis. J Pediatr ;59:473–496
31. Benirschke K, Clifford SH (1959) Intrauterine bacterial infection of the newborn infant. J Pediat;54:11–18
32. Driscoll SG (1965) Pathology and the developing fetus. Pediatr Clin North Am;12:493–514
33. Benirschke K (1965) Routes and types of infection in the fetus and the newborn. Am J Dis Child;28:714–721
34. Romero R, Fayek S, Avila C, et al. The prevalence and microbiology of intraamniotic infection in twin gestation with preterm labor. Am J Obstet Gynecol 1990;163:757–761

35. Galask RP, Varner MW, Petzold CR, et al. (1984) Bacterial attachment to the chorioamniotic membranes. Am J Obstet Gynecol;148:915–928
36. Schoonmaker JN, Lawellin DW, Lunt B, McGregor JA (1989) Bacteria and inflammatory cells reduce chorioamnionic membrane integrity and tensile strength. Obstet Gynecol;74(4):590–596
37. Leigh J, Garite TJ (1986) Amniocentesis and the management of premature labor. Obstet Gynecol;67:500–506
38. Altshuler G, Hyde S (1988) Clinicopathologic considerations of fusobacteria chorioamnionitis. Acta Obstet Gynecol Scand;67:513–517
39. Hillier SL, Martius J, Krohn M, Kiviat N, Holmes KK, Eschenbach DA (1988) A case-control study of chorioamnionic infection and histologic chorioamnionitis in prematurity. N Engl J Med;319:972–8
40. Thorp Jr JM, Katz VL, Fowler LJ, Kurtzman JT, Bowes Jr WA (1989) Fetal death from chlamydial infection across intact amniotic membranes. Am J Obstet Gynecol; 161:1245–6
41. Thomas GB, Jones J, Sbarra A, et al. (1990) Isolation of *Chlamydia trachomatis* from the amniotic fluid. Obstet Gynecol;76:519–20
42. Pao CC, Lao SM, Wang HC, et al. (1991) Intraamniotic detection of *Chlamydia trachomatis* deoxyribonucleic acid sequences by polymerase chain reaction. Am J Obstet Gynecol;164:1295–99
43. Arias F, Tomich P (1982) Etiology and outcome of low birth weight and preterm infants. Obstet Gynecol;60:277
44. Miller JM, Pupkin MJ, Hill GB (1980) Bacterial colonization of amniotic fluid from intact fetal membranes. Am J Obstet Gynecol;136:796–804
45. Bobbit JR, Hayslip CC, Damato JD (1981) Amniotic fluid infection as determined by transabdominal amniocentesis in patients with intact membranes in premature labor. Am J Obstet Gynecol;140:947–52
46. Wallace RL, Herrick CN (1981) Amniocentesis in the evaluation of premature labor. Obstet Gynecol;57:483–486
47. Hameed C, Tejani N, Verma UL, et al. (1984) Silent chorioamnionitis as a cause of preterm labor refractory to tocolytic therapy. Obstet Gynecol;149:726–730
48. Wahbeh CJ, Hill GB, Eden RD, et al. (1984) Intraamniotic bacterial colonization in premature labor. Am J Obstet Gynecol;148:739–743
49. Weible DR, Randall HW (1985) Evaluation of amniotic fluid in preterm labor with intact membranes. J Reprod Med;30:777–780
50. Gravett MG, Hummel D, Eschenbach DA, et al. (1986) Preterm labor associated with subclinical amniotic fluid infection and with bacterial vaginosis. Obstet Gynecol;67:229–237
51. Iams JD, Clapp DH, Contos DA, et al. (1987) Does extra-amniotic infection cause preterm labor? Gas-liquid chromatography studies of amniotic fluid in amnionitis, preterm labor, and normal controls. Obstet Gynecol;70:365–368

52. Duff P, Kopelman JN (1987) Subclinical intraamniotic infection in asymptomatic patients with refractory preterm labor. Obstet Gynecol;69:L756–759

53. Romero R, Emamian M, Quintero R, et al. (1988) The value and limitations of the gram stain examination in the diagnosis of intraamniotic infection. Am J Obstet Gynecol;159:114–119

54. Skoll MA, Moretti ML, Sibai BM (1989) The incidence of positive amniotic fluid cultures in patients in preterm labor with intact membranes. Am J Obstet Gynecol;161:813–816

55. Garite TJ, Freeman RK, Linzey EM, et al. (1979) The use of amniocentesis in patients with premature rupture of membranes. Obstet Gynecol;54:226–230

56. Garite TJ, Freeman RK (1982) Chorioamnionitis in the preterm gestation. Obstet Gynecol;59:539–545

57. Cotton DB, Hill LM, Strassner HT, et al. (1984) Use of amniocentesis in preterm gestation with ruptured membranes. Obstet Gynecol;63:38–43

58. Broekhuizen FF, Gilman M, Hamilton PR (1985) Amniocentesis for gram stain and culture in preterm premature rupture of the membranes. Obstet Gynecol;66:316–21

59. Vintzileos AM, Campbell WA, Nochimson DJ, et al. (1986) Qualitative amniotic fluid volume versus amniocentesis in predicting infection in preterm rupture of the membranes. Obstet Gynecol;67:579–83

60. Feinstein ST, Vintzileos AM, Lodeiro JG, et al. (1986) Amniocentesis with premature rupture of membranes. Am J Obstet Gynecol;68:147–52

61. Gonik B, Bottoms SF, Cotton DB (1985) Amniotic fluid volume as a risk factor in preterm premature rupture of the membranes. Obstet Gynecol;65:456–459

62. Romero R, Quintero R, Oyarzun E, et al. (1988) Intraamniotic infection and the onset of labor in preterm premature rupture of membranes. Am J Obstet Gynecol;159:661–666

63. Romero R, Avila C, Sepuleda W (1993) The role of systemic and intrauterine infection in preterm labor. In: Fuchs AR, Stubblefield PG (eds). Preterm Birth: Causes, Prevention, and Management. 2nd ed. New York: McGraw Hill;4:97–136

64. Romero R, Oyarzun E, Nores J, et al. (1993) The prevalence, microbiology and clinical significance of intraamniotic infection in spontaneous parturition at term. J Repro Med; 38:543–548

65. Romero R, Mazor M, Morrotti R, et al. Infection and labor. VII. Microbial invasion of the amniotic cavity in spontaneous rupture of membranes at term. Am J Obstet Gynecol (1992):166;129–33

66. Romero R, Mazor M, Brandt F, et al. (1992) Interleukin-1 and interleukin-1β in preterm and term human parturition. Am J Reprod Immunol;27:117–23

67. Guzick DS, Winn K (1985) The association of chorioamnionitis with preterm delivery. Obstet Gynecol;65:11–16

68. Romero R, Yoon BH, Gonzalez R, et al. (1993) The clinical significance of microbial invasion of the amniotic cavity with mycoplasmas in patients with preterm PROM. Presented at the 40th Annual Meeting of the Society for Gynecologic Investigation, March 31-April 3, 1993, Toronto, Ontario, Canada, p 70, Abstract S4.

69. Overbach AM, Daniel SJ, Cassady G (1970) The value of umbilical cord histology in the management of potential perinatal infection. J Pediatr;76:22–31

70. Maudsley RF, Brix GA, Hinton NA, et al. (1966) Placental inflammation and infection: A prospective bacteriologic and histologic study. Am J Obstet Gynecol;95:648–659

71. Driscoll SG (1973) The placenta and membranes. In: Charles D, Finland M (eds) Obstetric and perinatal infections. Philadelphia: Lea & Febiger, pp 529–539

72. Naeye RL, Peters EC (1980) Causes and consequences of premature rupture of fetal membranes. Lancet;1:(192

73. Pankuch GA, Appelbaum PC, Lorenz RP, et al. (1984) Placental microbiology and histology and the pathogenesis of chorioamnionitis. Obstet Gynecol;64:802–6

74. Aquino TI, Zhan J, Kraus FT, Knefel R, Taff T (1984) Subchorionic fibrin cultures for bacteriologic study of the placenta. Am J Clin Pathol;81:482–486

75. Hillier SL, Martius J, Krohn M, Kiviat N, Holmes KK, Eschenbach DA (1988) A case-control study of chorioamnionic infection and histologic chorioamnionitis in prematurity. N Engl J Med;319:972–978

76. Chellam VG, Rushton DI (1985) Chorioamnionitis and funiculitis in the placentas of 200 births weighing less than 2.5 kg. Br J Obstet Gynaecol;92:808–14

77. Romero R, Salafia CM, Athanassiadis AP, et al. (1992) The relationship between acute inflammatory lesions of the placenta and amniotic fluid microbiology. Am J Obstet Gynecol;166:1382–1388

78. Daikoku NH, Kaltreider DF, Khouzami VA, et al. (1982) Premature rupture of membranes and spontaneous preterm labor: Maternal endometritis risks. Obstet Gynecol;59:13–20

79. McCracken G, Shinefield H (1966) Changes in the pattern of neonatal septicemia and meningitis. Am J Dis Child;112:33–39

80. Buetow KC, Klein SW, Lane RB (1965) Septicemia in premature infants. Am J Dis Child;110:29–41

81. Mitchell MD(1984) The mechanism(s) of human parturition. J Dev Physiol;6:107–18

234   Roberto Romero et al.

82. Casey ML, MacDonald PC (1984) Endocrinology of preterm birth. Clin Obstet Gynecol;27:562–571
83. Casey ML, MacDonald PC(1988) Decidual activation: The role of prostaglandin in labor. In: McNellis D, Challis JRG, MacDonald PC, Nathanielsz PW, Roberts JM (eds) The onset of labor: Cellular and integrative mechanisms. Ithaca, New York: Perinatology Press, pp 141–156
84. Novy MJ, Liggins GC (1980) Role of prostaglandin, Prostacyclin, and thromboxanes in the physiologic control of the uterus and in parturition. In: Heymann MA (eds) Prostaglandins in the perinatal period: Their physiologic and clinical importance. New York: Grune & Stratton, pp 45–66
85. Garfield RE (1984) Control of myometrial function in preterm versus term labor. Clin Obstet Gyneol;27:572–901
86. Mitchell MD, Flint APF, Bibby JG (1978) Plasma concentrations of prostaglandin during late human pregnancy: Influence of normal and preterm labor. J Clin Endocrinol Metabol;46:947–951
87. Sellers SM, Mitchell MD, Bibby JG, et al. (1981) A comparison of plasma prostaglandin levels in term and preterm labour. Br J Obstet Gynaecol;88:362–366
88. Nieder J, Augustin W (1984) Concentrations of prostaglandin in amniotic fluid in premature labor. Z Geburtshilfe Perinatol;188:7–11
89. Weitz CM, Ghodgaonkar RB, Dubin NH, et al. (1986) Prostaglandin F metabolite concentration as a prognostic factor in preterm labor. Obstet Gynecol;67:496–99
90. TambyRaja RL, Salmon JA, Karim SM, et al. (1977) F prostaglandin levels in amniotic fluid in premature labor. Prostaglandin;13:339–48
91. Romero R, Wu YK, Mazor M, et al. (1988) Amniotic fluid prostaglandin E_2 in preterm labor. Prostaglandins Leukot Essent Fatty Acids;34:141–145
92. Romero R, Wu YK, Sirtori M, et al. (1989) Amniotic fluid concentrations of prostaglandin $F_{2\alpha}$, 13,14-dihydro-15-keto-11, 16-cyclo prostaglandin E2 (PGEM-II) in preterm labor. Prostaglandins;37:149–61
93. Romero R, Quintero R, Emamian M, Wan M, Hobbins JC, Mitchell MD (1987) Prostaglandin concentrations in amniotic fluid of women with intraamniotic infection and preterm labor. Am J Obstet Gynecol;157:1461–67
94. Lopez-Bernal A, Hansell DJ, Khong TY, et al. (1989) Prostaglandin E production by the fetal membranes in unexplained preterm labor and preterm labor associated with chorioamnionitis. Br J Obstet Gynaecol;96:1133–39
95. Walsh SW (1989) 5-Hydroxyeicosatetraenoic acid, leukotriene C, and prostaglandin $F_{2\alpha}$ in amniotic fluid before and during term and preterm labor. Am J Obstet Gynecol;161:1352–60

96. Romero R, Emamian M, Wan M, et al. (1987) Increased concentrations of arachidonic acid lipoxygenase metabolites in amniotic fluid during parturition. Obstet Gynecol;70:849–51

97. Romero R, Wu YK, Mazor M, Hobbins JC, Mitchell MD (1988) Increased amniotic fluid leukotriene C_4 concentration in term human parturition. Am J Obstet Gynecol;159:655–57

98. Bennett PR, Elder MG, Myatt L (1987) The effects of lipoxygenase metabolites of arachidonic acid on human myometrial contractility. Prostaglandins;33:837–844

99. Romero R, Mitchell MD (1988) Amniotic fluid 5-hydroxyeicosatetraenoic acid in term labor. Prostaglandins;36:179–187

100. Carraher R, Hahn DW, Ritchie DM, et al. (1983) Involvement of lipoxygenase products in myometrial contractions. Prostaglandins;26:23

101. Romero R, Wu YK, Mazor M, et al. (1988) Amniotic fluid 5-hydroxyeicosatetraenoic acid in preterm labor. Prostaglandins;36:179–87

102. Romero R, Wu YK, Mazor M, et al. (1988) Amniotic fluid arachidonate lipoxygenase metabolites in women with preterm labor. Prostaglandins Leukot Essent Fatty Acids;36:69–75

103. Romero R, Quintero R, Emamian M, et al. (1987) Arachidonate lipoxygenase metabolites in amniotic fluid of women with intraamniotic infection and preterm labor. Am J Obstet Gynecol;157:1454–60

104. Lopez-Bernal A, Hansell DJ, Canete Soler R, et al. (1987) Prostaglandin, chorioamnionitis and preterm labour. Br J Obstet Gynaecol;94:1156–58

105. Folco G, Hansson G, Granstrom E (1981) Leukotriene C_4 stimulates TXA_2 formation in isolated sensitized guinea pig lungs. Biochem Pharmacol;30:2491

106. Feuerstein, Foegh NM, Ramwell PW (1981) Leukotrienes C_4 and D_4 induce prostaglandin and thromboxane release from rate peritoneal macrophages. Br J Pharmacol;72:389

107. Serhan CN, Fridovich J, Goetzl EJ, et al. (1982) Leukotriene B_4 and phosphatidic acid are calcium ionophores: Studies employing arsenazo III in liposomes. J Biol Chem;257:4746

108. Bejar R, Curbelo V, Davis C, et al. (1981) Premature labor: Bacterial sources of phospholipase. Obstet Gynecol;57:479–482

109. Bennett PR, Rose MP, Myatt L (1987) Preterm labor: Stimulation of arachidonic acid metabolism in human amnion by bacterial products. Am J Obstet Gynecol; 156:649–655

110. Lamont RF, Rose M, Elder MG (1985) Effects of bacterial production prostaglandin E production by amnion cells. Lancet; 2:1131–1133

111. McGregor JA, Lawellin D, Franco-Buff A (1985) Phospholipase A2 activity of genital tract flora detected with two substrates. Society for Gynecologic Investigation, Phoenix, March 20–23

236   Roberto Romero et al.

112. Romero R, Kadar N, Hobbins JC, et al. (1987) Infection and labor: The detection of endotoxin in amniotic fluid. Am J Obstet Gynecol; 157:815–9
113. Romero R, Hobbins JC, Mitchell MD (1988) Endotoxin stimulates prostaglandin E_2 production by human amnion. Obstet Gynecol; 71:227–8
114. Romero R, Roslansky P, Oyarzun E, et al. (1988) Labor and infection. II. Bacterial endotoxin in amniotic fluid and its relationship to the onset of preterm labor. Am J Obstet Gynecol; 158:1044–1049
115. Lamont RF, Anthony F, Myatt L, Booth L, Furr PM, Taylor-Robinson D (1990) Production of prostaglandin E2 by human amnion in vitro in response to addition of media conditioned by microorganisms associated with chorioamnionitis and preterm labor. Am J Obstet Gynecol; 162(3):819–825
116. Romero R, Edwin S, Avila C, Foster J, Wu YK, Mitchell MD (1990) Prostaglandin production by amnion and decidual cells in response to bacterial products. Presented at the 36th Annual Meeting of the Society for Gynecologic Investigation, Houston, Texas, January 23–27
117. Dudley DJ, Chen CL, Hill RH, and Mitchell MD (1993) Effects of group B streptococci on prostaglandin production by fetal gestational tissues: Strain-specific differences. Presented at the 40th Annual Meeting of the Society for Gynecologic Investigation, March 31–April 3, Toronto, Ontario, Canada, p 279, Abstract P194
118. Dinarello CA (1984) Interleukin-1. Rev Infect Dis; 6:51–95
119. Romero R, Durum S, Dinarello C, et al. (1986) Interleukin-1: A signal for the initiation of labor in chorioamnionitis. Presented at the 33rd Annual Meeting of the Society for Gynecologic Investigation, Toronto, March 19–22
120. Romero R, LaFreniere D, Duff GW, et al. (1985) Human decidua: A potent source of interleukin-1-like activity. Presented at the 32nd Annual Meeting of the Society for Gynecologic Investigation, Phoenix, AZ, March 20–23
121. Romero R, Wu YK, Brody DT, Oyarzun E, Duff GW, Durum SK (1989) Human decidua: A source of Interleukin-1. Obstet Gynecol; 73:31–34
122. Romero R, Brody DT, Oyarzun E, et al. (1989) Infection and labor. III. Interleukin-1: A signal for the onset of parturition. Am J Obstet Gynecol; 160:1117–1123
123. Romero R, Mazor M, Brandt F, et al. (1992) Interleukin-1α and interleukin-1β in human preterm and term parturition. Am J Reprod Immunol;27:117–123
124. Romero R, Mazor M, Tartakovsky B. (1991) Systemic administration of interleukin-1 induces preterm parturition in mice. Am J Obstet Gynecol;165:969–971

125. Bukowski R, Scholz P, Hasan S, Chwalisz K (1993) Induction of preterm parturition with the interleukin-1β (IL-1β), tumor necrosis factor-α (TNF-α) and with LPS in guinea pigs. Presented at the 40th Annual Meeting of the Society for Gynecologic Investigation, March 31-April 3, Toronto, Ontario, Canada, p 81, Abstract S26

126. Kauma SW (1989) HLA-DR and interleukin-1β (IL-1β) mRNA expression in human decidua. Presented at the 36th Annual Meeting of the Society for Gynecologic Investigation, March 15–18, San Diego, California, p.333, Abstract 504

127. Paulesu L, King A, Loke YW, Cintorino M, Bellizzi E, Boraschi D (1991) Immunohistochemical localization of IL-1α and IL-1β in normal human placenta. Lymphokine Cytokine Res;10:443–48

128. Taniguchi T, Matsuzaki N, Kameda T, et al. (1991) The enhanced production of placental interleukin-1 during labor and intrauterine infection. Am J Obstet Gynecol;165;131–37

129. Hertelendy F, Todd H, Molnar M, Romero R (1993) Cytokine-initiated signal transduction in human myometrium. Presented at the 40th Annual Meeting of the Society for Gynecologic Investigation, March 31–April 3, Toronto, Ontario, Canada, p 131, Abstract S125

130. Bry K, Lappalainen U, Hallman M (1992) Interleukin-1 binding and prostaglandin E₂ synthesis by amnion cells in culture: Regulation by tumor necrosis factor-α, transforming growth factor-β, and interleukin-1 receptor antagonist. Biochimica et Biophysica Acta 1–6

131. Dinarello CA (1987) Clinical relevance of interleukin-1 and its multiple biological activities. Bull Inst Pasteur; 85:267

132. Romero R, Mazor M, Manogue K, et al. (1991) Human decidua: A source of tumor necrosis factor. Eur J Obstet Gynecol Reprod Biol;41:123–127

133. Casey ML, Cox SM, Beutler B, Milewich L, MacDonald PC (1989) Cachectin/tumor necrosis factor-formation in human decidua. J Clin Invest; 83:430–436

134. Gauldie J, Richards C, Harnish D, Lansdcorp P, Baumann H (1987) Interferon 2/B-cell hepatocyte-stimulatory factor type 2 shares identity with monocyte-derived hepatocyte stimulating factor and regulates the major acute phase protein response in liver cells. Proc Natl Acad Sci USA; 84:7251–7255

135. Romero R, Manogue KR, Murray MD, et al. (1989) Infection and labor. IV. Cachectin tumor necrosis factor in the amniotic fluid of women with intraamniotic infection and preterm labor. Am J Obstet Gynecol; 161:336–341

136. Romero R, Mazor M, Sepulveda W, et al. (1992) Tumor necrosis factor in term and preterm labor. Am J Obstet Gynecol;166:1576–1587

137. Silver RM, Lohner S, Chen CL, Mitchell MD, Branch DW (1993) Tumor necrosis factor-α (TNF-α) mediates LPS-induced abortion: Evi-

dence from the LPS-Resistant murine strain, C3H/HeJ. Presented at the
40th Annual Meeting of the Society for Gynecologic Investigation,
March 31-April 3, Toronto, Ontario, Canada, p 218, Abstract P71
138. Bry K, Hallman M (1991) Synergistic stimulation of amnion cell prostaglandin E_2 synthesis by interleukin-1, tumor necrosis factor and products
from activated human granulocytes. Prostaglandins Leukot Essent Fatty
Acids;44:241–45
139. Evans MI, Hajj SN, Devoe LD, et al. (1980) C-reactive protein as a predictor of infectious morbidity with premature rupture of membranes. Am
J Obstet Gynecol;138:648–52
140. Farb HF, Arnesen M Geistler P, et al. (1983) C-reactive protein with premature rupture of membranes and premature labor. Obstet Gynecol;62:49–51
141. Hawrylyshyn P, Bernstein P, Milligan JE, et al. (1983) Premature rupture
of membranes: The role of C-reactive protein in the prediction of chorioamnionitis. Am J Obstet Gynecol;147:240–246
142. Romem Y, Artal R (1984) C-reactive protein as a predictor for chorioamnionitis in cases of premature rupture of the membranes. Am J Obstet Gynecol;150:546–550
143. Handwerker SM, Tejani NA, Verma UL, et al. (1984) Correlation of
maternal serum C-reactive protein with outcome of tocolysis. Obstet
Gynecol;63:220–24
144. Potkul RK, Moawad AH, Ponto KL (1985) The association of subclinical infection with preterm labor: The role of C-reactive protein. Am J
Obstet Gynecol;153:642–45
145. Mitchell MD, Dudley DJ, Edwin SS, Lundin Schiller S (1991) Inter-
leukin-6 stimulates prostaglandin production by human amnion and decidual cells. Eur J Pharmacol;(192:189–91
146. Romero R, Avila C, Santhanam U, Sehgal P (1990) Amniotic fluid interleukin 6 in preterm labor: Association with Infection. J Clin Invest;85:1392–1400
147. Romero R, Sepulveda W, Kenney JS, Archer LE, Allison AC, Sehgal PB
(1992) Amniotic fluid interleukin-6 determination are of diagnostic and
prognostic value in premature labor. Presented at the 39th Annual Meeting of the Society for Gynecologic Investigation, March 18–21, San
Antonio, TX, p 334, Abstract 452
148. Matsuzaki N, Saji F, Kameda T, Yoshizaki K, Okada T, Sawai K, Tanizawa O (1990) In vitro and in vivo production of interleukin-6 by fetal
mononuclear cells. Placenta;11:205–213
149. Liechty KW and Christensen RD (1990) In vivo effect of IL-6 on cycling
status of hematopoietic progenitors from adults and neonates. Pediatr
Res;28:323–26
150. Romero R, Yoon BH, Sepulveda W, et al. (1993) Preterm labor and intact membranes: The diagnostic and prognostic value of amniotic fluid

white blood cell count, glucose determination, interleukin-6, and gram stain. Am J Obstet Gynecol;168:311 (A54)

151. Romero R, Yoon BH, Baumann P, et al. (1993) Which is the best rapid test for the evaluation of the patient with preterm PROM? A comparison of amniotic fluid (AF) glucose, AF-white blood cell count. Am J Obstet Gynecol;168:318 (A72)

152. Romero R, Ceska M, Avila C, Mazor M, Behnke E, Lindley I (1991) Neutrophil attractant/activating peptide-1/interleukin-8 in term and pre-term parturition. Am J Obstet Gynecol;165:813–30

153. Kelly RW, Leask R, Calder AA (1992) Choriodecidual production of in-terleukin-8 and mechanism of parturition. Lancet;339:776–77

154. Chwalisz K, Scholz P, Hegele-Hartung Ch, Roth G, Bukowski R (1993) Cervical ripening with the interleukin 1β (IL-1β) and tumor necrosis fac-tor-α (TNF-α) in pregnant guinea pigs. Presented at the 40th Annual Meeting of the Society for Gynecologic Investigation, March 31–April 3, Toronto, Ontario, Canada, p.82, Abstract S27

155. Romero R, Oyarzun E, Stanley ER (1989) Macrophage colony-stimulat-ing factor in amniotic fluid. Presented at the 36th Annual Meeting of the Society for Gynecologic Investigation, March 15–18, San Diego, Cali-fornia

156. Hoffman DR, Romero R, Johnston JM (1990) Detection of platelet-acti-vating factor in amniotic fluid of complicated pregnancies. Am J Obstet Gynecol;162:525–28

157. Dudley DJ, Collmer D, Mitchell MD, and Trautman MS (1993) Detec-tion of inflammatory cytokine mRNA from gestational tissues utilizing polymerase chain reaction (PCR). Presented at the 40th Annual Meeting of the Society for Gynecologic Investigation, March 31–April 3, To-ronto, Ontario, Canada, p 220, Abstract P76

158. Ito A, Hiro D, Ojima Y, et al. (1988) Spontaneous production of inter-leukin-1 like factors from pregnant rabbit uterine cervix. Am J Obstet Gynecol;159:261

159. Dinarello CA. Interleukin-1 and interleukin-1 antagonism. Blood;77:1627–1652

160. Dinarello CA, Thompson RC (1991) Blocking IL-1: Interleukin 1 recep-tor antagonist in vivo and in vitro. Immunol Today (1991);12:404–410

161. Arent WP (1991) Interleukin 1 receptor antagonist: A new member of the interleukin family. J Clin Invest;88:1445–1451

162. Balavoine JF, de Rochemonteix B, Williamson K, Seckinger P, Cru-chaud A, Dayer JM (1986) Prostaglandin E$_2$ and collagenase production by fibroblasts and synovial cells is regulated by urine-derived human in-terleukin 1 and inhibitor(s). J Clin Invest;78:1120–1124

163. Arend WP, Welgus HG, Thompson RC, Eisenberg SP (1990) Biological
     properties of recombinant human monocyte-derived interleukin 1 recep-
     tor antagonist. J Clin Invest;85:1694–7
164. Fidel PL, Romero R, Ramirez M, et al. (1993) Bacterial endotoxin stimu-
     lates the production of the natural IL-1 receptor antagonist by human de-
     cidua. Presented at the 40th Annual Meeting of the Society for Gyneco-
     logic Investigation, March 31–April 3, Toronto, Ontario, Canada, p 81,
     Abstract S25
165. Romero R, Sepulveda W, Mazor M, et al. (1992) The natural interleukin-
     1 receptor antagonist in term and preterm parturition. Am J Obstet Gyne-
     col;167:863–72
166. Granowitz EV, Santos AA, Poutsiaka DD, et al. (1991) Production of in-
     terleukin-1 receptor antagonist during experimental endotoxemia. Lan-
     cet;338:1423–24
167. Fischer E, Marano MA, van Zee KJ, et al. (1992) Interleukin-1 receptor
     blockade improves survival and hemodynamic performance in *Escheri-
     chia coli* septic shock, but fails to alter host responses to sublethal endo-
     toxemia. J Clin Invest;89:1551–57
168. Gavett MG, Witkin SS, Haluska GJ, Baggia S, Cook MJ, Novy MJ
     (1993) Stimulation of inteleukin-1 receptor antagonist protein (IRAP) by
     experimental intraamniotic infection or IL-1b infusion in Rhesus mon-
     keys. Presented at the 40th Annual Meeting of the Society for Gyneco-
     logic Investigation, March 31–April 3, Toronto, Ontario, Canada, p 180,
     Abstract S224.
169. Romero R, Tartakovsky B (1992) The natural interleukin-1 receptor an-
     tagonist prevents interleukin-1-induced preterm delivery in mice. Am J
     Obstet Gynecol;167:1041–45
170. Kauma S, Matt D, Strom S, Eierman D, Turner T (1990) Inteleukin-1β,
     human leukocyte antigen HLA-DRa, and transforming growth factor-β
     expression in endometrium, placenta, and placental membranes. Am J
     Obstet Gynecol;163:1430–37
171. Bry K, Hallman M (1992) Transforming growth factor-β opposes the
     stimulatory effects of interleukin-1 and tumor necrosis factor on amnion
     cell prostaglandin E$_2$ production: Implication for preterm labor. Am J
     Obstet Gynecol;167:222–26
172. Bry K, Hallman M (1993) Transforming growth factor-β2 prevents pre-
     term delivery inducted by interleukin-1a and tumor necrosis factor-α in
     the rabbit. Am J Obstet Gynecol; 168:1318–1322

9 Immune Control of Pregnancy

Gérard Chaouat, Elisabeth Menu, Valentino Djian,
Genevieve Delage, Duc-Can Dang, Aines Assal Meliani,
Sylvie Ropert, and Jacques Martal

9.1 Caveats

As Peter Johnson would say, "for every syndrome of unknown aetiology, there is always an immunological theory." First, it is my contention that immune events optimize and perhaps at some stage insure the success of pregnancy. Models of immune abortion do exist in animals and a subset of such probably exists in humans; immunological events are also likely involved in labor and threatened preterm delivery. However, it is *not* my contention that labor is an solely immune event, as maintained by some. Second, what is true for *rodents* does not necessarily hold for humans. Third, the reagents used in studies need to be of exquisite specificity and the patient series large. Unfortunately, these

Table 1. Cytokine production in the placenta an decidua

Uterus (second trimester decidua)	Mature placenta
IL-1 α and β +++++	IL-1s and IL-1-like ++++++
IL-2 (weak. passengers lymphos)	"IL-2-like"
IL-3 +++	(Source contradictory)
IL-6 ?	+++++ (Trophoblast + lymphos)
IL-8 ?	
IL 10 ++	++ Source ?
TNFs +++	++
Interferons ++ oTP (see text)	+ (α) γ
CSF 1 +++ to ++++++	++++
GM–CSF	++
TGFs +++ (α and β)	++ (β)

IL, interleukin; *TNF*, tumor necrosis factor; *oTP*, ovine trophoblast protein; *CSF*, colony stimulating factor; *TGF*, transforming growth factor; *GM–CSF*, granulocyte macrophage colony stimulating factor

Table 2. Timing of cytokine production during pregnancy

	day1	day3	PREIMPL	PERIMPL	POSTIMPL	LABOR
IL-1 α and β	++	0		+++	+	Increases ++
IL-3	+	0	++	++	+	?
IL-4	+	0	0	0	0	+ (?)
IL-6	++	0	+	+	+++	?
IL-8	+		+	+	+	?
IL-10	?	?	+?	++	++	?
TNFs	++	0	0		+	+++
Interferons	++	0	0			+++ (?)
oTPs	–	–	+	++++		?
CSF-1	++	+	++	+++	+++++	?
GM–CSF	+	0	0	++	+++	?
TGF α	+++	++	++	++	++	?
TGF β		0	++	++	++	?

For abbreviations see Table 1.

precautions are not always taken, and hence false conclusions are drawn.

9.2 Basic Problem

The fetus is viewed by many as nature's allograft, but before such an intimate connection is established, a whole process of development occurs from blastocyst to embryo, with variations in the antigenic status of the conceptus and preparation of the uterus. Table 1 shows this both for cytokine production and its timing.

9.2.1 Pre- and Peri-implantation Periods:
From Blastocyst to Placenta

In mice and rats, after mating plus fecundation (or at least after a factor present in seminal fluid issues the proper signal), T cells accumulate on day 1 in decidua and then decrease by day 3 (Noun et al. 1989), but not in the fallopian tubes of vasoligatured animals or those pseudopregnant after mating with vasectomized males. In rats, anti-PAF can prevent implantation, and if locally injected, implantation is only successful in the noninjected uterine horn.

Such an "inflammatory-like reaction" in response to spermatozoa depends on a signal in seminal fluid and is correlated by lymphokine secretion, already at this stage both from lymphocytes and cells of the reproductive tract (MacMaster et al. 1992; Sanford and de Wood 1992), e.g., decidual epithelial or glandular cells. The lymphokines involved are interleukin-1 (IL-1), tumor necrosis factor (TNF), IL-6, granulocyte colony stimulating factor (G-CSF), CSF-1, macrophage colony stimulating factor (M-CSF), and IL 3. Levels will, temporarily, return to normal by days 2–3, except for CSFs (see below), leukocyte inhibiting factor–human interleukin for leukocyte differentiation antigen (LIF-HILDA) (see below), and IL-6, traceable in blastocysts supernatants (S/N) in humans (unpublished data and P. Franchimont, University of Liége, and A. Hazout, Hop A. Béclére, personal communication).

The CSFs are important. CSF-1 first deserves mention (Arceci et al. 1989) but CSF-1 deficient oSP mutant mice have an apparently subnormal pregnancy (though there are some defects, pregnancy can pursue). An excess of CSF-1 is abortifacient (Tartakowsky et al. 1991). As for G-CSF, secreted in the pre-, peri-, and postimplantation uterus, its receptor, c kit, is expressed on gametes, repressed (or rather not expressed) on fertilized eggs, and then reexpressed on the two-cell stage, and then in the trophectoderm and the decidua. W locus mutant mice (c kit) are anemic and sterile with sometimes gross abnormalities in embryonic development (Pollard et al. 1991).

Similarly, GM–CSF is synthesized by early uterus and displays, at least in vitro, a key role in early trophoblast attachment and growth (Robertson and Seamark 1991; Tartakowsky et al. 1991).

For example, in a murine model of recurrent abortion, the CBA × DBA/2 (see below) system that we have treated by lymphokines (Chaouat et al. 1990), Tartakowsky has observed that in vitro cultured CBA × DBA/2 embryos had an abnormally high rate of morphologic anomalies, be they obtained from in vitro ferilization (IVF) or from flushed uterine horns. Culture of those embryos in GM–CSF prior to embryo transfer (ET) restored normal resorbtion rates. Such a maneuver also optimized embryo transfer in the more classic C57BL/6 strain.

Finally, lymphokines such as LIF-HILDA, whose use is well documented for culturing extrasomatic stem cells and allowing one to easily make transgenic mice from totipotent transfected cells, are detectable in the decidua at this period, along with the immunosuppressive lymphokine IL-10 (litter size and fetal weight are smaller in IL-10 gene-deficient mice).

Indeed, a very recent paper by Stewart et al. (1992) shows that rendering the mice deficient by "gene knockout" in LIF renders the mice sterile though they *are* fertile: the embryos are normal and can be transferred to normal CD1 recipient mice, where they implant and restoration of successful implantation of mutant embryo in LIF-HILDA-deficient mice can be obtained if LIF-HILDA is given via an osmotic pump. This is a programmed series of events in preparing for the implantation; disruption will cause early pregancy loss.

It has been demonstrated that such phenomena do not depend solely on estrogens since they cannot be mimicked by estradiol in adult ovariectomized animals, whereas they can be partly obtained by mating with vasectomized males, however, mechanical uterine stimuli failing to elicit such a response.

At this stage, the embryo itself is extremely resistant to an immune attack because of the presence of zona pellucida, a lack of major histocompatibility complex (MHC) expression (Dohr 1987), an intrinsic resistance to lysis (Croy and Rossant 1987), and expression, as for the gametes and later on placenta, of complement regulatory proteins (MCP, DAF, CD 56) (Hsi et al. 1991a).

Nothing even approaching consensus exists as regards the various "early immunosuppressive factors" that would predict or correlate with successful pregnancy such as early pregnancy factor (EPF), embryo associated suppressor factor (EASF), and blastocyst suppressor inducer factor.

Since the purpose of this meeting is to deal with labor and, I guess, somehow with early fetal wastage, I would summarize that treatment with a PAF antagonist, BN 52021, does prevent such cellular redistribution and the subsequent continuation of the inflammatory response correlates with early embryo loss (Kachkache et al. 1991) and that one can cause abortion at this stage in mice by γ-interferons, lipopolysaccharide (LPS), double-stranded RNAs (Chaouat et al. 1991; Kinsky et al. 1990).

Once or around when pregnancy is established, one of the main factors in its maintenance is the presence of the corpus luteum. Surprisingly enough, it has been shown that immunological molecules are involved in this. [In sheep oTP 1 trophoblastin is responsible for this, not chorionic gonadotrophin (Martal et al. 1990).] When purified to homogeneity, and to the surprise of many, both amino acid analysis and base comparison studies after cloning of the cDNA have shown that the material had strong analogies with α-interferons (Martal et al. 1991).

In fact, oTP-like materials have been found now in a variety of species and are remarkably conserved, exhibiting higher analogies between themselves across the species barrier than with α- or even ω interferons, defining a new family of interferons. According to a recent communication (Whaley et al. 1991) such materials could exist in human species, since analysis with an oTP probe of a genomic library from human term placenta has allowed detection of putative oTP coding sequences. That these intereferons perform classical functions is shown by their antiviral activities, as well as cytostatic activities on activated T lymphocytes. As such, we have demonstrated that they are immunosuppressive in vitro: recombinant oTP, produced in *Eschericia coli* or in yeast, is as active as natural one and in ovine species, both CD4 and CD8 proliferative functions are impaired.

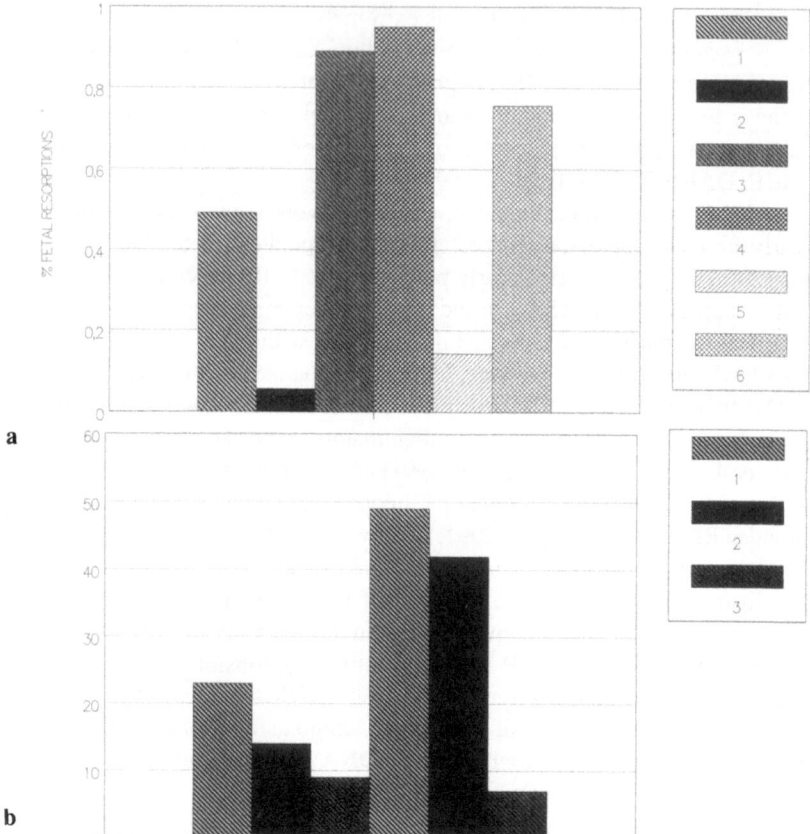

Fig. 1a,b. Enhancement of CBA DBA/2 fertility by r.oTP. **a** Prevention of re-sorbtion by r-oTP. *1,* Control CBA DBA/2, treated; *2,* r-oTP days 5,8; *3,* same but days 8,10 *4,* γ-interferon days 5,8; *5,* r-oTP day 5 only; *6,* α-interferon days 5,8. (α-, γ-, and τ-interferons are given at the same dose i.p.). **b** r-oTP acts mostly on fetal survival. *Left* group untreated; *right* group, treated. *1,* implantation sites; *2,* live fetuses; *3,* resorbed fetus

This suggests that oTP could, besides maintaining corpus luteum, and providing an antiviral barrier, act as an early safeguard against inflammatory events that would/could otherwise compromise a successful implantation, by abnormal uterine death or release of cytotoxic/cytostatic products such as TNF.

In fact, treatment in vivo by r-oTP blocks a local popliteal lymph node (PLN) graft versus host (GVH) assay. More importantly, we have recently tested r-oTP in the CBA × DBA/2 murine abortion model. Since oTP activates like interferons natural killer (NK) cells, although less than γ or α ones (γ more than α itself more than oTP), it is perhaps not surprising that r-oTP given i.p. on day 8 enhances resorbtions in this system. But even more importantly, we mated non-synchronized mice with DBA/2, and r-oTP was given exactly at the putative pre-/peri-implantation period. It apparently corrects resorption rates, whereas even at this period γ-interferon is clearly abortifacient.

In fact, more careful scrutiny of the data reveals that in the experiments the resorption rate itself was only a part of the modified parameters. There were more pregnant mice per group of mated mice and there were more implants (Fig. 1a,b) so that oTP acted more on implantation and perhaps uterus "preparation" than on embryo resorptions themselves.

If these effects were confirmed on a large series of mice in both resorption-prone and nonresorption-prone strains, they would further define the role of oTP in favoring implantation. Of course, since α-interferons may have 2-5-A synthethase-inducing activity and act on prostaglandins (PG) too, such definition requires controls with murine and human α-interferons, which are now being performed.

Having mentioned the role of oTP and its place in immunoendocrine networks, it is necessary to state that the production of human chorionic gonadotrophin (hCG) in humans also seems to be dependent on immune events. At least part of its production is known to be regulated by IL-6, secreted by trophoblast cells in a constitutive fashion (i.e., does not need T or B cell triggering, as is oTP in ovine species) and probably decidual lymphocytes. IL-6 production and hCG production are dependent on TNF (Neki et al. 1991).

It is important to state here that placental supernatant induces production of IL-6 by PBLs (Fig. 2) and that we have been able to trace IL-6 in supernatants of co-cultures of human IVF. P. Franchimont and A. Hazout

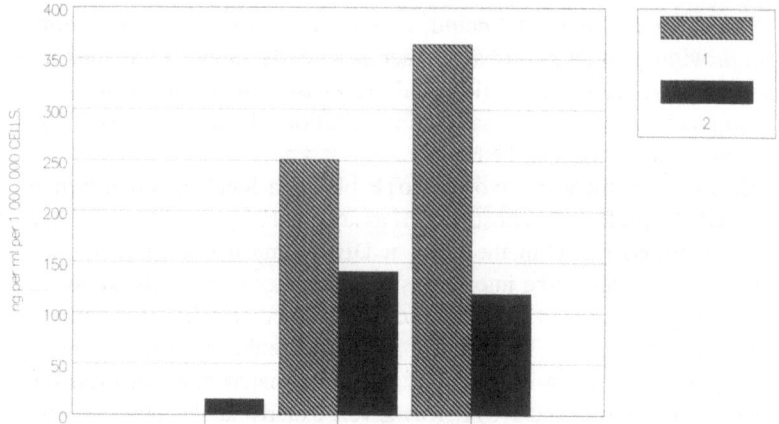

Fig. 2. Induction by a low molecular weight placental factor of IL-6 production by monocytes. *Column 1,* IL-6 control level; *columns 2 and 3,* with batch "C2" of placental material in serum-free *(column 2)* or fetal calf serum (FCS)-containing medium *(column 3); columns 4 and 5,* same in 10% FCS-containing medium

(personal communication) have correlated S/N levels of IL-6 with successful pregnancies, as stated above. We are rechecking these data.

9.3 Established Pregnancy

Let us examine the "alloantigenic" status of the placenta. There are differences between rodents and humans. In rodents, surprisingly, while the labyrinth does not express MHC antigens (as expected from cells which would be on the interface of an antigenically neutral barrier between the mother and the fetus), the outer layer of the placenta, the spongiotrophoblast, expresses (as do the annexes) MHC class I (H-2 K, D, L in mice) in detectable amounts (Singh et al. 1983). The material can be detected at membrane level in situ or on isolated trophoblasts, and proper mRNA transcripts are readily detectable.

In the rat, the same is true: spongiotrophoblasts do express class I MHC, the RT1a antigens, but monomorphic MHC antigens, named Pa1 and Pa2 (allelic forms), are also expressed on the rat placenta, and there is evidence from elegant ET experiments that their expression is influenced by genomic imprinting, biased towards MHC expression of maternally derived class I only being repressed (Kanbour-Sharir et al. 1990).

In mice and rats, those class I MHCs are in direct contact with maternal circulation and are accessible to monoclonal antibodies of the relevant antipaternal specificity, quickly degraded once bound to the placenta in smaller peptidic fragments. This has been called the "paternal strain immunoadsorbant status of the placenta" (Singh et al. 1983).

No class II (Ia antigens) whatsoever are expressed on the placental trophoblast. In fact, abnormal expression of these under the effects of azacytidine always induces abortion. This effect is not due to embryotoxic effects, since anti-Ia monoclonal antibody treatment with antibodies of the relevant specificity protects against such an embryo loss (Athanassakis et al. 1991).

Let us now deal with primates, and, more precisely, humans. The villous placenta has two layers. The outer one, the syncytiotrophoblast, is clearly MHC class I negative. The underlying cytorophoblast is weakly but, for some reports, definitely MHC class I positive, for others, class I negative, too. As in rodents, class II MHCs are never expressed.

The detection of reactivity with the monoclonal antibody W6 32 led first to postulate in the 1980s that HLA-A, -B, -C were expressed in the so-called extravillous placenta. In fact, W6 32 reacts with a conserved portion of classical MHC molecules, and immunoprecipitation studies have shown that the molecule does not show migration variations between individuals when analyzed by the sodium dodecyl sulfate polyacrylamide gel electrophoresis (SDS PAGE) technique. Therefore, the molecule was suspected to be monomorphic and to have some domains deleted, since its size was inappropriate for a complete HLA molecule. It is indeed the case: the molecule has been cloned from choriocarcinomas or trophoblast by two groups and christened HLA-G (Ellis 1990; Kovatts et al. 1990, 1991).

It seems to have evolved distinctly from HLA-A, -B, -C (see Geraghty et al. 1992a,b) as determined by "genetic tree" analysis of HLA-G and its pseudogene companion HLA-J. Its expression is *absolutely* restricted to placental trophoblasts. In baboons, however, it is expressed on villous rather than on extravillous trophoblasts.

This is a typical example of variations between species that should caution us about the general protective function of HLA-G as generally ascribed, and, in general, as discussed in the first lines of this review, about generalization of the immune situations from one animal model to the other.

The same localization problem for HLA-G is seen when observing subspecies of *Macaca fascicularis* vs *Macaca cynolmongus*, for example, to cite widely used animal models.

Yet, HLA-G shows conservation in evolution (it has analogies with Pa). This, and such restricted expression suggests, but does not prove, a role in defense of the fetoplacental unit. Indeed, L cells transfected with HLA-G are much less sensitive to NK cell-mediated lysis than HLA-A, -B or -C transfected controls (Kovatts et al. 1991).

However, recent data by other groups presented at the recent Immunology of Reproduction 5th World Congress in Rome conflict with those data, but this could be due to the cell lineage used, transfection techniques, etc.

If confirmed, the mechanisms of such anti-NK action is not yet known in detail: HLA-G might confuse, trap, or defuse the T cell–receptor complex or present an abnormal, immunoregulatory set of (self?) peptides. Alternatively, HLA-G might play a role in placental functions such as peptide or hormone transport at the interface. At present, there is no convincing proof that disruption of HLA-G or, in rats, of a monomorphic class I gene (Pa Ag) would lead to an immune abortion.

Without getting into anatomical details, the antigenic (MHC) status of porcine and equine placenta is that they also suppress class II and express class I MHC (swine lymphocyte antigens, SLA, or equine lymphocyte antigens, ELA).

The regulation of polymorphic MHC expression in trophoblasts relies upon hypermethylation of DNA and, possibly, action of specific regulatory elements acting on the interferon consensus sequence and NF Kappa b (KBF1) or KBF1-like enhancers interactions (Le Bouteiller et al. 1991; see also Boucraut 1992).

Such is the antigenic status of the established pregnancy.

The threat could come from activated NKs cells or lymphokine-activated killer cells (LAKCs) to which trophoblast is sensitive, or to the breach of NK cells to layers underneath trophoblasts, since embryonic fibroblasts are targets for and lysable by NK cells, *do express* MHC class I antigens, and are a target for putative maternal antipaternal cytolytic T lymphocytes (CTLs).

This requires local immunoregulatory mechanisms. It is now convincingly demonstrated that systemic active suppression is *not* necess-

ary for successful pregnancy, whether mediated by enhancing antibodies or suppressor cells (which both do exist, but are not required for
successful pregnancy outcome).

Indeed, at peri-implantation, the future placenta (the ecto placental
cone, EPC) is MHC negative and, as stated, intrinsically resistant to
NK- and CTL-mediated lysis, as indeed were the gametes themselves
or, as stated above, blastocysts. But it attracts in a short period of time
NK cells at the implantation site.

Among them are, in mice, cells bearing the Asialo GM1 marker, a
subset of cells in the NK lineage proved (see below) to be involved in
immune abortion. Premature upregulation of such cells by double-
stranded (DS) RNAs induce if given at day 3 abortion/nonimplantation
(see below). These could also be inappropriately activated by γ-interferons, TNF, and LPS (a bacterial wall product), as a consequence of
local infection (or, possibly, dysregulation in MHC expression, such as
Class II induction), again leading to immune abortion (see below).
Under normal conditions, they most probably play an " immunotrophic
role" (see below for role of lymphokines) as well as controlling via
TNF at low doses placental growth and invasion.

Indeed, in humans, high doses of placental supernatant induce TNF
production by NKs (whose cytotoxic activity it suppresses) at high
doses, suggesting that trophoblasts emit both protective signals and
signals initiating a feedback negative control of its eventual excess
growth (TNF being cytostatic for trophoblasts) (Fig. 3).

Those NK cells quickly see their number reduced once the fully
mature placenta develops. They are then replaced by decidua-associated suppressor cells which are *not* of the same lineage and secrete in
the decidua TGF-β_2, a very potent immunosuppressor (Clark et al.
1990,1991). Murine abortion models decidua are TGF-β_2 deficient at
implantation sites and recruit less efficiently decidua-associated suppressor cells.

So far, however, it has not yet been published that TGF-β *corrects*
resorptions in such models, but, conversely, monoclonal antibodies of
carefully checked specificity against TGF-β enhance abortion in resorption-prone systems, such as the CBA × DBA/2 one. The few residual decidual CD8+ T cells are then also suppressor cells and are paternal strain antigen specific, as are systemic cells (Clark et al. 1991).
Their action depends, however, on a proper previous hormonal preacti-

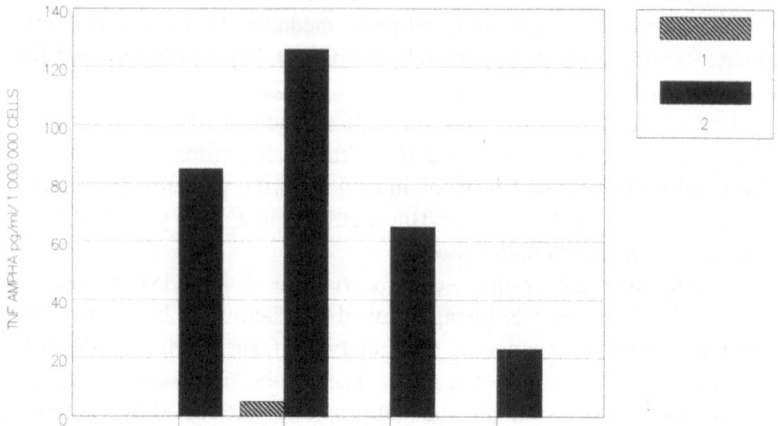

Fig. 3. Induction by a low molecular weight placental factor of tumor necrosis factor (TNF) production by purified human NK cells. *Black columns,* TNF at 4.5, 12.5, 18, and 24 h; *dotted bar,* control value at 12.5 h (it remained the same throughout) Bioassay on L929 cells

vation, as well as on an uncharacterized, but demonstrated, trophoblast-derived soluble factor.

One has still to explain the (statistical) decrease in maternal NK immunity observed earlier by Barret et al. (1982). In peripheral blood lymphocytes of *healthy* pregnant women, which has been attributed to a serum factor as well as to the total lack of antipaternal CTLs in a normal pregnancy, and the immunity of placenta to CTLs even after deliberate maternal antipaternal immunization. One has also to propose an explanation for the enhancement of placental weight in such a situation. This involves placental properties, factors, and immunoendocrine networks.

9.3.1 Placental and Decidual Factors

Indeed, the placenta downregulates maternal cell-mediated cellular immunity at the afferent, central, and effector stages. The main active molecule is one of low molecular weight (> 5000), as purified in cooperation with Bruce Acres and Hanno Kolbe (Transgene S.A., Stras-

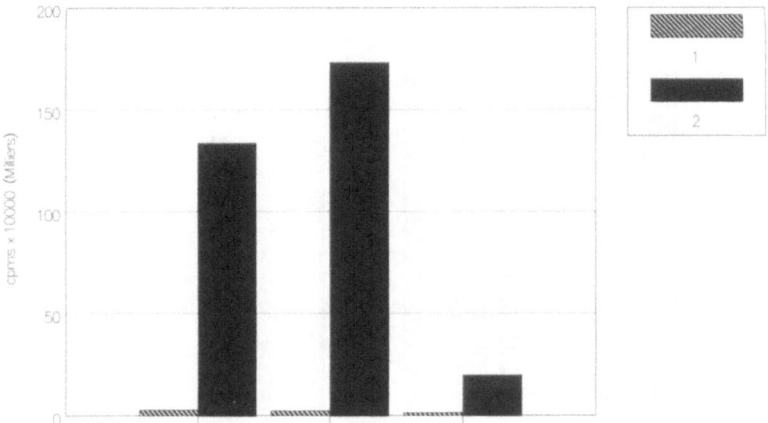

Fig. 4. Suppression of a phytohemagglutinin (PHA) assay by a factor from JAR S/N treated cells. Peripheral blood lymphocytes (PBL) were treated for 24 h in control (RPMI medium) or JAR choriocarcinoma S/Ns. The resulting supernatants were assayed in a PHA-driven T-cell lymphoblastogenesis assay. Group 1 *(left)*,control values (*1*,without PHA: background proliferatin levels; *2*, with PhA). Group 2 *(middle)*, PBLs assayed in control supernatant. Group 3 *right)*, PBLs assayed in supernatant of JAR-treated regulatory cells

bourg, France). It blocks IL-2, IL-4, and other lymphokine-induced dependent lymphocyte proliferation (Menu et al. 1991 a,b).

In vivo, the active moiety corrects abnormal resorbtion rates in normal and artificially induced abortion models and prevents local (popliteal lymph node assay) and general GVH (Fig. 4).

At the effector stage, CTL activity is inhibited by a 60-kDa material, distinct from the above reported lymphocyte proliferation inhibiting activity (LPIA), though this action is presumably seen only on $\alpha\beta$ T cell receptor bearing T cells, not γ ones.

Similarly, the placenta secretes in both mice and humans T suppressor inducer factors. These can be obtained in large amounts from the supernatants of choriocarcinomas such as JAR, JEG, and BeW0. The activated suppressor T cells (Ts) are active in vivo (GVH assays) as well as in vitro (Krishnan et al. 1991). In vitro, they block the expansion and activation of antipaternal CD8+ CTLs as well as proliferation of CD4 helper T cells and block IL-1 production from peripheral

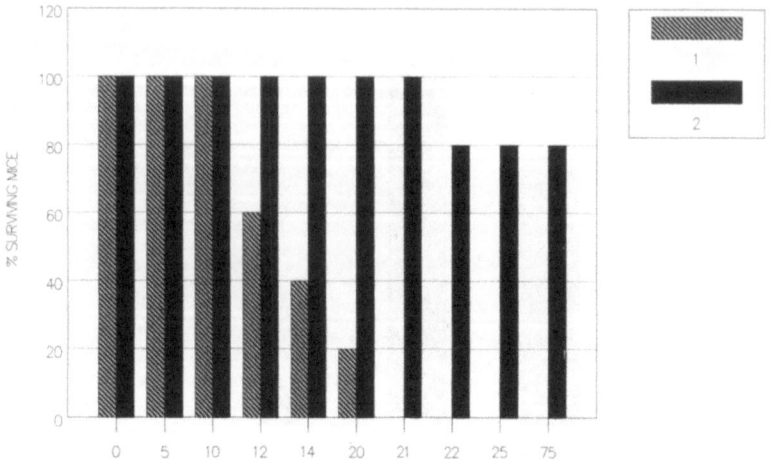

Fig. 5. Prevention of lethal GVH in C3H × BALB/c F1s injected with C3H cells after lethal irradiation. Percentage of surviving animals as function of number of days. *1*, untreated; *2*, cells injected with placental material

blood lymphocytes (PBLs). They act via a nonspecific, non-MHC-restricted souble factor (M. Doric, in preparation, and O. Heine, in preparation, and Doric and Heine, 8th Congress of Immunology, Budapest, Hungary, August 1993) (Fig. 5).

Finally, freshly explanted murine trophoblasts, human cytotrophoblasts, and some weakly class I-positive choriocarcinoma are remarkably "intrinsically" resistant to cell mediated lysis (CML), from conventional CTLs or NKs (Zuckerman and Head 1987, 1988), probably because the apoptosis program is "shut off."

CD46 and CD 55, CD 56 (MCP, DAF, CD 56) probably also suppress lytic activity at the cellular level, especially antibody-dependent cell-mediated cytotoxicity (ADCC). It ensures that a wide variety of suppressors of cellular lytic processes are operational at the maternofetal interface.

Soluble factors as described above are likely involved. This resistance, however, is not absolute, since trophoblast cells are sensitive to activated NK cells or LAKCs in vitro. They are also sensitive to CTLs which have NK–LAKC activity cultured in Gibco Opti MEM medium, a medium which elicits a stronger CTL activity than that obtained using normal culture medium (Drake and Head 1989), endows them with some LAKC-like killing characteristics, though this requires that the murine trophoblast be cultured for 24–48 h, a still unexplained phenomenon (Drake and Head 1988,1989).

9.3.2 Steroids

Two steroids. deserve special attention. The 1–25 dihydroxy cholecalciferol (1,25,-(OH)2D3) secreted by the placenta is immunosuppresive by acting on the IL-1/IL-2 pathway. It exerts such an activity at concentrations compatible with a local immunosuppressive role during pregnancy.

Progesterone also plays an immunosuppressive role. It was first discarded as in vitro artifact, since it was immunosuppressive in vitro on normal lymphocytes, but no progesterone receptors have been demonstrated on resting T lymphocytes. However, J. Szekeres-Bartho et al. (1989a,b) and Hoversland and Beaman (1991) have observed that, upon progesterone treatment at physiological doses, alloactivated or pregnancy lymphocytes released at least two suppressor factors.

The first one, still not purified or cloned at present, which blocks NK cell-mediated lysis is progesterone-induced blocking factor (PIBF). It blocks NK action not only on K 562, a classic NK target, but also on human embryonic fibroblasts. This is physiologically relevant since in immune abortion in mice fibroblasts are present in the placenta!

The factor is also endowed with other pleiotropic immunoregulatory activities. It blocks a mixed lymphocyte reaction (MLR) and activates/enhances the absolute number of CD8+, D44–T cells (Ts subset in humans) and the CD4 2H4+, TQ1+ (Ts inducer subset in humans) in MLR.

The second material is a T suppressor inducer factor (TSIF), named J6B7, recently purified and cloned (Lee et al. 1990). It plays an integral role in pregnancy, since well-defined monoclonal antibodies raised against it are abortifacient in mice (Beaman and Hoversland 1988) *without* being teratogenic (a *very important* control indeed, one not employed in some studies using polyclonal antibodies of unchecked cross-reactivity and performed on a very low number of mice; Hoversland and Beaman 1991). With monoclonal antibodies there are titers in lymphoid organs of pregnant mice.

Interestingly, this monoclonal antibody also cross-reacts with lipo-modulin and the not yet well-characterized PIBF shows lipomodulin like activities. The release of such PIBF (and probably of J6B7) is blocked if one uses the antiprogestin RU486, and similarly transcription of mRNAs for J6B7 is a progesterone-dependent event (ASIR 1991). The block of secretion of PIBF after progesterone action on activated T cells is not seen with the RU 43044 glucocorticoid blocker specific for these, not acting on progesterone receptors.

Monoclonal antibodies specific for the N-terminal part of the conventional progesterone receptor exist, developed by the Milgrom team and commercially available from Transbio (Paris, France), Abbott (Paris, France; EIA PR kit). They react with activated T cells, whether from phytohemagglutinin (PHA) or concanavaline (ConA) stimulation – an MLR which mimiks allorecognition as is the case in the peripheral blood in pregnant women or when in contact with class I positive cells or decidual antigen-pressenting cells (APC) processed presented peptides from paternal MHC molecules (as far as in vitro assays are concerned). In vivo, this is indeed the case of pregnancy, but, more importantly, of allotransplanted patients.

The presence of such receptors can be ascertained by immunohisto-chemical studies, enzyme-linked immunosorbent assay (ELISA), fluorescence-activated cell sorter (FACS) analysis (which requires previous lysolecithin or saponin membrane permeation), a tricky technique but necessary since the receptors are nuclear ones and in rats their presence has been found by appropriate binding studies with the relevant labelled materials (D. Philibert, Roussel Uclaf), However, levels at first trimester are quite low (below 10 fmol/mg protein, the limit of sensitivity of Abbott EIA being 1 fmol/mg protein, but it requires much care) and the standard deviation is, as expected, due to HLA variation between individuals in a randomly selected sample of an outbred wild primate population such as humans, too high for this assay be applied as a diagnostic tool to distinguish between aborter and normal first trimester populations (for example, a cell such as T 47 D or MCF7 display 100–400 fmol/mg protein).

The number of such receptors increases as a function of gestational age, and it was confirmed by an independent study at ASIR, which also reports a decrease in preeclampsia, not only in labor (Jenkins et al. 1992).

The PIBF blocks TNF secretion by activated lymphocytes and also can correct/prevent the effect of low doses of RU486 in mice, suggesting an immune component in RU486 activity. This is now being explored in humans in our laboratory in Clamart. Preliminary polymerase chain reaction (PCR) evidence has recently been obtained by Andras Paldi, in cooperation with L. d' Auriol and M. Misrahi, suggesting it is a classical progesterone receptor (PR). The bands obtained with the appropriate primers have now been sequenced and are those of a classical progesterone receptor (A. Paldi, J. Szekeres Bartho, G. Chaouat et al., unpublished data and submitted for publication).

This is of importance since the placenta is the only place where T cells could be simultaneously activated and subjected to progesterone in high enough amount to secrete immunomodulatory materials. These PRs abruptly disappear during parturition or at the onset of threatened preterm delivery. They persist if the onset of labor is delayed. Therefore, it is possible that this event is associated with the induction of labor, by allowing the release of TNF or TNF-like materials in the placental bed. Indeed, smooth muscle is a target and once triggered, a further source of TNF, which most likely increases the motoricity of the smooth muscle, also promotes vasoconstriction. In addition, TNF is a stop signal for placental growth and is cytotoxic for placenta at high doses. Besides, TNF or TNF-like cytostatic activity is produced by uterine granulated metrial glands cell. *It is, therefore, possible that immunoendocrine events are ONE of the (many) pathways involved in parturition.*

In addition, stress-mediated abortion or premature delivery can be prevented by alloimmunization as in normal or artificial models of abortion (Chaouat et al., 8th International Congress of Immunology; Clark et al., ASIR Meeting, Charleston, 1992, and Clark and Chaouat 1993) (Fig. 3).

Finally, although there are many different PGs in the uterus, reports that PGE_2 plays a cardinal role in pregnancy, as claimed to be shown by the demonstration that indomethacin alone or in association with IL-2 was abortifacient in mice, have not been substantiated by independent investigators. Indeed, a wide number of patients affected with rheumatoid disease have taken indomethacin inconspicuously without any effect statistically on reproduction.

9.4 Immune Control of Trophoblast Growth and Functions (Immunotropism)

Let us now examine the stimulation of trophoblast growth (and function) by immune events already alluded to earlier on. Trophoblasts cells rely for optimal growth on autocrine and paracrine growth factors. They contain a lot of retroviral sequences and express a high number of oncogenes. Some are receptors for paracrine growth factors, such as c-fms, expressed on spongiotrophoblast cells which are the receptors for the colony stimulating factor 1 (CSF-1) (see above). CSF-1 levels drastically increase in the uterus during pregnancy, up to a 1000-fold increment. *Nothing is known (yet!) at parturition.*

CSF-1 is a direct growth factor for placenta (autocrine pathway) but is also produced in large amounts by uterine epithelial cells.

In vitro, trophoblasts respond positively to members of the CSF family of growth factors, such as the aforementioned CSF-1, a product of activated T cells but also of uterine epithelium and implantation sites, and IL-3 (or CSF-β_2) and granulocyte macrophage colony stimulating factor (GM–CSF)). The response of trophoblast cells to such stimuli is enhanced proliferation, enhanced phagocytosis, a phenomenon seen as optimal when one deals with EPC-derived cell outgrowth (Athanassakis et al. 1987; Armstrong and Chaouat 1989). A similar role is ascribed to G-CSF, and, as stated, c-kit (M-CSF receptor) or M-CSF-deficient mice are sterile.

Conversely, lymphokines such as TNF, γ-interferon, are stop signals for placental growth, and, indeed, are abortifacient in mice, the latter also dysregulating in such a process MHC class I regulation, labyrinthine cells becoming MHC class I positive. TNF might come, as discussed above, from granular uterine metrial gland cells, at least in part. It is tempting to speculate that when intrauterine infection occurs, high levels of bacterial wall degradation products accumulate, such as LPS. Such a product, at various doses according to gestational age, is abortifacient in mice at every stage of pregnancy via a TNF-mediated mechanism (Chaouat et al. 1990).

These data explain why placenta is larger in allopregnancy than in syngeneic pregnancy, larger with successful pregnancies, and smaller in T cell-depleted animals or mice made tolerant (unresponsive to) paternal alloantigens and the "hybrid vigor" effect. They also offer an explanation for treatment of natural or artificially induced abortions in mice by alloimunization. As such, they gave rise to the "immunotrophic theory" (Wegmann 1984).

9.5 Abortion and Immunity

The first abortion putatively due to immune malfunction was the caroli/musculus system: transfer of *Mus caroli* embryos into the uterine horns of the laboratory mouse *Mus musculus* fail. The abortion/resorbtion process starts by day 9.5--10, and by day 11–12 all *M caroli* embryos are dead/resorbed, whereas cotransferred *M musculus* embryos develop further successfully until delivery.

M caroli embryos are prone to massive lymphocytic infiltration. By making embryonic injection chimeras, Janet Rossant has shown in a series of elegant experiments that even though the ICM/future embryo could be *M caroli*, if the ECM/ECM derived placenta was of *M musculus* genotype, a sucessful pregnancy across the species barrier was possible.

This demonstrates a crucial role for trophoblast decidua interactions. Since then, similar injection chimeras as well as aggregation chimeras have been used in the sheep/goat system, also giving birth to chimeras across the species barrier, the quite unique and spectacular "unicornlike" offspring which were shown on the front cover of *Nature*.

The *M caroli* model, however, has its weaknesses. In fact, it is now admitted that rejection of *M caroli* embryos in *M musculus* uterus is not immunologically initiated, though later on immunological mechanisms are involved in embryonic rejection (Croy et al. 1987).

On the other hand, there is a xenogenic pregnancy system which seems to be influenced by immunological manipulations. Elegant experiments by Allen et al. (1986) used the equine endometrial cup, where one can see a gross lymphocytic infiltrate, almost at the limit of rejection. They have shown that in donkey mare and horse donkey pregnancy, the equine endometrial cup, e.g., the placentation in equids, was even more largely infiltrated by lymphocytes, with large necrotic and hemorrhagic areas (Allen et al. 1986).

When Allen transferred donkey × donkey embryos into foster mare mothers, pregnancy failed, with two precise characteristics: The trophoblast outgrowth stopped at the peri-implantation phase, and in a quite reproducible time span there was then no secretion of equine choriononic gonadotrophin (ECG) and the embryos were massively infiltrated by lymphocytes.

Inspired by the CBA × DBA/2 system (see below), Allen et al. then immunized the mare against donkeys or proceeded to transfer anti-donkey horse serum into donkey embryos bearing mares. Remarkably, whereas there was almost 90–100% failure of such ET into untreated mothers, pregnancy was successful in 75–85% of the passively or actively immunized mothers range. Though one can discuss environmental components in this system, as indeed for mice, there are few doubts that an immunological component was involved in protection (transfer of serum), i.e., an immunological maneuver was an immunological treatment of pregnancy loss. In fact, attempts have been made to define in the equine species decidua-associated suppressor cells and factors, but have been hampered by the small number of experimental animals available.

There exist, fortunately, models of intraspecies pregancy failure and their treatment: When one mates CBA/J (H-2^k) with DBA/2 males (H-2^d), this mating leads to a high rate of spontaneous resorbtions (about 40%). This is peculiar to this combination (Chaouat et al. 1988). It is not seen in other H-2^k × H-2^d combinations, such as the C3H × DBA/2, nor in the C3H × BALB/c combination, or, more interestingly, in the CBA × BALB/c (also H-2^k × H-2^d). High resorption rates do not appear in CBA × CBA, DBA/2 × DBA/2 either, and DBA/2 × CBA/J behave "normally." The "abortifacient trait" is recessive, e.g., CBA × (BALB/c × DBA/2) F1s are not prone to high rates of resorbtion.

The model involves minor loci differences in the same MHC (H-2^k × H-2^d) combination since immunization against BALB/c cells corrects CBA DBA/2 resorbtion rates, BALB/C being H-2^d as DBA/2. In the same vein, B10 mated with B10.A exhibit an abnormally high resorption rate which is not seen in B10 × B10.Br and B10 × B10.D2 matings (Chaouat et al. 1988).

This second model "mirror images" CBA × DBA/2, involving subtle MHC disparity on the same (B10) genetic background. The "MHC effect" certainly involves presentation/interactions and self-peptide competition, since B10.A is H-2^a , H-2^a being a recombinant haplotype between H-2^d (B10.D2) and H-2^k (B10.Br).

The two models share several characteristics: resorptions increase with aging, repeated pregnancies with the "wrong" father enhance this, and there is a local deficiency of decidua-associated suppressor cells near "would be resorbed" embryos and the process involves NK cells. The models are environment dependent: Gnotobiotic CBA/J do not abort.

The resorbing embryos are infiltrated first by NK lineage cells, probably coming from the early local accumulation around embryos at the early implantation period in mice. Modulation of NK activity, and NK accumulation at implantation sites, correlates with abortion (De Fougerolles and Baines 1988; Chaouat et al. 1989): In vivo elimination of Asialo GM1 positive cells reduce resorptions. PIBF is curative of these abortion systems, as it is of DS RNAs induced ones (Szekeres Bartho and Chaouat 1990). Similarly, the anti-TNF-α drug pentoxifylline reduces the abortion rates. Conversely, activation of NK cells, be it by POLY I C12 U or other DS RNAs (Kinsky et al. 1990) or interferon inducer or γ-interferon which also induces class I MHC expression in trophoblasts in vivo, results in enhancement of abortion. In fact, interferon and DS RNA systems work in *any* strain of mice, as do LPS and recombinant TNF, to induce abortion (Chaouat et al. 1990).

Decidua of "natural" aborting combinations contain more γ-interferon and TNF-α than normal matings (Chaouat et al. 1987; and Raghupathy et al., submitted). LAKCs being involved, it is tempting to use IL-2. Reports of very low doses of IL-2 as abortifacient or combinations of IL-2 plus indomethacin have not been confirmed.

Immunization of CBA/J prior to DBA/2 mating against BALB/c splenocytes, but not with DBA/2 ones, corrects resorption rates.

The effect is transferrable by cells or serum, correlates with increased systemic Ts activity and increased alloantibody production and could be investigated genetically using recombinants between BALB/c and DBA/2 (Chaouat et al. 1988).

After alloimmunization, one sees an increase in placental weight – an observation which led to the immunotropism theory (Wegmann 1984) – suggesting that lymphokines were involved in optimal placental growth. Indeed, anti-L3T4 plus anti-CD8 treatment reduced placental size and weight. Then it was shown in vitro that factors of the CSF family of lymphokines were indeed growth factors for murine trophoblast in vitro (see above). MRL/lpr mice are an autoimmune strain in which there is too much CSF secretion and antibodies develop with anti-idiotypic activity to IL-3 receptor and are endowed with IL-3-like activity as a consequence of autoimmunity. In this strain, anti-CD4 plus anti-CD8 treatment normalizes the excess placental weight seen habitually as a correlative of the lpr trait (in other strains, it reduces litter size and mean placental weight and trophoblast phagocytic activity, even in congenic inbred matings, though less so than in semiallogenic ones). The same treatment enhances drastically CBA × DBA/2 fetal loss and prevents fetal protection by alloimmunization. MRL mice being H-2^k as CBA/J, we transferred cells from lpr mice into CBA/J without obtaining immediate GVH or HVG, and this was active in reducing resorption. In view of in vitro data (Athanassakis et al. 1987; Armstrong and Chaouat 1989), we attempted to treat CBA/J with IL-3 and GM–CSF (Chaouat et al. 1990).

Other models of induced abortion relevant to clinics are induction of abortion by syngeneic tumor grafts as seen in C57BL/6 mice: When one injects a regressor tumor, a real abortion (with bleeding and expulsion, not merely a resorption) is observed (Tartakowsky 1986; Tartakowsky and Gorelik 1988). In humans, the syndrome known as recurrent idiopathic abortion, putatively thought to be immunologically mediated, is still highly controversial. The immune abortions induced by autoimmunity are, of course, indisputable.

It has long been recognized, however, that there exist in humans a subset of women (couples, in fact) who experience recurring spontaneous abortions in first term. These are called recurrent spontaneous abortions (RSA) and occur with their regular partner. One of the proposed therapies for RSA was artificial insemination with sperm from an anonymous third party donor. Such RSAs occur without obvious endocrinological, mechanical, or infectious processes, at least according to our present knowledge and technology. They are classed as primary (no previous child) or secondary (one or several previous children). The diagnosis proceeds at present per exclusion *only*. There is no consensus on the immunological parameters involved and, hence, no consensus at all on the syndrome; in fact there is no existing positive diagnosis assay, though attempts in this direction have been made.

Based initially on the "TLX" theory, immunization of the mother against paternal or third party lymphocytes were boldly attempted, yielding impressive success rates (80%). These data were confirmed by a double-blind study of James Mowbray (Mowbray et al. 1985). This treatment is now routinely performed in the world, *but* a controversy is raging about the placebo or spontaneous remission rate and the efficiency of the therapy and its inocuity. Other studies failed to observe so big a difference between control and treated patients, some finding in fact showed no difference at all. The most important aspect of the controversy is pending on the background remission rate (for both sides of the controversy, refer to Hsi et al. 1991a, "Immune abortion, the controversy" and several issues of the *American Journal of Reproductive Immunology*). The American Society for Immunology of Reproduction has started a large scale analysis.

9.6 Conclusion

Finally, we have moved from the 1960 concept of an enhanced allograft to the notion of a complex symbiosis using immunoendocrine and cytokine networks at the fetomaternal interface. The relevance to labor and early pregnancy loss has been stressed throughout the text. Nevertheless, I would like to point out the following points: Though a great difference obviously exists between abortion (leading to death of the conceptus) and labor, TNF and IL-1 levels, and perhaps IL-6, should be monitored during labor and following RU486 treatments in order to assess their role in the events leading to delivery. TNF levels in particular should be monitored in preeclampsia and threatened preterm delivery. Treatment with pentoxifylline could be envisaged in threatened preterm deliveries and preeclampsia.

Monitoring of the human equivalent of Beaman, Lee, and Hoversland's factor (Beaman and Hoversland 1988), well determined because purified, cloned, and for which monoclonal antibodies of well-defined specifity are available, should be started, with the caveat of antilipomodulin reaction. The effects of neutralizing monoclonal antibodies against TNF and IL-1 should be monitored in animal models.

The reason why I do not believe that immune events are an *integral* part of labor is simple. Alloimmunization corrects resorptions. It enhances progesterone receptor bearing lymphocytes. It decreases *decidual* and circulating TNF levels. Yet, it does *not* promote enhanced pregancy duration. On the other hand, changing daylight cycle can, in a variety of species. This does not mean that understanding why progesterone receptors in lymphocytes, TNF levels, etc. are modified during labor should not be studied; they must. But the regulatory pathways must be well defined. Otherwise, if reproductive immunologists who were performing experiments with reagents of undefined specificities and low numbers of animals claimed to have the key to parturition events, I am pretty sure everyone in this audience and in the field of immunology as well would quote Voltaire's famous: "tout ce qui est excessif est insignifiant" (everything excessive is insignificant).

References

Allen WR, Kydd JH, Antczack DF (1986) Successful application of immunotherapy to a model of pregnancy failure in equids. In: Clark DA, Croy BA (eds) Reproductive immunology. Elsevier, Amsterdam, pp 253–261

Arceci B, Shanahan F, Stanley ER, Pollard JW (1989) The temporal expression and localisation of colony stimulating factor (CSF-1) and its receptor in the female reproductive tract are consistent with CSF-1 regulated placental development. Proc Natl Acad Sci 86:8811–8818

Armstrong DTA, Chaouat G (1989) Effects of lymphokines and immune complexes on murine placental cell growth in vitro. Biol Reprod 401:460–475

Athanassakis I, Bleackley RC, Paetkau V, Guilbert L, Barrp J, Wegmann TG (1987) The immunostimulatory effects of T cells and T cell lymphokines on murine fetally derived placental cells. J Immunol 138:37–44

Athanassakis Vassiliadis I, Papamettheakis J (1991) Modulation of class II antigens on fetal placenta lads to fetal abortion. In: Chaouat G, Mowbray J (eds) Biologie cellulaire et moléculaire de la relation materno fetale. Editions INSERM, Libbey, Paris, pp 69–81

Barret DS, Rayfield LS, Brent L (1982) Suppression of natural cell mediated cytotoxicity in man by maternal and neonatal serum. Clin Exp Immunol 47:742–748

Beaman KD, Hoversland RC (1988) Induction of "spontaneous" abortion by blocking antigen specific suppression. J Reprod Fert 82:135–139

Boucraut J (1992) Regulation de l'expression des antigenes du CMN dans le placenta human. Ph.D. Thesis, University of Marseille Luminy

Chaouat G, Lankar D, Kolb JP, Clark DA (1987) 2 Modèles d'avortements d'origine immunitaire chez la souris de laboratoire: mécanismes abortifs, modalites, et mécanismes du traitement par l'immunisation contre un male relié ou non relié suivant les differences antigeniques pere-mere.In: Chaouat G (ed) Colloque INSERM CNRS Immunologie de la relation feto-maternelle. Editions INSERM, John Libbey, pp 243255

Chaouat G, Clark DA, Wegmann TG (1988) Genetics aspects of the CBA/J DBA/2 J and B10 B10.A models of murine spontaneous abortions and prevention by leukocyte immunisation.18th R.C.O.G. study group. Early Pregnancy loss. Mechanisms and treatment. R.C.O.G. Allen WR, Clark DA, Gill III TJ, Mowbray JF, Robertson WR (eds) RCOG Press, pp 89–105

Chaouat G, Szekeres Bartho J, Menu E, Kinsky R, Thang MN, Dy M, Minkowski M, Wegmann TG, Clark DA (1989) Abortion and maternal immunity. (Placenta interacts with the maternal immune system). Fourth Meeting of the Latin American Society of Obstetrics and Gynecology (FLASOG). Santiago de Chile, 30 November 1989

Chaouat G, Menu E, Dy M, Minkowski M, Clark DA, Wegmann TG (1990) Control of fetal survival in CBA DBA/2 mice by lymphokine therapy. J Fert Steril 89:447–458

Chaouat G, Menu E, Wegmann TJ (1991) Role des lymphokines de la famille du CSF, et du TNF, de l'interféron gamma, et de l'IL-2 sur la survie fetale et la croissance placentaire étudiées in vivo dans 2 modèles d'avortements immunitaires spontanés murins. Biologie cellulaire et moléculaire de la relation materno fetale. Editions INSERM, John Lebbey, pp 91–101

Clark DA, Chaouat G (1993) J Immunol (in press)

Clark DA, Flanders KC, Banwatt D, Millar-Book W, Manuel J, Stedronska-Clark, Rowley B (1990) Murine pregnancy decidua produces a unique immunosuppressive molecule related to transforming growth factor beta-2. J Immunol 144:12.3008–3004

Clark DA, Lea RG, Denburg J, Barwatt D, Manuel J, Daari N, Underwood J, Michel M, Mowbray J, Daya S, Chaouat G (1991) Transforming growth factor beta related factor in mammalian pregnancy decidua: homologies between the mouse and human in successful pregnancy and in recurrent unexplained abortion. Biologie cellulaire et moléculaire de la relation materno fetale. Editions INSERM, John Libbey, pp 131–141

Croy BA, Rossant J (1987) Mouse embryonic cells become susceptible to CTL mediated lysis after midgestation. Cell Immunol 104:355–365

Croy BA, Crepeau M, Yamashiro S, Clark DA (1987) Further studies on the transfer of Mus Caroli embryos to immunodeficient Mus Musculus. Colloque Inserm 154. Immunologiede la reproduction: relation materno fetale. Chaouat G (ed) Editions Inserm, Paris, pp 101–111

De Fougerolles R, Baines M (1988)Modulation of Natural Killer activity influences resorbtion rates in CBA X DBA/2 matings. J Reprod Immunol 11:147–153

Dohr G (1987) HLA and TLX expression on the human ovocyte, zona pellucida and granulosa cells. Human Reprod 2:657–661

Drake BL, Head JR (1988) Murine trophoblast cells are susceptible to Lymphokine Activated Killer (LAK) cell lysis. Am J Reprod Immunol 16:114

Drake BL, Head JR (1989) Murine trophoblast cells can be killed by allospecific cytotoxic T lymphocytes generated in Gibco OPTI MEM medium. J Reprod Immunol 15:71

Ellis SA (1990) HLA-G: at the interface. Am J Reprod Immunol 23:84–86

Geraghty DE, Keller BH, Hauser JA, Orr HT (1992a) The HLA class I gene family includes at least 6 genes and 12 pseudogenes and gene fragments. J Immunol 149:1934–1946

Geraghty DE, Keller BH, Per J, Hannsen JA (1992b) Examination of 4 HLA class I genes. Common events in the evolution of class I genes and pseudogenes. J Immunol 149:1947–1956

Hoversland RC, Beaman KD (1991) The lack of effect of a monoclonal antibody against murine T cell suppressor factor on murine embryo development in vitro. Am J Reprod Immunol 26:84–88

Hsi BL, Fenichel P, Cervoni F (1991a) Expression of complement regulatory proteins on human gametes and trophoblast. Immune abortion: the controversy. Biologie cellulaire et moléculaire de la relation materno fetale. Editions INSERM, John Libbey, pp 3–13

Hsi BL, Fenichel P, Cervoni F (1991b) In Biologie cellulaire et moléculaire de la relation materno fetale. Editions INSERM John Libbey.1991. pp 261–307

Jenkins DM, Shidchru N, Dobson A (1992) Lymphocyte progesterone receptors (LPR) in mid pregnancy predict pre-eclampsia. ASRI meeting 27, 1,2:39 (abstract 39)

Kachkache M, Acker GM, Chaouat G, Noun A, Garabedian M (1991) Hormonal and local factors control the immunohistochemical distribution of immunocytes in the rat uterus before conceptus implantation: Effects of ovariectomy, fallopian tube section and RU 486 injection. Biol Reprod 45:860–868

Kanbour-Sharir A, Zhang X, Rouleau A, Armstrong DT, Kunz HW, Macpherson TA, Gill III TJ (1990) Gene imprinting and major histocompatibility complex class I antigen expression in the rat placenta. Proc Natl Acad Sci 87:444–448

Krishnan L, Menu E, Chaouat G, Talwar GP, Raghupathy R (1991) In vitro and in vivo immunosuppressive effects of supernatants from human choriocarcinoma. Cell Immunol 138:313–326

Kinsky R, Delage G, Rosin N, Thang MN, Hoffmann M, Chaouat G (1990) A murine model of NK cell mediated resorption. Am J Reprod Immunol 23:73

Kovatts S, Main EK, Libbrach C, Stublebline M, Fischer SJ, De Mars R (1990) A class I antigen, HLA-G, expressed in human trophoblasts. Science 248:220–223

Kovatts S, Librach C, Fisch P, Main EK, Sondel PM, Fischer SJ, De Mars R (1991) The role of nonclassical MHC class I on human trophoblast. In:

Chaouat G, Mowbray J (eds) Biologie cellulaire et moléculaire de la relation
meterno fetale. Editions INSERM, John Libbey, Paris, pp 13–21 and 41–51

Le Bouteiller P, Boucraut J, Fauchet R, Pontarotti P (1991) Transfected HLA
class I genes human cell line escape the negative cis regulatory control
exerted on endogenous class I genes. In: Chaouat G, Mowbray J (eds) Biologie cellulaire et moléculaire de la relation materno fetale. Editions IN-
SERM, John Libbey, Paris, pp 41–51

Lee CK, Ghoshal K, Beaman KD (1990) Cloning of a cDNA coding for a T cell
molecule with putative immunoregulatory role. Mol Immunol 27:1137–1134

MacMaster MT, Newton RC, Sudhanski K, Dey, Andrews GK (1992) Activation and distribution of inflammatory cells in the mouse uterus during the
preimplantation period. J Immunol 148:1699–1705

Martal J, Chene G, Charlier M, Guillomot M, Reinaud P, Bertin J, Danet G,
Zouari K, Charpigny G (1990) Trophoblastin, oTP, embryonic interferons.
In: Placental communications: biochemical, morphological and cellular as-
pects. Cedard L et al. (eds) Colloque INSERM/John Libbey, p 125

Martal J, Charpigny C, Fillion C, Assal Meliani A, Chaouat G (1991) Purifica-
tion et clonage de la trophoblastine et de ses 5 isoformes. Activités lutéoly-
tiques, antivirales, immunologiques. Biologie cellulaire et moléculaire de
la relation materno fetale. Editions INSERM, John Libbey, pp 317–325

Menu E, Jankovic DL, Theze J, David V, Chaouat G (1991) Immunoactive
products of human placenta, III- Characterization of an inhibitor affecting
lymphocyte proliferation. Regional Immunol 3:254–259

Menu E, Djian V, Kinsky R, Jankovic D, Delage G, Rosin N, Chaouat G (1991)
Immunoactive products of human placenta. Biologie cellulaire et moléculaire
de la relation materno fetale. Editions INSERM, John Libbey, pp 197–205

Neki R, Matsusaki N, Masuhiro K, Taniguchi T, Shimoya K, Jo T, Li Y, Ta-
kagi T, Saji F, Tanizawa T (1991) Analysis of the machanism of Tumor
Necrosis Factor alpha induced release of human chorionic gonadotropohin
from normal human trophoblasts. 5th Annual meeting of Japan Society for
Basic Reproductive Immunology and Japan Society for Medical Reproductive Immunology. Joint meeting 1990. Edited by K Honjo, S Kasahura.
JSBR and JSMR. Editions Mure Printing, pp 108–114

Noun A, Acker G, Chaouat G, Antoine JC, Garabedian M (1989) Macrophages
and T lymphocyte bearing antigens bearing cells in the uterus before and during ovum implantation in the rat. Clin Exp Immunol 78:434–438

Pollard JW, Pampfer S, Arceci RJ (1991) Class III tyrosine kinase receptors at
materno fetal interface. Biologie cellulaire et moléculaire de la relation
materno fetale. Editions INSERM, John Libbey, pp 81–91

Robertson SA, Seamark RF (1991) Uterine granulocyte macrophage colony
stimulating factor (GM CSF) in early pregnancy. Cellular origin and poten-

tial regulators. In Biologie cellulaire et moléculaire de la relation materno fetale. Editions INSERM, John Libbey, pp 113–123

Sanford TR, De M, Wood G (1992) Expression of colony stimulating factors and inflammatory cytokines in the uterus if CD1 mice during days 1 to days 3 of pregnancy. J Reprod Fert 94:213–220

Singh B, Raghupathy R, Anderson DJ, Wegman TG (1983) The murine placenta as an immunological barrier between the mother and the fetus. In: Wegmann TJ, Gill III (eds) Immunology of reproduction. Oxford University Press, New York, pp 229–251

Stewart C, Kaspar P, Brunet LJ, Bhatt H, Gadi I, Köntgen F, Abbondanzo S (1992) Blastocyst implantation depends on maternal expression of leukemia inhibitory factor. Nature 359:76–79

Szekeres Bartho J, Autran B, Debre P, Andreu G, Denver L, Blot P, Chaouat G (1989) Immunoregulatory effects of a suppressor factor from healthy pregnant women's lymphocytes after progesterone induction. Cell Immunol 122:281

Szekeres Bartho J, Reznikoff Etievant M, Varga P, Pichon MF, Varga Z, Chaouat G (1989) Lymphocytic progesterone receptors in human pregnancy. J Reprod Immunol 16:239

Szekeres Bartho J, Chaouat G (1990) A T cell derived progesterone induced blocking factor correct resorbtions in a murine abortion system. Am J Reprod Immunol 23:26–29

Tartakowsky B (1986) Murine abortion models. In: Wegmann TG, Gill III TJ (eds) The placenta and the survival of the fetal allograft. Oxford University Press, Oxford (Reproductive immunology, vol II)

Tartakowsky B, Gorelik E (1988) Immunisation with asyngeneic regressor tumor causes resorptions in allopregnant mice. J Reprod Immunol 13:113

Tartakowsky B, Goldstein O, Ben Yair B (1991) In vivo modulation of pre embryonic development by cytokines. Biologie cellulaire et moléculaire de la relation materno fetale. Editions INSERM, John Libbey, pp 239–245

Wegmann TG (1984) Fetal protection against abortion: is it immunosuppression or immunostimulation? Ann Immunol Inst Pasteur 135 D. pp 309–311

Whaley AE, Caroll RS, Nephew KP, Imakawa (1991) Molecular cloning on unique interferons from human placenta. Workshop of the annual meeting of the International Society for Interferon Research, Nice. J Inf Res 11 [Suppl] (Abstract 69)

Zuckerman FA, Head JR (1987) Murine trophoblast resist cell mediated lysis.I-Resistance to allospecific cytotoxic T lymphocytes. J Immunol 139:2856–2865

Zuckerman FA, Head JR (1988) Murine trophoblast resist cell mediated lysis. II-Resistance to Natural cell mediated cytotoxicity. Cell Immunol 116:274

10 Molecular Mechanisms Regulating Contractility of the Pregnant Uterus

Colin T. Jones, Ahmad Khouja, Daniel Wichelhaus, and Helen Warsop

10.1 Introduction

The changes in uterine contractility that take place several days or hours before birth, depending upon the species, are amongst the most dramatic seen in mammalian physiology [1,2]. The shift from the innervated nonpregnant uterus dominated by the effects of neurotransmitters, neuropeptides and sex steroids to the pregnant uterus potentially dominated by circulating factors, such as the sex steroids and locally produced agents from submyometrial tissues, is a particularly striking change [3–6]. This has been variously hypothesized as the explanation for the control of uterine contraction during labour residing with the fetus and intrauterine tissues and removed from the mother [7,8]. However, the reappearance of oxytocin, both circulating and of endometrial origin, as a potentially important stimulating of myometrial contraction, directly and indirectly, during labour implies that some maternal influence may be present in the later stages of birth [9]. Despite some evidence to the contrary, the labour-related production of prostaglandins, whether oxytocin stimulated or not, still appears to be the primary mechanism for initiating and sustaining uterine contraction through labour [10].

There are a number of candidates for the stimulation of prostaglandin production oxytocin, platelet-activating factor, interleukins and others [11–15] and there is evidence for supression of their production [16,17]. However, the precise agents responsible for their production in association labour is yet to be confirmed with certainty.

It is also unclear what are the precise mechanisms by which prostaglandins stimulate uterine contraction. They have been shown to increase inositol phosphate production [18] and at high concentrations to release directly Ca^{2+} from intracellular stores [19,20]. However, the quantitative contribution of each of these pathways is yet to be defined.

In smooth muscle as elsewhere contraction is the result of an increase in intracellular Ca^{2+} [21,22]. This is caused by transient opening of receptor- or voltage-operated calcium channels initiating a rapid but brief influx of calcium across the cell membrane [23,24] and by release from intracellular stores. Release from intracellular stores is through of inositol 1,4,5-trisphosphate (IP_3) [25,26]. It is clear that IP_3 is responsible for the initial phasic contraction of smooth muscle. Tonic contraction may occur subsequent to this, resulting from a secondary influx of Ca^{2+} across the plasma membrane and through phosphorylation of myosin-light-chain kinase (MLCK) by activation of protein kinase C (PKC). Although there is clear evidence for the existence of these pathways within uterine smooth muscle, their contribution to regulation of contraction at the time of birth is far from clear. Prostaglandins have the ability both to initiate contraction through increase in IP_3 production and phosphorylation of MLCK through diacylglycerol activation of PKC and to modulate these effects through cAMP production. Hence the degree of contraction induced by different prostanoids varies considerably from tissue to tissue.

In uterine smooth muscle approaching birth key questions about molecular events underlying contractile changes are:

1. Are there changes in receptor performance that are central to enhancements of contractility?
2. Are there changes in the pathway for uterine IP_3 production that are important to the increase in uterine contractility?
3. What are the primary mechanisms responsible for enhanced prostanoid production in the near-term uterus?

These questions will be addressed.

10.2 Changes in Receptor Function

The production of inositol polyphosphates by uterine cells in response to hormonal stimulation show a broad range of or responses. In smooth muscle cells from the nonpregnant uterus there is little IP response to a number of hormones, including oxytocin, vasopressin and histamine, while responses to endothelin-1, bradykinin and interleukin-1 are substantial (Fig. 1). In pregnancy the response to all agents except interleukin falls. Such a decline in uterine responsiveness is a common feature of pregnancy. For instance the relative lack of response to oxytocin stimulation is found commonly and only very late in pregnancy when there are sharp increases in oxytocin receptor number is significant and substantial response to the peptide apparent [9,27,28]. Throughout much of pregnancy perhaps the most striking receptor changes relate to the denervation of the uterus leading to almost complete loss of sympathetic nerve terminals from the body of the uterus [4,29,30]. Despite the denervation α_1 and α_2 receptors retain relatively high levels and broad distribution across the uterus in a wide range of species [31,32](Fig. 2). Moreover responsiveness of the uterus to stimulation by noradrenaline, although low, is significant throughout pregnancy (Fig. 1). These responses exhibit a particularly important facet of the bulk of the pregnant uterus, that is, a poor hormonal response relative to receptor density and by comparison to the responses observed in vascular smooth muscle. This poor responsiveness to catecholamines occurs despite the fact that in rat, rabbit and monkey use of α-adrenergic blocking agents disrupts normal birth [33,34]. In nonpregnant uterus contractile responses are much enhanced by estrogens [35].

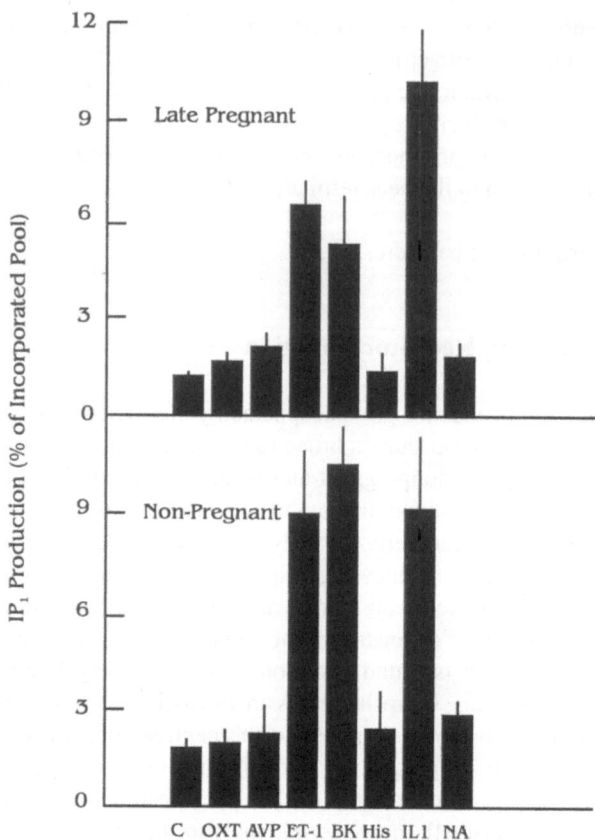

Fig. 1. Production of [³H]inositol monophosphate from guinea pig myometrial cells prelabelled with [³H]inositol. The cells were incubated for 30 min either in the absence (*C*, control) or presence (*OXT*, 1 μM oxytocin; *AVP*, 1 μM arginine vasopressin; *ET-1*, 10 mM endothelin-1; *BK*, 1 μM bradykinin; *His*, 1 μM histamine; *IL1*, 10 nM interleukin-1; *NA*, 10 μM noradrenaline) of a range of hormones and inositol phosphate separated by ion-exchange chromatography

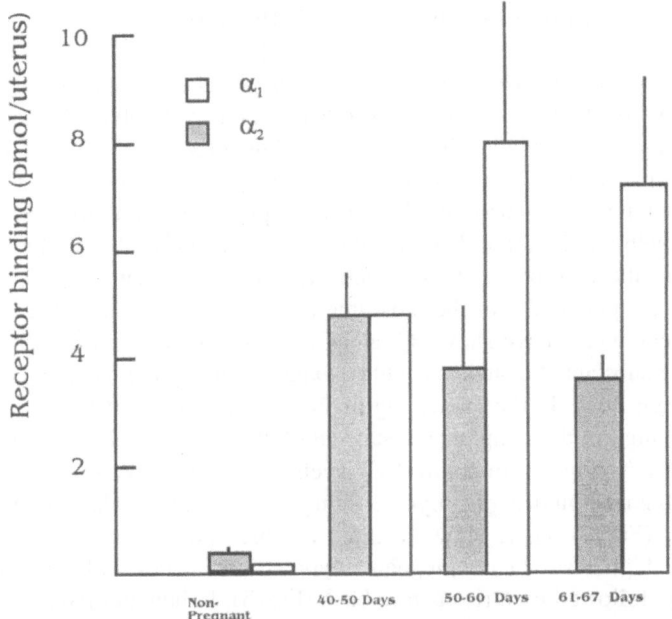

Fig. 2. Changes in α_1 *(open bars)* and α_2 *(stippled bars)* receptor binding in guinea pig uterus during pregnancy. Binding to α_1 receptor was determined with [3H]Prazosin and to α_2 receptor with [3H]Rauwolscine as described previously [31,32]

Summarizing, it appears that the pregnant uterus has relatively limited sensitivity to a range of normal contractile agents and that, with the exception of oxytocin, there is no clear evidence for increased responses at the time of birth. It is therefore important to question whether the transmembrane pathway for generation of IP3 is in a fully functional state in the pregnant uterus and whether this changes as birth approaches. If the pathway is well developed that implies that local paracrine rather than endocrine changes may be important in mediating contraction and therefore the development of the pathway controlling prostanoid production has been compared.

10.3 Development of the Uterine G Proteins

Coupling of cell surface receptors to ion channels of second messenger systems mediating contraction or other responses involves the action of guanosine-5'-triphosphate (GTP)-binding proteins [36,37]. These are trimeric proteins with well-conserved β- and γ-subunits and a broad range of α-subunits. The specific properties conferred on the G protein are therefore determined by the α-subunit (Table 1). In relation to the control of uterine contraction the G proteins G_s, G_i and $G_{q/11}$ are potentially of the most interest. Previous studies in the nonpregnant uterus have shown G_s and G_i under estrogen control [35,38]. Hence substantial changes would be expected in G protein content during pregnancy. In the guinea pig myometrium G_s content is comparatively high in the nonpregnant state and falls progressively throughout pregnancy (Fig. 3). In contrast G_i levels are comparatively low in the nonpregnant guinea pig myometrium and rise sharply late in pregnancy (Fig. 3). $G_{q/11}$ shows a more complex pattern with comparatively high level in nonpregnant myometrium, a fall early in pregnancy, followed by a rise before birth (Fig. 3). In human myometrium similar changes in G protein content were observed with a fall in G_s (Fig. 4), a rise in G_i (Fig. 5) and a rise and fall in $G_{q/11}$ (Fig. 6).

Table 1. Functions of major G proteins

α-Subunit	Function	M_r (kDa)	Properties
G_s	Activates Ca^{2+} channels, activates adenylate cyclase	44.5–46	CT sensitive
G_i	Inhibits adenylate cyclase, PT-sensitive PI-PLC/PLA$_2$? Stimulates Ca^{2+} channels?	40.4–40.5	Pertussis toxin sensitive
G_o	Closes Ca^{2+} channels, K^+ and Na^+/H^+ antiporter regulator?	40	PT sensitive
G_t	Activates cGMP phosphodiesterase	40.5	CT and PT sensitive
$G_{q/11}$	Activates β-PI-PLC	40–41	CT and PT insensitive
G_z	Unknown	41	PT sensitive

PI-PLC, phosphoinositide phospholipase C; CT, cholera toxin; PT pertussis toxin; PLA$_2$, phospholipase A$_2$

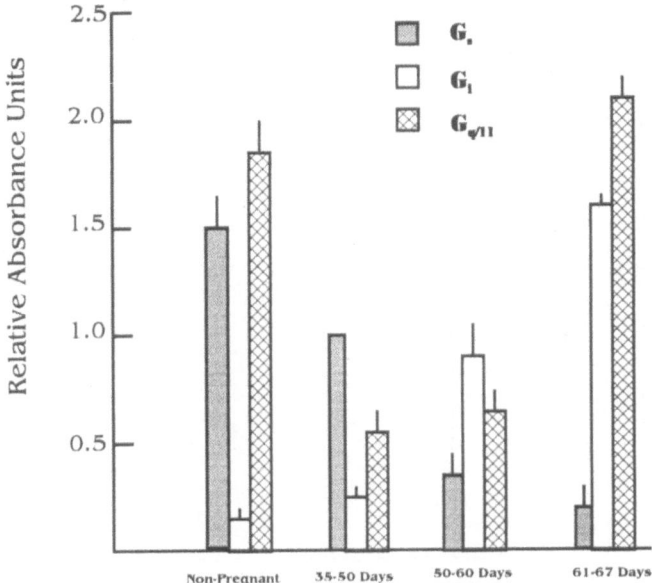

Fig. 3. Pregancy-related changes in G protein content of guinea pig myometrium. G protein content was determined by western blotting using specific antibodies against α-subunits and quantitfied by scanning laser densitometry. *Open bars,* G$_s$; *stippled bars,* G$_i$; *cross-hatched bars,*G$_{q/11}$

G$_s$ and G$_i$ function to control adenylate cyclase and their coupling to adrenergic receptors in the uterus has been demonstrated [39]. The changes are consistent with a decline in the capacity for cAMP production in the pregnant myometrium and therefore in cAMP-mediated relaxation as term approaches. G$_i$ may also have a role in regulating phosphoinositide phospholipase C (PI-PLC) and phospholilpase A$_2$ (PLA$_2$) [40–42], although current evidence indicates that G$_{q/11}$ is the G protein primarily responsible for coupling to PI-PLC [40,43,44]. Hence the myometrial changes would be consistent with an increase in the capacity for PI-PLC, and possibly PLA$_2$ activation and therefore IP$_3$-induced contraction as term approaches. However, previous studies, in the guinea pig at least, have indicated that in the myometrial cell membranes the capacity for GTP-stimulated activation of PI-PLC declines sharply in pregnancy and close to birth [45].

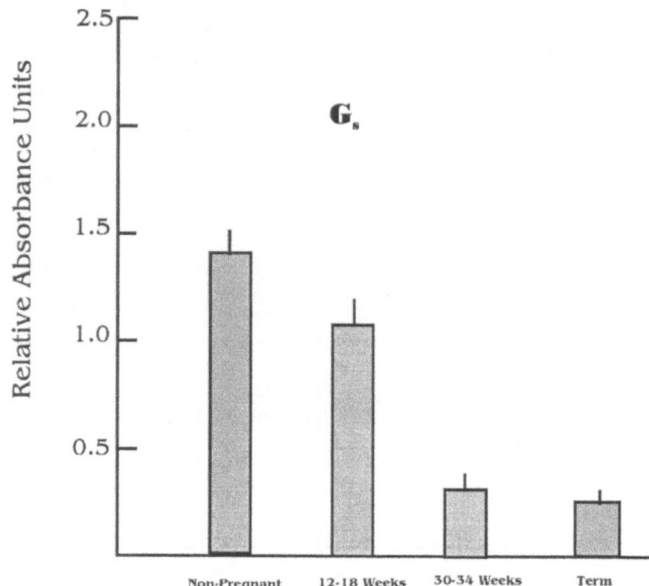

Fig. 4.

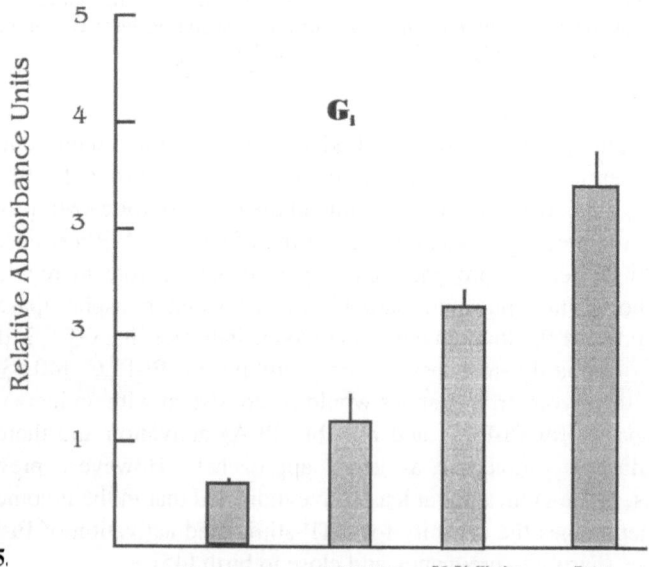

Fig. 5.

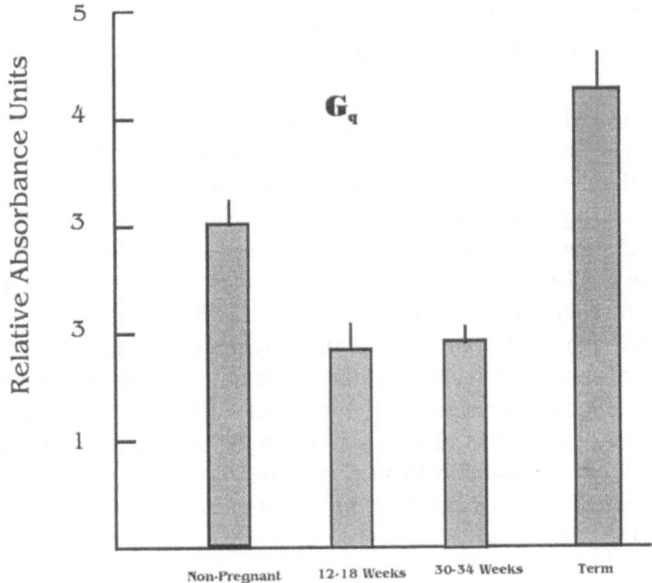

Fig. 6. Pregancy-related changes in G_q content of human myometrium. $G_{q/11}$ content was determined by western blotting using specific anitbodies against α-subunits and quantitfied by scanning laser densitometry

Fig. 4. Pregancy-related changes in G_s content of human myometrium. G_s content was determined by western blotting using specific anitbodies against α-subunits and quantitfied by scanning laser densitometry

Fig. 5. Pregancy-related changes in $G_{i\alpha2}$ content of human myometrium. $G_{i\alpha2}$ content was determined by western blotting using specific anitbodies against α-subunits and quantitfied by scanning laser densitometry

10.4 Pregnancy-Related Changes
in Myometrial Phosphoinositide Phospholipase C

There are at least four accepted, possibly five, different forms of PI-PLC, one of which, β-PLC, has been demonstrated to be firmly associated with the GTP-binding protein $G_{q/11}$ and activated by it [43,44]. In the guinea pig and rat uterus GTPγS has been shown to activate PI-PLC and thereby stimulate IP production [45–47]. Myometrium contains two PI-PLC isoenzymes α-PLC and δ-PLC (Fig. 7). Both isoenzymes are present in the cytosol and membrane fractions with at least 80% apparently cytosolic. As the enzymes readily leave the membrane on isolation it is not possible to predict confidently how much enzyme is membrane associated in the intact cell. However, cytosolic levels are a good index of total activity in the myometrial cell. Hence in the guinea pig uterus α-PI-PLC activity and content rises progressively throughout pregnancy (Fig. 8), whilst that of the δ-PLC rises then falls (Fig. 9). Similar changes in human myometrial PI-PLC

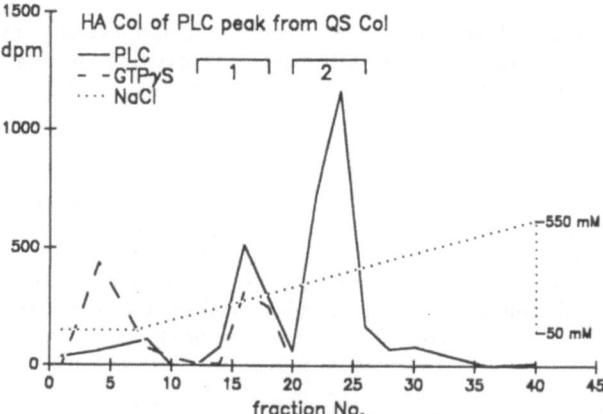

Fig. 7. Heparin–Agarose chromatography of cytosolic forms of phosphoinositide phospholipase C (PI-PLC) from guinea pig uterus at mid-pregnancy. A 70 000 *g* uterine cytsolic fraction was first separted on Q-Sepharose, then the PLC fractions applied to the heparin-agarose column and subsequently eluted with a linear NaCl 50–500 mM gradient. PLC activity is expressed as liberated [3H]inositol phosphate *(continuous line),* GTPγ35S binding was measured in all fractions *(discontinous line).* Peak 1 is α-PLC and *peak 2 is δ-PLC. Dotted line,* NaCl

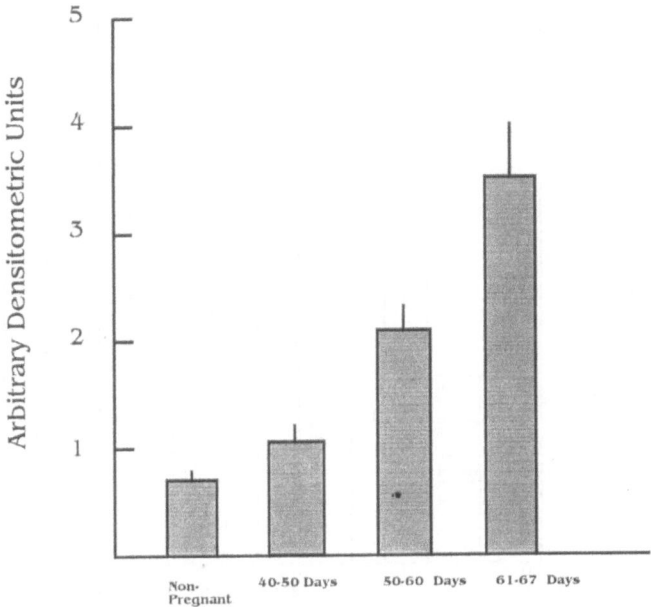

Fig. 8. Changes in the amount of cytosolic α-phosphoinositide phospholipase C (PI-PLC) in guinea pig uterus during pregnancy. Enzyme amounts were determined by western blotting using specific antibodies against α-PI-PLC and quantified by scanning laser densitometry

isoenzymes are also observed (Fig. 10). The previous reports of a decline in membrane PI-PLC and GTPγS stimulation of the enzyme probably reflects loss of the enzyme to the cytosolic fraction in preparation as total activity increases [45,46]. The α-PI-PLC form in the uterus, as in other tissues, has GTP-binding activity associated with it, and in the guinea pig myometrium this increases sharply close to term such that about 30–40% of the total cell GTP-binding activity is associated with this enzyme [48]. The GTP-binding activity is firmly associated with the enzyme and immunological studies indicate that this is G_i and possibly $G_{q/11}$. In the guinea pig myometrium, as has been shown for the rat [46–48], GTPγS will stimulate isolated partially purified α-PI-PLC that has GTP-binding activity associated with it (Fig. 10). However, the interaction is complex as the isolated enzyme

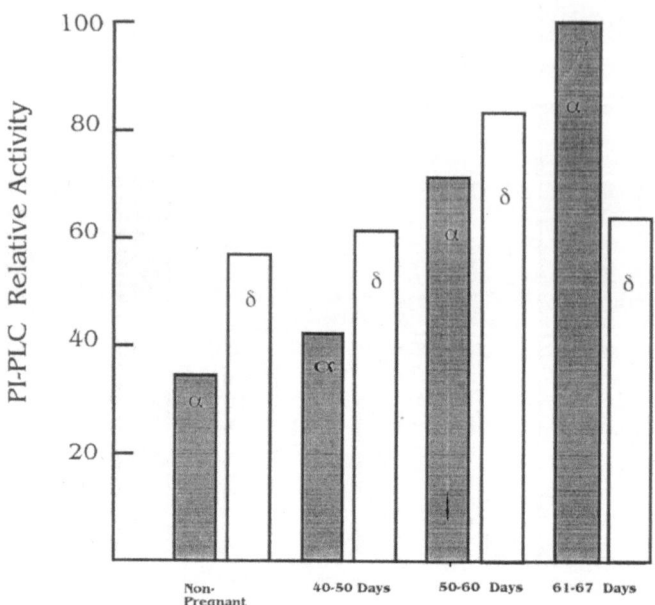

Fig. 9.

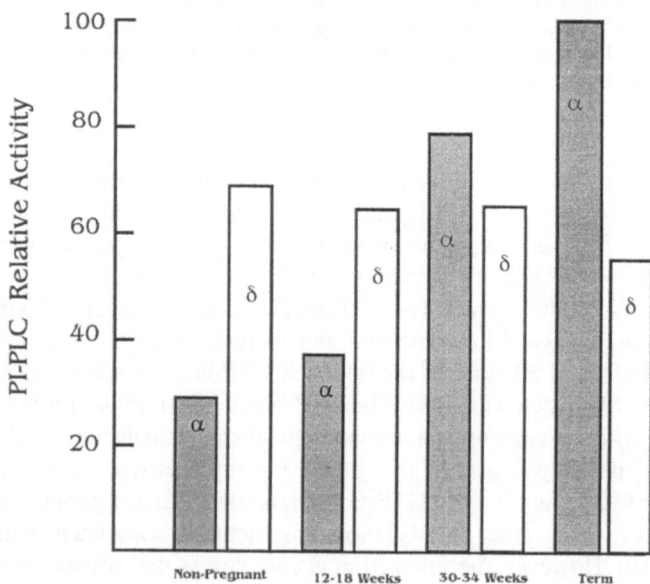

Fig. 10.

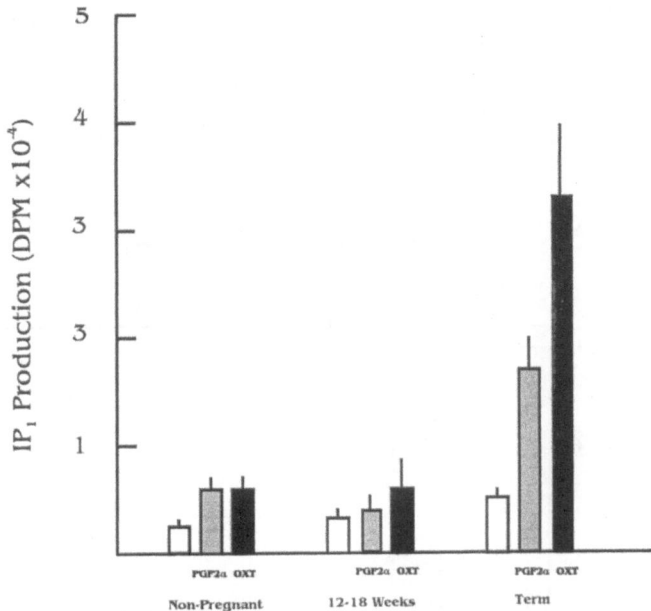

Fig. 11. Production of IP₁ by human myometrial cells. Cells were incubated for 30 min during which [3H]IP₁ production was measured. The stimulants used were either 10 μM PG$_{F2\alpha}$ or 1 μM oxytocin *(OXT)*. Other details are as for Fig. 1

is inhibited at midpregnancy and activated in late pregnancy, implying that factors in addition to G$_i$ or G$_{q/11}$ are probably involved, but are as yet unidentified (Fig. 10). Studies with permeabilized guinea pig myometrial cells demonstrate clearly that GTPγS has a marked stimulatory action on IP production. In such cells IP₁ and IP₂ are used as indices of IP₃ production, the latter being rapidly broken down [46].

Fig. 9. Relative changes in the activity of α- and δ-PI-PLC in uterine cytosol of guinea pigs during pregnancy. Isoenzymes were separated by chromatography as described in Fig. 7. The activity of α-PI-PLC in the near-term guinea pig uterus was set at 100%

Fig. 10. Relative changes in the activity of α- and δ-PI-PLC in human uterine cytosol during pregnancy. Details as for Fig. 9

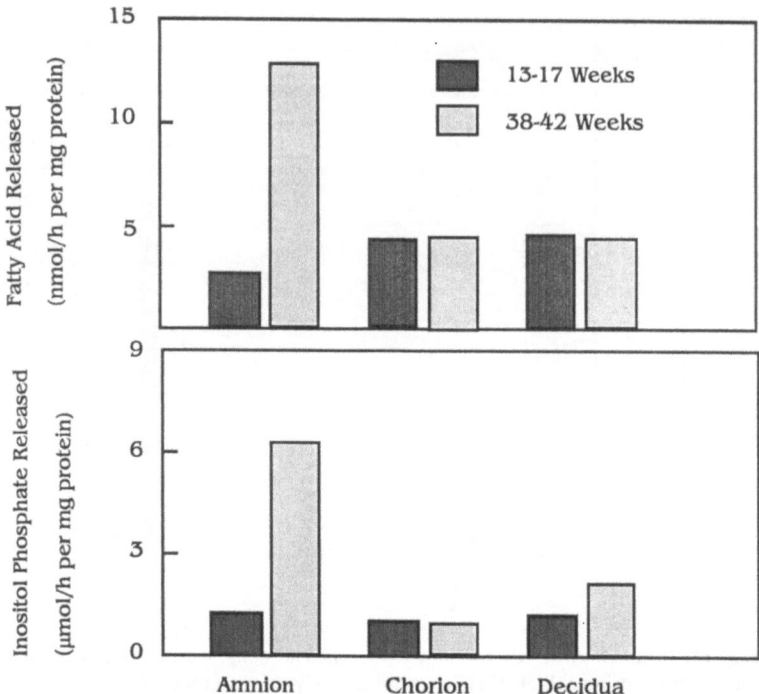

Fig. 12. Release of inositol phosphate and arachodonic acid from human amnion, chorion or decidua. Data is redrawn from Angle and Johnson [12]

These data indicate that as birth approaches, both in terms of PI-PLC activity and capacity for activation, there is a sharp increase in the potential for IP production and hence for IP-induced myometrial contraction. This is illustrated for $PG_{F2\alpha}$-induced IP_1 production in human myometrial cells which is sharply higher from term them mid-pregnant or non-pregnant tissue (Fig. 11). Similar observations have been made for IP production in amnion cells, but not chorion or decidua (Fig. 12).

10.5 Changes in Phospholipase A₂ (PLA₂) Activity in Myometrium During Pregnancy

Consistent with the increase in prostaglandin production late in pregnancy both in guinea pig and human myometrium there is a rise in amount of phospholipase A₂ (Figs. 13,14). Functional activity can be measured by following release of arachidonic acid from a prelabelled membrane phospholipid pool. Although the arachidonic acid can arise from the action of both PLC and PLA₂, the guinea pig myometrial cells more than 50% of the released arachidonic acid appears to arise from PLA₂ action. Hence it is the rise in arachidonic acid release as term approaches in cells from guinea pig and human myometrium is consistent with the changes in PLA₂ activity (Figs. 15,16). Similar increases have been reported for arachidonic acid release from amnion (Fig. 12).

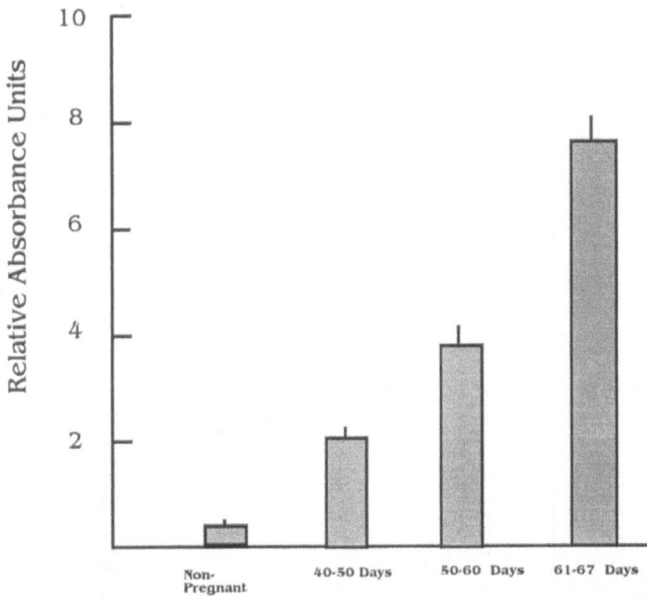

Fig. 13. Changes in the amount of phospholipase A₂ (34 kDa) in guinea pig uterus during pregnancy. Enzyme anount was determined by western blotting using a specific antibody against the 110-kDa sequence and quantitfied by scanning laser densitometry

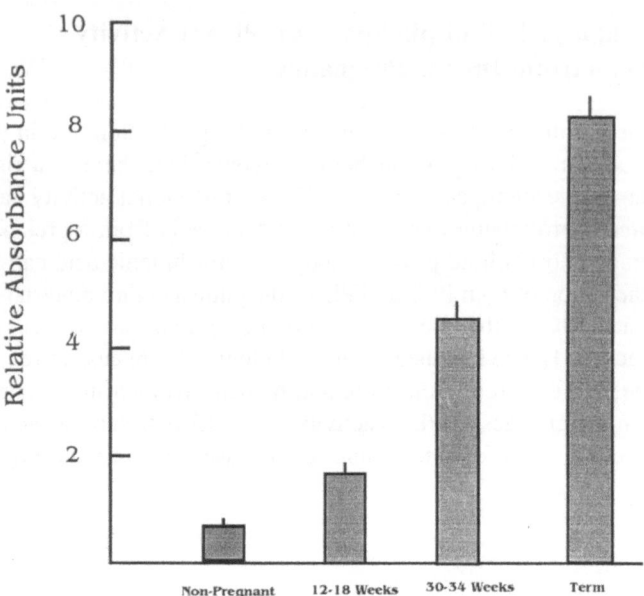

Fig. 14.

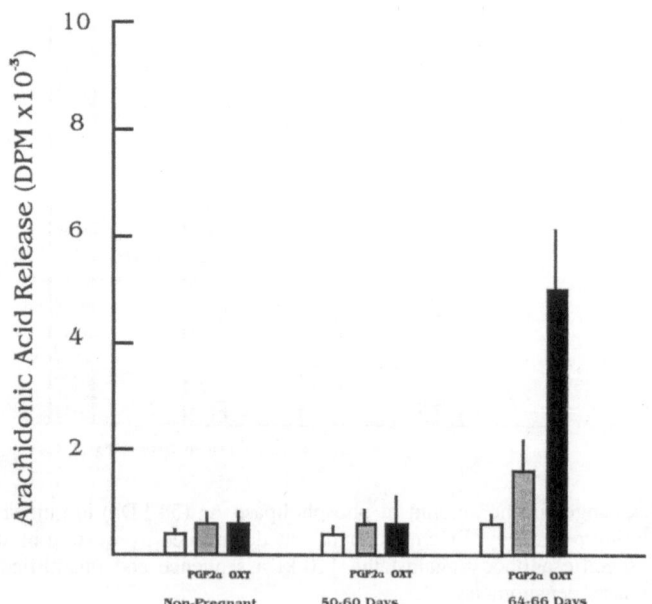

Fig. 15.

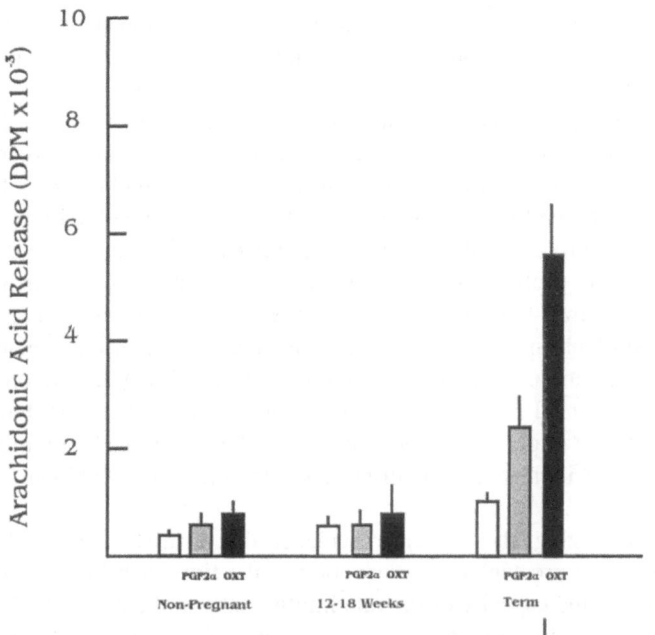

Fig. 16.

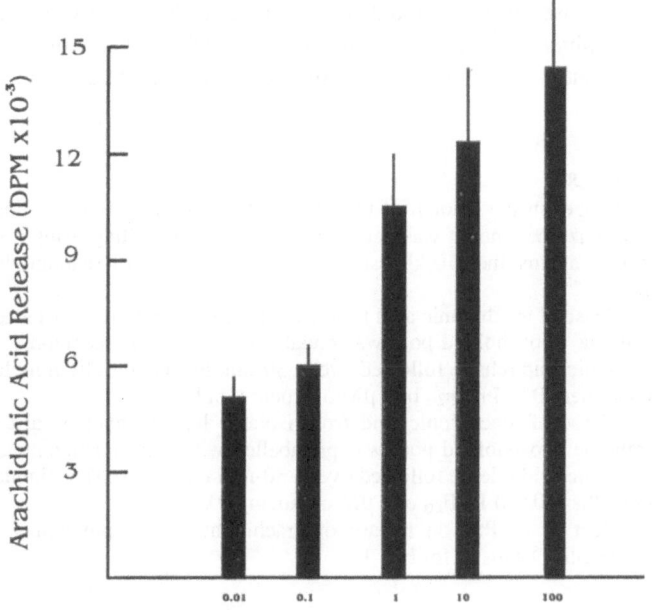

Fig. 17.

These observations indicate that increased production of prostaglandins at birth occurs against a background of a sharp rise in the capacity for production.

Moreover, the rise in content of G_i and possibly $G_{q/11}$ reported above could contribute a sharp rise in coupling efficiency between receptors and PLA_2. Arachidonic acid release in a number of systems has been shown to be G protein-mediated and may involve more than one G protein [49], although the most important G protein for enzyme activation is probably G_i [50]. However, in guinea pig myometrium [51], in contrast to rat [52], it appears that G_i is not involved in release of inositol phosphate, although it is unclear how this may relate to PLA_2 regulation. In guinea pig myometrial cells PLA_2 activation and arachidonic acid release [53] is G protein activated (Fig. 17), but the nature of the G protein involved has yet to be determined, preliminary experiments indicate that pertussis is not inhibitory to arachidonic acid release.

Summarizing, there is clear evidence that in association with the sharp rise in prostaglandin production at the time of birth there is an increase in the capacity of the rate limiting step in the pathway, PLA_2. This is not only through a rise in oxytocin receptor number, but also through increases in the amount of enzyme and the activity of the G protein coupling pathway. As to the latter, much further work is required to identify the components of the PLA_2 activation pathway.

Figs. pp 284/285
Fig. 14. Changes in the amount of phospholipase A_2 in human uterus during pregnancy. Enzyme amount was determined by western blotting using a specific antibody against the 110-kDa sequence and quantitfied by scanning laser densitometry
Fig. 15. Release of arachidonic acid from isolated guinea pig myometrial cells. The membrane phospholipid pool was prelabelled with [^{3}H] arachidonic acid and arachidonic acid release followed over a 30-min incubation.The stimulants used were either 10 μM $PG_{F2\alpha}$ or 1 μM oxytocin *(OXT)*
Fig. 16. Release of arachidonic acid from isolated human myometrial cells. The membrane phospholipid pool was prelabelled with [^{3}H] arachidonic acid and arachidonic acid release followed over a 30-min incubation.The stimulants used were either 10 μM $PGF2\alpha$ or 1 μM oxytocin *(OXT)*
Fig. 17. Effect of GTPγS on release of arachidonic acid from guinea pig myometrial cells. Details as for Fig. 15

Table 2. Important molecular changes in the pregnant uterus

Increases	Decreases
Oxytocin receptors	Denervation
α-Adrenergic receptors	β-Adrenergic receptors
PI-PLC	G protein G_s
IP production	
PLA$_2$	
Arachidonic acid production	
G proteins G_i and G_q	
Gap junction proteins	
Contractile proteins	
Cyclo-oxegenase	

PI-PLC, phosphoinositide phospholipase C; PLA$_2$, phospholipase A$_2$; IP, inositol phosphate

10.6 Summary

The picture presented by this and other studies is of a striking increase in the activity of the pathways potentially central to the activation of uterine smooth muscle contraction. It is particularly apparent that substantial changes occur in the pathways for arachidonic acid and inositol phosphate release and the G protein systems regulating them. This together with the main changes in the pregnant uterus summarized in Table 2 suggest that at the time of birth the uterus, although relatively insensitive to a wide range of hormones, may have its contractile state dictated largely by oxytocin and prostaglandins. The capacity to respond to these appears to rise sharply around the time of birth to ensure explosive contraction of the uterus. This is a potentially attractive outline of the significance of some of the major still leaves a number of important questions to be answered (Table 3).

Table 3. Questions for molecular mechanisms of labour

What is the mechanism of suppression of prostanoid production throughout pregnancy?

To what extent is oxytocin or are other agents responsible for enhanced prostanoid production?

Is the increase in uterine contractility at birth mediated by the action of inositol polyphosphates?

What is the role of external calcium in sustaining uterine contraction in labour?

Are the submyometrial layers the primary sites for uterotonic agent production?

Are the changes in uterine membrane transduction pathways important to contraction?

Are there any phosphorylation changes causing sustained tonic contraction?

Acknowledgements. We are grateful to Birthright, Action Research for the Crippled Child and the British Heart Foundation for financial support for these studies, to Dr S.J. Arkinstall and Dr R.M. Haigh for helpful discussions, to Prof. A.C.T. Turnbull for assistance in obtaining human myometrium and to R. Green, R. James and P. Williams for expert technical assistance.

References

1. Germain G, Gabrol D, Visser A, Sureau C (1982) Am J Obstet Gynecol 142:513–519
2. Harding R, Poore ER, Bailey A, Thorburn GD, Jansen CA, Nathanielsz PW (1982) Am J Obstet Gynecol 142:448–457
3. Thorbert G (1978) Acta Obstet Gynecol Scand Suppl 79:1–32
4. Arkinstall SJ, Jones CT (1985) J Reprod Fertil 73:547–557
5. Thorburn G.D (1979) Anim Reprod 2:1–27
6. Liggins GC (1983) Clin Obstet Gynecol 26:47–55
7. Casey ML, MacDonald PC (1988) In Fetal and Neonatal Development (Jones CT, ed) Perinatology Press, Ithaca, New York, pp 521–532
8. Liggins GC, Fairclough RJ, Grieves SA, Forster CS, Knox BS (1977) In The Fetus and Birth, CIBA Foundation Symposium (Knight J, O'Connor M) Elsevier North Holland, Amsterdam, pp 5–30
9. Soloff MS (1990) In Uterine Function: Molecular and Cellular Aspects (Carsten M.E, Miller J.D.) Plenum Press, New York, pp 373–392
10. Bleasdale JE, Johnston JM (1984) Rev Perinatal Med 5:151–191

11. Billah MM, DiRenzo GC, Ban C, Troung CT, Hoffman DR, Anceschi MM, Bleasdale JE, Johnston, JM (1985) Prostaglandins 30:841–850
12. Angle MJ, Johnsston JM (1990) In Uterine Function: Molecular and Cellular Aspects (Carsten ME, Miller JD. eds.) Plenum Press, New York, pp 471–500
13. Abel MH, Baird DT (1980) Endocrinology 106:1599–1606
14. Schrey MP. Holt JR, Cornford PA, Monaghan H, Al-Ubaidi F (1992) J Clin Endocrinol Metab 74:426–435
15. Reimer RK, Goldfien AC, Goldfien A, Roberts, JM (1986) Endocrinol 119:699–709
16. Wilson T, Liggins GC, Aimer GP, Skinner SJM (1985) Biochem Biophys Res Comm 131:22–29
17. Saeed SA, Mitchell MD (1982) Prostaglandins 8:635–640
18. Ruzycky AL, Crankshaw DJ (1988) Can J Physiol Pharmacol 66:10–17
19. Carsten ME (1974) Prostaglandins 5:33–40
20. Fukuo K, Morimoto S, Koh E, Yukawa S, Tsuchiya H, Imanaka S, Yamamoto H, Onishi T, Kumashara Y (1986) Biochem Biophys Res Commun 136:247–252
21. Bulbring E, Tomita T (1987) Pharmacol Rev 39:49–96
22. Kamm KE, Stull JT (1989) Ann Rev Physiol 51:299–313
23. Bean BP (1989) Ann Rev Physiol 51:367–384
24. Hallam TJ, Rink TJ (1989) TIPS 11:8–10
25. Somlyo AV, Bond M, Somlyo AP, Scarpa A (1985) PNAS 82:5231–5235
26. Berridge MJ (1990) Ann Rev Biochem 56:159–193
27. Soloff MS, Alexandrova M, Fernstrom MJ (1979) Science 204:1313–1315
28. Fuchs AR, Fuchs F, Husslein P, Soloff MS (1984) Am J Obstet Gynecol 150:734–741
29. Cha KS, Lee WC, Rudzik A, Miller JW (1965) J Pharmacol Exp Ther 148:9–13
30. Owman C, Alm P, Rosengren E, Sjoberg NO, Thorbert G (1975) Am J Obstet Gynecol 122:961–964
31. Arkinstall SJ, Jones CT (1988) Am J Physiol 255:E272–E279
32. Arkinstall SJ, Jones CT (1989) Am J Physiol 256:E215–E220
33. Harbert GM, Spisso KR (1981) Am J Obstet Gynecol 139:767–777
34. Legrand C, Maltier JP (1986) J Reprod Fertil 76:415–424
35. Riemer RK, Goldfien A, Roberts JM (1987) J Pharmacol Exptl Ther 240:44–50
36. Bourne HR, Sanders DA, McCormick F (1990) Nature 349:125–132
37. Hepler JR, Gilman AG (1992) TIBS 17:383–387

38. Roberts JM, Riemer RK, Bottari SP, Wu YY, Goldfien A (1989) J Devel Physiol 11:125–134
39. Arkinstall SJ, Jones CT (1990) J Endocrinol 127:15–21
40. Sternweis PC, Smrcka AV (1992) TIBS 17:502–506
41. Nakashima S, Hattori H, Shirato L, Takenaka A, Nozawa Y (1987) Biochem Biophys Res Commun 148:971–978
42. Teitelbaum I (1990) J Biol Chem 265:4218–4222
43. Smrcka AV, Hepler JR, Brown KO, Sternweis PC (1991) Science 251:804–807
44. Blank JL, Ross AH, Exton JH (1991) J Biol Chem 266:18206–18216
45. Arkinstall SJ, Jones CT (1990) Am J Physiol 259:E57–E65
46. Wichelhaus DP, Khouja A, Jones CT (1992) J Devel Physiol 18:179–186
47. Wen Y, Anwer K, Singh SP, Sanborn BM (1992) Endocrinol 131:1377–1382
48. Wichelhaus DP, Jones CT (1992) J Devel Physiol 18:49–58
49. Kajiyama Y, Murayama T, Kitamura Y, Imai SI, Nomura Y (1990) Biochem J 270:69–75
50. Gupta SK, Diez E, Heasley LE, Osawa S, Johnson GL (1990) Science 249:662–666
51. Marc S, Leiber D, Harbon S (1988) Biochem J 255:705–713
52. Liebmann C, Offermans S, Spicher K, Hinsch KD, Schnittler M, Morgat JL, Reissmann S, Schultz G, Rosenthal W (1990) Eur J Pharmacol 207:67–71
53. Khouja A, Jones CT (1993) J Devel Physiol 19:1–7

11 Pros and Cons
of Preterm Labor Treatment

Joachim Dudenhausen

In 1991 the *New England Journal of Medicine* reported on the survival of a newborn weighing only 280 g at birth (Muraskas et al. 1991). A 36-year-old woman pregnant for the first time after induced ovulation and homologous insemination achieved a triplet pregnancy. Two embryos were destroyed transabdominally using KCl solution in the tenth week of gestation. Growth and karyogram of the surviving fetus were normal. Severe preeclampsia developed. At 26 + 6 weeks the Doppler blood flow examination showed an enddiastolic stop. Due to this indication a cesarean section was performed. The newborn female showed signs of severe symmetric intrauterine retardation with a weight of 280 g and a body length of 25 cm; the Apgar score was 6 at 1 min and 9 at 5 min. After spending 120 days in the neonatal intensive care unit, the baby was sent home weighing 1900 g and with a length of 41 cm. After a year had passed she weighed 4390 g and was 61 cm in length;

Table 1. Survival rate depending on the gestational age at birth and on the birthweight (after Goldenberg et al. 1984)

Gestational age (weeks)	Weight (g)	Survival rate	Increase (%) per week	per day
24	650	17	13	1.9
25	775	30	21	3.0
26	900	51	13	1.9
27	1025	64	11	1.6
28	1150	75	6	0.9

all values below the fifth percentile; at the neuropediatric follow-up examination she was found to have a mild developmental delay relative to her adjusted age. This example illustrates a large number of the legal and ethical problems of perinatal medicine which today influence the work in obstetrical and neonatological departments. It is impossible for me to formulate principles of how this case should have been managed, giving ethical argumentation to which all sides of a pluralistic society would agree. But in any event this is an example of how an extremely underweight and immature baby survived. It must be made clear that survival is dependent on the gestational age at birth and on the birthweight. This connection can be seen when looking at this case. And it is also apparent that prolongation of the pregnancy by 1 week, for example from 25 to 26 weeks brings about an increase of the survival rate of 13% (Table 1).

11.1 Risk Factors and Diagnosis of Preterm Labor

Now we have arrived at the central question, namely, to what extent are morbidity and mortality of premature babies reduced when pregnancy is prolonged. Here the success of the therapy is not to be seen in the rate of prematurity but rather in the prolongation of the pregnancies and in the improvement achieved in the survival rate. For example, if one compares the use of tocolytic substances with the rate of babies born prematurely, as was done in various parts of Bavaria within the scope of the Bavarian perinatal census, then in spite of the broad span

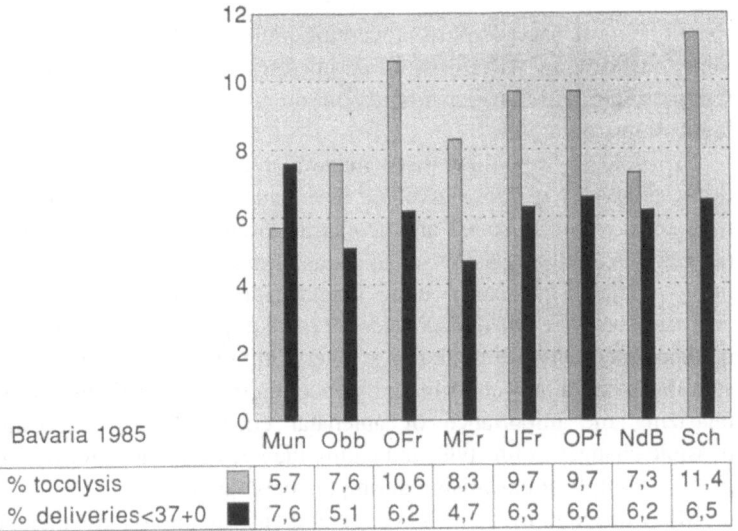

Bavaria 1985	Mun	Obb	OFr	MFr	UFr	OPf	NdB	Sch
% tocolysis	5,7	7,6	10,6	8,3	9,7	9,7	7,3	11,4
% deliveries<37+0	7,6	5,1	6,2	4,7	6,3	6,6	6,2	6,5

Fig. 1. Composition of the use of tocolytics with the rate of prematures in various parts of Bavaria (after Dudenhausen 1986)

of oral tocolytic substances prescribed, no great differences can be seen in the rate of premature births (Fig. 1; Dudenhausen 1986). Therefore it is important to point out that, when considering tocolytic substances, the therapeutic goal is not so much to judge the rate of premature births in a region, but rather to judge the effect of tocolytic substances in prolongation of the pregnancy and improvement in the survival rate.

There has been hardly any change in the frequency of premature births since the 1950s. At the present time the rate of premature births in western countries is mostly between 4% and 9% (Riegel et al. 1991). Premature babies account for two thirds of the infant mortality, and about 60% of the morbidity in newborns can be traced back to prematurity.

Various risk factors are known from the literature, including pregnancy with a higher risk of prematurity. Medical and surgical measures for the treatment of the symptoms of threatened premature delivery are well known. With this therapeutic arsenal one would expect that we would be more successful in preventing premature births. The fact remains that despite our efforts, we do not know enough about why contractions start and that, for this reason, no great progress has yet been made in this field.

Manifold attempts have been made to prevent premature contractions, and many of those involved in antenatal counselling think that they are a result of social and economic circumstances. As a consequence it is thought that the social status of pregnant women should be improved, such that social status would have less effect on them. In any case social treatment is considered to be of greater importance than medical intervention. I think this is a very important point to which obstetricians could also contribute: by encouraging their patients, by emphasizing the importance of antenatal check-ups, by suggesting possible changes in lifestyle, and taking the necessary steps for treating medical complications arising during the pregnancy.

Many different measures have found their way into the treatment of threatened premature birth; unfortunately their value is not undisputed and quite often the expectations aroused by the originator are not in any way fulfilled. Additionally when weighing up the pros and cons of some of these measures, we have to bear in mind that not only is the success questionable, but also that dangerous situations for mother and child can arise.

One of the main reasons why the various methods of management are not uniformly accepted can be seen in the fact that we still do not have a definite and widely established opinion about the causes and progress of premature birth. We do not even have a widely accepted definition as to what is meant by "threatened premature birth" or "premature contractions." Are cervical effacement or cervical dilatation or a prefixed number of uterine contractions the causative symptoms? There is no consensus at present as to the number of uterine contractions necessary within a certain period of time to be considered premature contractions and thus as a prerequisite for therapy.

For some time ultrasound has been playing a role in diagnosing premature contractions. The visualization of the effacement of the cervical canal is certainly a very important criterion, as is the protrusion of the membranes into the cervical canal with a funnel-shaped widening of the os uteri.

There is no doubt whatsoever that many studies on the importance of tocolytic substances, for example, have come up with such excellent results because women who in fact did not have premature contractions requiring treatment were included in their study cohort.

Postgraduate training of doctors and midwives in the early detection of premature contractions is a very important approach. Unfortunately, however, even with these risk factors and this training, it is very difficult to distinguish the patients who are really at risk of giving birth prematurely from those who are not. So, some women are sent to a high-risk antenatal care unit and receive intensive prenatal care who in fact do not need it. Of course it is very important that pregnant women be taught how to recognize early contractions themselves. On the other hand, women in the so-called high risk group can be helped today by using a commercially available tocometer. They can attach this device themselves and register the uterine contractions; these are then passed on by telephone to a central exchange and if necssary it can be arranged that a midwife visit the patient daily. It is quite clear that the value of these measures has *not* yet been proved in prospective studies.

11.2 Success of Tocolytic Therapy

Tocolytic substances are used very often in the western world. The high expectations which were placed on tocolytics for the treatment of premature contractions have not been completely fulfilled. Even if the rate of premature birth has not been decreased, some authors have reported that pregnancies have been prolonged. Today we can certainly say that delivery can be inhibited for up to 2 weeks by intravenous administration of β-sympathomimetics, whereas long-term use of this medication is questionable (Caritis et al. 1984; Downey and Martin 1983; Graber 1992). Accompanying the short-term effect of the tocolytics there has also been a change in the rate of cases of respiratory distress syndrome and a change in morbidity and mortality. Some authors are doubtful about the effect of the tocolytics and suggest that only in cases where delivery would not have taken place anyway was an effect to be seen. On the other hand, however, enough randomized studies can substantiate a short-term effect. The success rate of the various tocolytics is very similar (Beall et al. 1983). Today it is recognized that tocolytic treatment enables us to prolong pregnancies for the promotion of lung maturity with glucocorticoids.

It is questionable how many pregnant women with an indication for tocolytic therapy should really be given this medication. We are convinced that there are few contraindications for tocolytics. So premature rupture of the membranes or preeclampsia are not ipso facto a contraindication for tocolytic therapy. However, the amnion infection syndrome is a very important contraindication. When there are signs of infection the pregnancy should be terminated. This also applies to the time of 25–30 weeks. Certainly from the point of view of neonatal care it is preferable to have a slightly immature infant without infection than a less immature infant with an infection. This is one of the reasons why today's interest is focused on finding parameters for early diagnosis of intrauterine infection. The value of tocolysis is undisputed in cases of external version of the fetus from breech presentation to vertex presentation or when situations of acute risk occur during labor.

11.3 Side Effects of Premature Labor Treatment

Feneterol is the β-sympathomimetic substance used most in Germany, while in the United States ritodrine is more widespread. Its side effects, particularly on the heart circulatory system and also on metabolic functions, have been widely described. Some guidelines for its use have been established, such as having an ECG before starting tocolytic therapy and also taking blood sugar controls into consideration and the increased insulin requirement for diabetic patients. There is no doubt at all that magnesium sulfate has an inhibiting effect on contractions, but only to a slight extent. Magnesium sulfate is mostly used today as an adjuvant to β-sympathomimetics. Some other substances are also used in addition as for example indomethacin and acetylsaliaphic acid. The negative effects of these medicines were discussed, for example, to affect on the ductus arteriosus; it is crucial that they are not dangerous to the child.

Intravenous administration of β-mimetics stimulates the β-receptors in many organ systems and is responsible for various clinical side effects that are connected with this form of therapy. Maternal cardiovascular side effects are seen most frequently from the β-adrenergic medications, particularly hypotension, tachycardia, and arrhythmia.

The stimulation of the vascular β-2 receptors leads to vasodilatation and thus to hypotension. The result is a reflectory compensation with an increase in heart rate, in the beat volume, and in the cardiac output. Cardiac output during tocolytic therapy with β-mimetics can be up to over 50% above the original level. This is why β-mimetic therapy should not be performed on women known to have a heart insufficiency.

Cardiac arrhythmia is also described in β-mimetic therapy; here it is mostly supraventricular tachycardia, but atrial fibrillation and ventricular ectopy have also been described. Due to these side effects it is necessary that a detailed case history be drawn up, paying special attention to heart diseases, and that an ECG should be performed before the β-mimetic therapy is started.

In some cases myocardial ischemias have been described during β-adrenergic therapy. However, changes in the ECG which show the presence of myocardial ischemia were only observed by a few authors. More frequently temporary ST segment depressions are described which are dose dependent and which regress after the dosage has been reduced or after the therapy has been discontinued (Michalak et al. 1983). Observations have been made which suggest that coronary artery perfusion is reduced during tocolytic therapy. Also, the fact that observations have been made about the increased cardiac enzyme and myocardium-specific myoglobin has led to the view that subclinical myocardial impairments can occur. Clinical interest has focused on cases of maternal lung edemas that were reported during β-adrenergic therapy. Up to 1992, 80 such cases have been reported (Besinger and Niebyl 1990). Regularly when the therapy starts pulmonary edema occurs, often associated with glucocorticoid therapy parallel to β-adrenergic therapy. It is often the combination of glucocorticoid therapy and β-adrenergic therapy in multiple pregnancy. Also anemia and blood transfusions create a risk factor for lung edema.

The pathophysiology of this particularly severe complication has not yet been definitely explained. At the center of the explanation we have excessive stress to the heart during β-adrenergic therapy and a cardiac reaction to present impairments.

The large and broad range of β-adrenergic complications relates to metabolic side effects as, for example, the effect on the glucose metabolism and the potassium metabolism. Here it should be emphazised that an abnormal glucose tolerance can occur during β-mimetic therapy. Furthermore there is wide discussion whether β-mimetics are contraindicated in insulin-dependent diabetes. We do not think this is the case. The insulin dosage should be changed corresponding to the daily profile.

The maximal fall in serum potassium levels normally occurs a few hours after the start of the therapy. After about 1 day the serum potassium concentration is normal again, just as the serum glucose and insulin levels stabilize again.

The results of blood flow examinations in animals and humans during β-mimetic therapy are contradictory. Some authors have found a drop in the uteroplacental blood flow, others report an increase. In a third group -- as so often happens in medical studies – no changes were found at all. It may be that these differences are to be found in the various β-mimetics or in the duration of their usage or in the form of the adjuvant medication. Nonetheless this question has not been definitely answered. This question is really important because the therapy of limited uteroplacental perfusion, for example, with intrauterine growth retardation is seen by many as an indication for tocolysis. There are not enough results to prove this.

11.4 Side Effects on the Fetus

The effects of tocolytic therapy with β-sympathomimetics on the fetus or on the newborn have not been examined in such detail as the side effects on the mother. The maturation of the sympathetic system and the development of α- and β-receptors in the various organ systems enables the fetus to react to corresponding pharmacological stimuli. Thus immediately after intravenous administration of β-sympathomimetics fetal tachycardia occurs, which is interpreted as a result of a direct stimulation of the β-receptors on the fetal myocardium. Sodha and Schneider (1983) have reported that β-mimetics with a molecular weight of 300–500 can pass through the placenta and that due to their positive load and to an extensive insolubility in organic solvents are dependent for their entry on the pores of the placental membrane (Table 2). The placenta takes over a certain protective function. When

Table 2. Transplacental diffusion of four β-mimetics after bolus injection into intervillous space (after Sodha and Schneider 1983)

	Venous return (Maternal circuit)	Venous return (Fetal circuit)
^{3}H-fenoterol	87	2.3
^{3}H-ritodrine	92	2.4
^{14}C-hexoprenaline	91	1.1
^{3}H-salbutamol	88	2.8
^{14}C-albumin	94	0.2

the β-mimetics enter through the placenta they are partially inactivated by a conjugate formation. It has been shown in experiments that after Feneterol infusion a relatively higher portion is found as conjugate – therefore pharmacologically inactive – in fetal plasma than in maternal plasma. The enzyme edge of the placenta for conjugate formation of various β-mimetics has been proved by Schneider's working unit. In this connection results by Siimes et al. (1978) are interesting; they observed that in the sheep fetus after infusion of ritodrine directly into the fetal circulation there was a clear increase in the heart frequency,

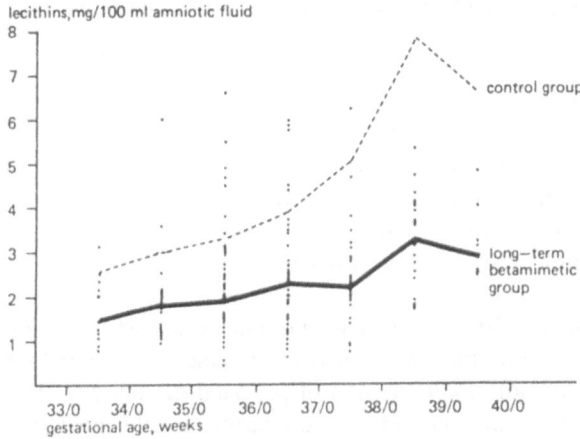

Fig. 2. Medians of the lecithin contents in amniotic fluid in long-term β-mimetic therapy group and in the control group dependent on the pregnancy week (after Dudenhausen et al. 1978)

whereas the dose given to the mother animal did not have any effect on the action of the fetal heart.

The effect on the surfactant production of the unborn child seems to be an important secondary function. While some authors found a drop in the membrane syndrome rate after short term treatment with β-adrenergic substances, some years ago we pointed out that even in long-term administration of β-mimetics limited lung maturity occurs (Fig. 2), (Dudenhausen et al. 1978). For management this means that before long-term tocolysis is terminated, the lung maturity in the fetus must be established.

An important side effect in the infant is hypoglycemia in the new-born after long-term tocolysis. This phenomenon seems to be caused by a hyperinsulinism and an increased growth hormone level which is due to a β stimulation of the fetal pancreas.

11.5 Conclusion

After all that has been said, it is clear that the current situation in treating threatened premature birth and delivery is not satisfactory. We would certainly be further advanced in this if we had more exact information about the pathophysiology of premature onset of labor. What is premature onset of labour? Is it

- An immunological problem
- A problem to do with infection
- A question of the oxytocin receptors
- A question of the prostaglandin synthesis
- A PO_2 problem of the fetus
- A question of the fetus as a trigger?

During the past 20 years we have not discovered why the fire started, only been treating the smoke. It is of extreme importance for the drecrease of infantile mortality and of perinatal mortality in the western world to reduce the rate of premature births, and this goal can only be achieved when we have more precise information about the course of premature labor and the reason for its happening.

References

Beall, MH, BW Edgar, RH Paul, T Smith-Wallace (1985) A comparison of ritodrine, terbutaline, and magnesium sulfate for the suppression of preterm labor. Am J Obstet Gynecol 153:834–839

Besinger RE, JR Niebyl (1990) The safety and efficacy of tocolytic agents for the treatment of preterm labor. Obstet Gynecol Surv 45:415–440

Caritis SN, G Toig, LA Heddinger, G Ashmead (1984) A double-blind study comparing ritodrine and terbutaline in the treatment of preterm labor. Am J Obstet Gynecol 150:7–14

Downey LJ, Martin AJ (1983) Ritodrine in the treatment of preterm labour: a study of 213 patients. Br J Obstet Gynaecol 90:1046–1053

Dudenhausen JW (1986) Kontroverse über die orale Tokolyse. In: JW Dudenhausen, E Saling (Hrsg): Perinatal Medizin. Band XI. Thieme, Stuttgart-New York, S 55–57

Dudenhausen JW, Kynast G, Lange-Lindberg AM, Saling E (1978) Influence of long-term beta-mimetic therapy on the lecithin content of amniotic fluid. Gynecol Obstet Invest 9:205–209

Goldenberg RL, Nelson KG, Davis R (1984) Delay in delivery: influence of gestational age and the duration of delay on perinatal outcome. Obstet Gynecol 64:480–492

Graber EA (1992) Prematurity 1992. Obstet Gynecol Surv 47:521–524

Michalak D, Klein V, Marquetti GP (1983) Myocardial ischemia: a complication of ritodrine tocolysis. Am J Obstet Gynecol 146:861–869

Muraskas JK, Carlson NJ, Halsey C, Frederiksen MC, Sabbagha RE (1991) Survival of a 280 g infant. N Engl J Med 324:1598–1599

Riegel K, Ohrt B, Brandmaier R (1991) Prognose von Früh- und Mangelgeborenen. Gynäkol prax 15:1–11

Siimes ASJ, Creasy RK, Heymann MA (1978) Cardiac output and its distribution and organ blood flow in the fetal lamb during ritodrine administration. Am J Obstet Gynecol 132:42–49

Sodha RJ, Schneider H (1983) Transplacental transfer of beta-adrenergic drugs studied by an in vitro perfusion method of an isolated human placental lobule. Am J Obstet Gynecol 147:303–310

12 Therapies for Starting and Stopping Labour

Andrew A. Calder

The clinical objective of exercising control over the human uterus arises when it becomes desirable either to interrupt the pregnancy or to inhibit parturition. The most obvious examples of these requirements are when the appropriate timing of parturition has been disturbed either by failing to occur at term of pregnancy or by commencing or threatening to commence before term. In both of these circumstances, the welfare of the fetus may be in jeopardy and may ultimately depend on the clinician's ability to cause it to remain within the uterus while it matures or to bring about its safe expulsion from a hostile environment.

In addition to complications which interfere with the appropriate timing of parturition and which call for interventions which aim to restore delivery at term of pregnancy, there are other clinical circumstances in which it may be appropriate to effect preterm delivery (because continuation of the pregnancy exposes the mother, offspring or both to dangers) or even occasionally to inhibit labour at term (for instance, if uterine contractions appear to be causing fetal hypoxia).

In common with the pattern of research in human parturition during recent decades, the proceedings of this Schering Foundation Workshop have focused principally on physiological mechanisms relating to contractility of the myometrium. It is gratifying to note that due recognition is also now being accorded to that other component of human parturition, namely, the behaviour of the uterine cervix. Although long considered to represent two component parts of a single organ (namely, the uterus) the corpus and cervix are so fundamentally different in their tissue composition, function and control as almost to deserve the status of separate organs. In essence, the corpus consists almost entirely of smooth muscle while the cervix contains very little myometrium and is predominantly made up of fibrous connective tissue. Furthermore, the corpus and cervix might be regarded as fulfilling passive and active roles, respectively, which change to active and passive, respectively, during parturition. Thus, the corpus remains passively quiescent through the course of pregnancy and becomes actively contractile during labour; the cervix must fulfil an active role of remaining closed and unyielding during pregnancy, changing to a more passive yielding and dilating role during labour. Clinical control of parturition must, therefore, take account of both uterine components, and those agents and procedures which may be applied for such purposes must be considered in the context of both corpus and cervix. The therapeutic weapons available to the clinician for these purposes may be considered in a number of different categories, depending on whether they are, for instance (a) physiological or unphysiological, (b) proparturient or antiparturient, and (c) in current clinical use, under investigation, potentially exciting for future exploration, or long since abandoned. The catalogue of therapies and potential therapies is extensive (Fig. 1) and in the interests of simplicity will be considered under the following headings:

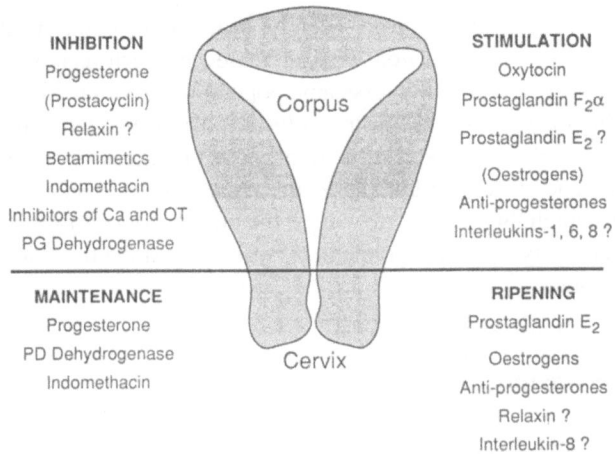

Fig. 1. Some of the main biological and pharmacological substances known or postulated to influence myometrial contractility and cervical ripening

- Oxytocin and its inhibitors
- Prostaglandins and their inhibitors
- Placental steroids and their antagonists
- Relaxin
- Cytokines
- Miscellaneous pharmacological tocolytic drugs

Before considering these categories individually, however, mention should be made of physical interventions which may influence the continuation or interruption of pregnancy. The most time honoured of these is amniotomy. It has been recognized for several centuries that physical interference to rupture the fetal membranes will usually result in delivery of the fetus. This procedure was recommended by Thomas Denham of the Middlesex Hospital, London, in the mid-eighteenth century to advance the timing of delivery with the object of avoiding cephalopelvic disproportion. As a result it became known in Europe as "the English method" of labour induction. Other forms of mechanical interference such as the passage of uterine bougies or intracervical balloons, were popularized in the mid-nineteenth century by Tarnier and de Ribes (Speert 1958). More recently, with the recognition that the risks of intrauterine infection were increased by such techniques and particularly by amniotomy, it became popular to carry out a process of "sweeping the membranes" from the lower pole of the uterus. It is now recognized that such effects as were provoked by these physical forms of interference were attributable to the release of endogenous prostaglandins which they stimulated within the uterine compartment (Keirse 1979). The only "mechanical method" which has stood the test of time is amniotomy.

12.1 Oxytocin

Oxytocin, this decapeptide hormone emanating from the posterior pituitary, has been the substance most widely used for labour induction and augmentation and remains in many respects the most important. It was discovered by Sir Henry Dale (Dale 1906) and first applied in clinical practice by William Blair Bell (Bell 1909). Ever since then oxytocin has been the subject of fierce controversy. At first the preparations were crude extracts of pituitary and were of widely varying potency. Following intramuscular administration (which was initially favoured) absorption was unpredictable. In consequence, such preparations often led to either poor success or violent and uncontrolled uterine stimulation. The latter could lead to asphyxia and death of the fetus, rupture of the uterus and even, in some instances, death of the mother. It is thus hardly surprising that it attracted violent opposition.

In spite of this, oxytocin has evolved into a safe therapeutic weapon. Three principal developments have brought this about. The first was the isolation of pure oxytocin and its chemical characterization (Du Vigneaud et al. 1953) and its subsequent commercial synthesis (Boissonas et al. 1955). The second was the demonstration and acceptance that in contrast to other routes of administration (such as intranasal, sublingual and buccal) the optimal route of administration from the standpoint of safety and reliability is intravenous. The third was the recognition that the sensitivity of the pregnant uterus to oxytocin varies widely between individuals and at different stages of gestation. From this emerged the important principle of "oxytocin titration" (Turnbull and Anderson 1968) with the use of sophisticated apparatus to control the rate of infusion. The infusion of oxytocin is commenced at a low rate but this is rapidly increased until uterine contractility becomes established. The dose rate is thus titrated against the myometrial response, and it is crucial that the dose rate should not be mindlessly increased beyond the level producing effective uterine contractility. Indeed, it may be appropriate to reverse the increase of dose rate once labour becomes well established.

Clinical experience has demonstrated that the effectiveness of oxytocin is greatly increased if it is combined with performance of amniotomy. Unless the fetal membranes are ruptured at the outset or early in the attempt at labour induction, oxytocin is of little value.

Amniotomy stimulates the release of endogenous prostaglandins within the uterine compartment and this may be responsible for sensitizing the myometrium to the action of oxytocin.

On a world scale, oxytocin (as synthetic "Syntocinon") in combination with amniotomy, represents the most popular and widely employed procedure for labour induction. It is generally highly successful but carries a number of drawbacks. The most important of these is a widely varying record of success, which is particularly related to the condition of the cervix (Calder 1981). When the cervix is unripe, particularly in primigravid subjects, the ability of oxytocin to advance the labour effectively is seriously diminished. This almost certainly derives from the fact that oxytocin plays no part in the process of cervical ripening and while it may produce powerful uterine contractility in such circumstances, this may be counterproductive if all that results is a prolonged nonprogressive labour, which may be harmful to the fetal condition (Calder 1979). Thus, if the clinician strives to perform amniotomy and infuse oxytocin when the cervix is unripe, the likely result is a prolonged labour with an increased incidence of intrauterine infection, fetal hypoxia and the need to resort to delivery by caesarean section. Other deficiencies of oxytocin as a labour-inducing agent arise from its association with neonatal hyperbilirubinaemia (Ghosh and Hudson 1972), an increased association with post-partum haemorrhage (Kennedy et al. 1982) and a clear dislike of the method on the part of the mothers concerned (Moran-Ellis 1991). In spite of these drawbacks, if applied in properly selected cases, oxytocin remains highly effective and is still the agent of choice for improving the quality of labours of spontaneous onset which fail to show normal patterns of progress.

12.2 Oxytocin Antagonists

A number of substances have been developed which appear to oppose the action of endogenous oxytocin and their ability to influence uterine contractility has been explored (Akerlund et al. 1985). These have not been used widely in clinical practice although early claims have been made regarding their potential usefulness as inhibitors of premature labour (Akerlund et al. 1987). Whether these agents will become important in the therapeutic armamentarium for the inhibition of premature labour remains to be seen and may ultimately depend on the extent to which oxytocin itself is crucial for human parturition.

12.3 Prostaglandins

As has been discussed elsewhere in this volume, the prostaglandins are of special importance in respect of uterine contractility. $PGF_{2\alpha}$ appears to be invariably associated with stimulation of myometrial contractility. Prostacyclin (PGI_2) is almost certainly invariably inhibiting to the myometrium. PGE_2 on the other hand remains controversial. In many circumstances, especially the nonpregnant myometrium, PGE_2 appears to inhibit uterine contractility, while at some stages of pregnancy it appears to be stimulatory (Horton 1972). Although its widely applied success in pharmacological induction of labour points to a direct stimulant effect on the myometrium at term, we have grounds for believing that, in some circumstances, PGE_2 acts via a provocation of endogenous $PGF_{2\alpha}$ production (Greer et al. 1990). Perhaps the most important aspect of PGE_2, however, lies in the fact that it appears to represent the pre-eminent prostanoid for the process of cervical ripening (Calder and Greer 1991, 1992a).

Prostacyclin has not been investigated or applied to any extent in the control of parturition and while $PGF_{2\alpha}$ is undoubtedly a highly potent uterine stimulant, its use has been virtually abandoned because of the unacceptable level of side effects (mostly gastrointestinal) which its use provokes. The 20 years or so during which prostaglandins have been available for clinical application have gradually seen the establishment of PGE_2 as the agent of choice for cervical ripening and as an effective and acceptable labour-inducing agent where the cervix is al-

ready ripe (Calder 1983). Much of the development work of prostaglandins for these purposes has concerned the exploration of a variety
of routes of administration and vehicles within which the prostaglandins may be delivered. It has long been obvious that local administration within the genital tract is preferable to systemic administration
either orally or intravenously, and the best vehicles at the moment appear to be the water-based gels such as methyl cellulose or triacetin
(Calder and Elder 1993).

The simplest and most acceptable local route of administration is
via the posterior vaginal fornix and this is effective for most purposes.
Where the cervix is particularly unripe, there may be advantages from
the use of more invasive routes such as the extra-amniotic or intracervical route and these will require a lower dose of PGE_2 (Calder 1987).

The use of PGE_2 for labour induction, therefore, requires that the
state of the cervix should be assessed, employing a scoring system such
as that of Bishop (1964). If the cervix is already ripe (a Bishop score of
perhaps 5 or greater), labour will be effectively induced by the administration of a small dose of PGE_2 into the posterior vaginal fornix. Amniotomy should be deferred until the cervix has completed the process
of effacement and reached a dilatation of 3 or 4 cm. To rupture them
earlier is to risk imparing completion of the effacement process while
deferring rupture much later may be to lose the clinical advantages amniotomy brings in the form of optimal labour efficiency and fetal sur-
veillance (Calder 1987). Employed in this way, PGE_2 has been found to
improve the success and reduce the complications associated with labour induction when the cervix is unripe (Calder and Greer 1992b) and
to reduce the incidence of postpartum haemorrhage and patient dissatis-
faction in circumstances where the cervix is already ripe compared to
the use of amniotomy and intravenous oxytocin (Kennedy et al. 1982).

12.4 Inhibitors of Prostaglandin Synthesis

Because the prostaglandins are considered crucial in the control of
myometrial contractility and cervical ripening, it is hardly surprising
that inhibitors of their synthesis should represent important therapies
for inhibition and suppression of labour. The agent most widely explored for this purpose is indomethacin. Its clinical application has

been cautious because of the capacity of this agent to bring about premature closure of the fetal ductus arteriosus (Moise et al. 1988) but there has been a gradual acceptance of this risk, leading to a wider use of indomethacin following the recognition that (a) the theoretical risk of ductus closure is often heavily outweighed by the benefit of efficient tocolysis, and (b) there has been an increasing recognition that conventional tocolytic therapy is of limited value (see below). Thus, indomethacin may be exhibited with appropriate safeguards and has been found to bring significant benefit in both inhibition of threatened premature delivery (Novy et al. 1990) and in the suppression of established preterm labour (Wiqvist 1979; Keirse 1992).

12.5 Placental Steroids

The principal placental steroids, oestrogens and progesterone, although recognized as central players in the dramatic events of pregnancy and parturition, have not been widely applied to modify them clinically, perhaps as a result of a surprising lack of clinical efficacy. Thus, progesterone, which on theoretical grounds might have been thought likely to represent a potent tocolytic agent, has never established a clinical role for this purpose. Similarly, oestrogens, although implicated in the process of cervical ripening and generally thought to be an important link in the progression from pregnancy to parturition, have not established any important or regular place in the pharmacopoeia of labour-inducing agents. While oestradiol given either intravenously (Pinto et al. 1964) or locally within the genital tract (Gordon and Calder 1977) appears to enhance cervical ripening, this effect has never established a place in clinical practice. This is perhaps partly because of anxieties about the exposure of the fetus to oestrogens in pregnancy but also because the objective of cervical ripening can be much more efficiently realized with the use of prostaglandins.

An alternative approach to cervical ripening which has been extensively explored in Japan has been to administer dehydroepiandrosterone sulphate by repeated intravenous injections (Mochizuki and Tojo 1980). This substance is converted in the placenta to oestradiol with apparent benefits in the form of cervical softening, but this concept has not been widely adopted beyond the boundaries of Japan.

12.6 Steroid Antagonists

Although progesterone itself has never found a therapeutic place in the clinical control of uterine contractility, the antiprogesterone drugs currently under development represent one of the most exciting and potentially important innovations in obstetric practice. The first of this new generation of compounds to be widely investigated is mifepristone and this drug and its inevitable successors seem destined to alter the pattern ˙of clinical practice very significantly. Mifepristone has been extensively investigated and applied as an agent for the interruption of pregnancy at early gestations and has been found to have major benefits in the first and second trimesters (Bygdeman and Swahn 1985; Rodger and Baird 1990), benefits which derive from effects on both uterine contractility and cervical softening. To date there has been little published regarding the use of mifepristone in late pregnancy but such studies as have been conducted (Frydman et al. 1991) raise the prospect of an important clinical benefit from this new therapy. As in early gestation, it seems likely that progesterone receptor blockade with mifepristone may render the pregnancy more vulnerable to the effects of prostaglandins on both the myometrium and the cervix, perhaps by interfering with the prostaglandin-metabolizing enzyme prostaglandin deyhydrogenase (Kelly and Buckman 1990). An early imperative in the application of progesterone receptor blockers lies in the need to exclude any risk to the fetus from such therapy. The theoretical possibility of adverse effects on glucose homeostasis because of the antiglucocorticoid action of mifepristone has so far not materialized (Frydman et al. 1991).

Although a variety of drugs are available which are known to interfere with the action of oestrogenic steroids, none of these has so far found any clinical application in late pregnancy and it seems unlikely that any will do so in the foreseeable future.

12.7 Relaxin

Relaxin, a hormone principally produced by the corpus luteum of pregnancy, has been recognized to exist since its discovery by Hisaw in 1926 (Hisaw 1926). During the intervening years, a variety of biological roles have been ascribed to this hormone. With regard to late pregnancy, these raise the intriguing possibility that it might be involved in the process of cervical softening (thus favouring delivery) while at the same time restraining uterine contractility (Sherwood et al. 1990). In human clinical practice the 1960s saw a brief period of excitement concerning relaxin as a tocolytic agent but human relaxin was at that time not characterized or available and research stopped following a number of unhappy complications associated with the use of porcine relaxin. Interest was re-aroused with the suggestion from MacLennan et al. (1980) that purified porcine relaxin might have a clinically beneficial effect on cervical ripening and labour induction. More recently, with the characterization and synthesis of recombinant human relaxin it has been possible to embark on clinical studies using this hormone, although a multicentre study of human relaxin administered vaginally in gel in an effort to ripen the cervix in late pregnancy has demonstrated no clinical benefit (J.E. Brennand et al., unpublished observations) and further development of this therapeutic approach has, for the time being, been suspended.

12.8 Cytokines

The group of compounds which are currently generating the greatest excitement in relation to parturition are the cytokines. Interest in these substances (especially interleukin-1, IL-1, IL-6 and tumour necrosis factor) arose from investigation of the mechanisms whereby intrauterine infection may provoke preterm labour (Romero et al. 1990). These substances are produced by macrophages within the decidua as well as fetal placental tissues in response to stimulation by microbial products. Although IL-1 may be important in the process of cervical ripening (Ito et al. 1988), this is a fairly nonspecific mediator of inflammatory change. Our own interest centres on IL-8 which we have found to be active in intrauterine tissues in association with parturition (Kelly et al. 1992). Recognized as an important substance in the attraction and activation of neutrophils, we postulate that this substance may be particularly important in the process of cervical ripening, where the main source of collagenase may be an influx of neutrophils which are attracted into the tissue and stimulated to degranulate thereby releasing their lytic enzymes under the influence of IL-8 (Barclay et al. 1993). The potential therapeutic application of these observations remains to be explored.

12.9 Tocolytic Drugs

The objective of suppressing or inhibiting parturition has been addressed by the administration of a wide variety of pharmacological agents with very mixed success. Historically, sedative agents including barbiturates were used for this purpose but have long since been abandoned. Ethanol was advocated (Steer and Petrie 1977). It has a mildly inhibitory effect on the uterus, perhaps partly direct and partly mediated by the inhibition of oxytocin secretion, but its use is confounded by adverse maternal and fetal side effects. The pleasurable and amusing effects of alcohol on the mother were in practice quickly replaced by deeply unpleasant side effects of vomiting and headache and while these might have been tolerated had the agent proved of clinical benefit, the use of ethanol was abandoned when it was demonstrated that it produced no useful extension of the length of gestation and carried serious adverse effects to the fetus.

The agents which have dominated pharmacological tocolysis for the past two decades have been the β-mimetics such as ritodrine and terbutaline (King et al. 1988; Lyrenas et al. 1986). These agents cause a rise of intracellular cyclic AMP production inhibiting the action of myosin light chain kinase and thereby the coupling of actin and myosin to produce myometrial contractility (Berg et al. 1982). In clinical practice, however, their efficacy has been disappointing. These agents can undoubtedly induce short term inhibition of uterine contractility and delay in labour for a matter of hours, but it is increasingly apparent that if labour has become properly established, the β-mimetic drugs will not produce any useful extension of gestation other than for a matter of hours (King et al. 1988). This effect may, however, be clinically useful in allowing the fetus to be transferred in utero to tertiary neonatal centre or to allow the exhibition of glucocorticoids such as β-methasone with the object of enhancing fetal pulmonary maturation. Against such benefits, however, must be balanced the risk to the mother from these powerful agents and, in particular, the dangers of cardiovascular disturbance. A significant number of maternal deaths have followed the use of these agents and the maternal risks must not be underestimated (Benedetti 1983).

Calcium channel blockers such as nifedipine (Read and Wellby 1986) have likewise been explored although so far these agents have not established a regular place in the management of preterm labour. It seems likely that, as with β-mimetic agents, their clinical effectiveness may be shortlived. This may yet be clinically useful to allow in utero transfer of the fetus and enhance lung maturity and perhaps also for acute uterine inhibition in the face of fetal hypoxia due to excessive uterine contractility.

12.10 Summary

The catalogue of therapies which can influence human parturition, either to bring it about or to inhibit it, is extensive and bewildering. Ever increasing knowledge concerning the biological control of parturition, far from simplifying therapeutic approaches, seems to make them ever more complex.

Where labour requires to be induced, the challenge is made much greater if the uterine cervix is unripe. If this is the case, local application of PGE_2 currently represents the best available therapy to bring about cervical ripening and thereby improve the success of labour induction. Oxytocin, while a potent stimulant of myometrial contractility, has no beneficial effect on the cervix and, indeed, is contraindicated in circumstances in which the cervix remains unripe. Its role should be reserved for the later stages of the induction process when the cervix has already been ripened and the fetal membranes ruptured. In future, relaxin and (more probably) the antiprogesterone drugs may be destined to alter clinical practice for the better and the cytokines, particularly IL-8 may also establish a therapeutic role for these purposes.

Inhibition and suppression of labour in the face of threatened preterm delivery, remains a less successful area of clinical practice. Although widely used in current practice, the β-mimetic agents are of dubious value. Prostaglandin synthesis inhibitors are clearly much more effective but require to be used with considerable caution. There remains plenty of scope for the exploration of other therapeutic strategies, including oxytocin inhibitors and possibly also relaxin, but clinical advances especially in the vital area of prevention of preterm delivery will only come from further advances in our knowledge of the biological control of myometrial contractility and cervical ripening and of how these two phenomena are synchronized.

References

Akerlund M et al. (1985) The effect on the human uterus of two newly developed competitive inhibitors of oxytocin and vasopressin. Acta Obstet Gynaecol Scand 64:499–504

Akerlund M et al. (1987) Inhibition of uterine contractions of premature labour with an oxytocin analogue. Results from a pilot study. Br J Obstet Gynaecol 94: 1040–1044

Barclay CG, Brennand JE, Kelly RW, Calder AA (1993) Interleukin-8 production by the human cervix: Another factor in cervical ripening. Am J Obstet Gynecol 169:625–632

Bell WB (1909) The pituitary body and the therapeutic value of infundibular extract in shock, uterine atony and intestinal paresis. BMJ 2:1609–1613

Benedetti TJ (1983) Maternal complications of parenteral β-sympathomimetic therapy for premature labor. Am J Obstet Gynecol 145: 1

Berg G et al. (1982) β-adrenergic receptors in human myometrium during pregnancy: changes in the number of receptors after β-mimetic treatment. Am J Obstet Gynecol 151: 392–396

Bishop EH (1964) Pelvic scoring for elective induction. Obstet Gynecol 24: 266–268

Boissonas RA et al. (1955) A new synthesis of oxytocin. Helvetica Chimica Acta 38:1491–1495

Bygdeman M, Swahn ML (1985) Progesterone receptor blockade: Effect on uterine contractility in early pregnancy. Contraception 32: 45–51

Calder AA (1979) Management of the unripe cervix. In: Anderson ABM, Keirse MJNC (eds) Human parturition: new concepts and developments. Leiden University Press, Leiden

Calder AA (1981) The human cervix in pregnancy: a clinical perspective. In: Ellwood DA, Anderson ABM (eds) The Cervix in Pregnancy and Labour. Clinical and Biochemical Investigations, Edinburgh, Churchill Livingstone, pp 103–122

Calder AA (1983) Methods of induction of labour. In: Studd J (ed) Progress in Obstet Gynaecol, Volume 3: Churchill Livingstone, Edinburgh, pp 145–264

Calder AA (1987) Prostaglandins – their role in labour and labour induction. In: Keirse MJNC, de Koning HJ (eds) Priming and induction of labour by prostaglandins. "A State of the Art". Leiden, Boerhaave Cursus, pp 79–86

316   Andrew A. Calder

Calder AA, Greer IA (1991) Pharmacological Modulation of Cervical Compliance in the First and Second Trimesters of Pregnancy. Semin Perinatol 15 (2):162-172

Calder AA, Greer IA (1992a) Prostaglandins and the cervix. Bailliere's Clinical Obstet Gynaecol 6:771–786

Calder AA, Greer IA (1992b) Cervical physiology and induction of labour. In: Bonnar J (ed) Recent advances in obstetrics and gynaecology, vol 17. Churchill Livingstone, Edinburgh, pp 33–56

Calder AA, Elder MG (1993) Prostaglandins for cervical ripening and labour induction. In: Vane JR and Grady J (eds) Clinical Application of Prostaglandins. Sevenoaks, Arnold, pp 92–104

Dale HH (1906) On some physiological aspects of ergot. J Physiol 34:163

DuVigneaud V et al. (1953) The sequence of amino acid in oxytocin with a proposal for the structure of oxytocin. J Biol Chem 205:949–955

Frydman R et al. (1991) Mifepristone for induction of labour. Lancet 337:488–489

Ghosh A, Hudson FP (1972) Oxytocic agents and neonatal hyperbilirubinaemia. Lancet ii:823

Gordon AJ, Calder AA (1977) Oestradiol applied locally to ripen the unfavourable cervix. Lancet 2:1319–1321

Greer IA, McLaren M, Calder AA (1990) Plasma prostaglandin E_2 and prostaglandin $F_{2\alpha}$ metabolite levels following vaginal administration of prostaglandin E_2 for induction of labor. Acta Obstet Gynecol Scand 69:621–625

Hisaw FL (1926) Experimental relaxation of the pubic ligament of the guinea pig. Proceedings of the Society of Exploratory Biological Medicine 23:661–663

Horton EW (1972) Prostaglandins. William Heinemann Medical Books Ltd., London, Springer, Berlin Heidelberg New York

Ito A et al. (1988) Spontaneous production of interleukin-1-like factors from pregnant rabbit uterine cervix. Am J Obstet Gynecol 160:1117–1123

Keirse MJNC (1979) Endogenous prostaglandins in human parturition. In: Keirse MJNC, Anderson A, Bennebroek Gravenhorst J (eds) Human Parturition. Leiden University Press, Leiden, pp 101–142

Keirse MJNC (1992) Inhibitors of prostaglandin synthesis for treatment of preterm labour. In: JO Drife, AA Calder (eds) Prostaglandins and the Uterus. Springer, London, pp 277–296

Kelly RW, Buckman A (1990) Antiprogestagenic inhibition of uterine prostaglandin inactivation: a permissive mechanism for uterine stimulation. J Steroid Biochem Mol Biol 37:97–101

Kelly RW et al. (1992) Choriodecidual production of interleukin-8 and mechanism of parturition. Lancet 339:776–777

Kennedy JH et al. (1982) Induction of labour: a comparison of a single prostaglandin E2 vaginal tablet with amniotomy and intravenous oxytocin. Br J
Obstet Gynaecol 89:704–707

King JF et al (1988). Betamimetics in preterm labour: an overview of the randomized controlled trials. Br J Obstet Gynaecol 95:211–222

Lyrenas S et al. (1986) Pharmacokinetics of terbutaline during pregnancy. Eur
J Clin Pharmacol 29:619–623

MacLennan AH et al. (1980) Ripening of the human cervix and induction of
labour with purified porcine relaxin. Lancet i:220

Mochizuki M, Tojo S (1980) Effect of DHA sulphate on softening and dilatation
of the uterine cervix in pregnant women. In: Naftolin F, Stubblefield PG (eds)
Dilatation of the uterine cervix. New York, Raven Press, pp 267–286

Moise KJ et al. (1988) Indomethacin in the treatment of premature labour: effects on the fetal ductus arteriosus. New Engl J Med 319 (6):327–331

Moran-Ellis J (1991) Rupture of the membranes in labour. J Obstet Gynaecol
11(Suppl 1):S6–S10

Novy et al. (1990) Shirodkar cerclage in a multifactorial approach to the patient
with advanced cervical changes. Am J Obstet Gynecol 162:1412–1420

Pinto RM et al. (1964) Action of oestradiol 17β on the activity of the pregnant
human uterus. Am J Obstet Gynecol 88:759–769

Read MD, Wellby DE (1986) The use of calcium antagonist (nifedipine) to
suppress preterm labour. Br J Obstet Gynaecol 93:933–937

Rodger MW, Baird DT (1990) Pretreatment with mifepristone (RU 486) reduces interval between prostaglandin administration and expulsion in second trimester abortion. Br J Obstet Gynaecol 97:41–45

Romero R et al. (1990) The role of systemic and intrauterine infection in preterm parturition. In: Garfield RE (ed) Uterine Contractility. New York,
Raven Press, pp 319–353

Sherwood OD et al. (1990) Relaxin promotes diverse physiological processes
in the pregnant rat. In: Garfield RE (ed) Uterine Contractility. New York,
Raven Press, pp 237–252

Speert H (1958) Obstetric and Gynecologic Milestones. New York, Macmil-
lan, pp 518–524

Steer CM, Petrie RH (1977) A comparison of magnesium sulfate and alcohol
for the prevention of premature labor. Am J Obstet Gynecol 129:1–4

Turnbull AC, Anderson ABM (1968) Induction of labour; results with amniotomy and oxytocin titration. J Obstet Gynaecol of the Br Commonwealth
75:32–41

Wiqvist N (1979) The use of inhibitors of prostaglandin synthesis in obstetrics.
In: Keirse MJNC, Anderson ABM, Bennebroek Gravenhorst J (eds)
Human parturition. The Hague, Leiden University Press, pp 189–200